I0461624

TRANSITIONS IN LEADERSHIP, VOLUME 2

Veteran Stories *of* Transition *and* Transformation

Twenty-five personal narratives of military service, higher education, and career success

Robert B. Turrill, Ph.D.
and Heriberto Arámbula, Ph.D.

pine peak press

www.pinepeakpress.com

Copyright © 2025 by Pine Peak Press
First Edition — 2025

All rights reserved.

No part of this publication may be reproduced in any form, or by any means, electronic or mechanical, including photocopying, recording, or any information browsing, storage, or retrieval systems, without permission from Pine Peak Press.

The views expressed in this publication are solely those of the authors and do not necessarily represent the views of the University of Southern California or any other entity or person.

ISBN
979-8-9925868-0-0 (Hardcover)
979-8-9925868-2-4 (Paperback)
979-8-9925868-3-1 (eBook)

1. Business & Economics, Leadership
2. Biography & Autobiography, Military
3. Biography & Autobiography, Educators

Distributed to the trade by The Ingram Book Company

Dedicated to the men and women of
the armed forces of the United States

Table of Contents

Acknowledgements

A project of this nature always requires a team effort with a common goal and common commitment. Our first major acknowledgement goes to the veterans who committed to the project and persisted when it was not easy nor comfortable. Their collaboration and commitment continued strongly throughout the two years of writing and production. Many who started did not finish for a variety of reasons, but throughout the process these twenty-five writers stayed intentional about reaching the finish line.

We would also like to acknowledge each veteran's family and support network that made it possible for them to continue writing and drafting until completion, often sacrificing other activities.

Two veteran alumnae included in this volume volunteered their time and organizational skills to foster communication about the project and to plan social events with the other veteran co-authors: Kathy Takayama, Cohort VI, and Jessica Felix-Bradshaw, Cohort IV, were immensely helpful. Ty Smith, Cohort III, inspired the project through his enthusiastic approval.

The Master of Business for Veterans program staff does so much to keep the program running. They recruit new students, assist in career transitions, and lead the rigorous ten-month academic program that results in an MBV degree. The authors wish to specifically thank Program Director Lt. Col. James Bogle, Ret., Academic Director Kevin Fields, Associate Director Jamie Saure, and Director of Career Services Michael Phillips.

Steven Turrill, MFA, edited the manuscript and designed the book and covers.

Finally, grateful appreciation is owed to our families who supported our efforts throughout the entire process. Among the Arámbulas, Cindy, Eddie Jr., Celeste, and Rosie are cornerstones of resilience. Their unique contributions to our lives remind us of what truly matters. They have been pillars of love and family that have sustained us. For the Turrill family, we appreciate the constant support and encouragement of Peggy and the support of Michael, Brendan, Allison, and a special note of appreciation to Steven, our editor, for his contribution to the book.

Preface

A Unique Project for Veterans

The first volume of *Transitions in Leadership* describes the Master of Business for Veterans (MBV) program at the University of Southern California. Published by Pine Peak Press in 2020, it included forty-four personal stories from students who had completed their MBV degree. The book focused on the uniqueness of the program, its design and process, and its intention to provide a graduate business education that assisted veterans in their civilian transition from military service, whether that transition be from active duty or some other prior status.

The key characteristic of the program's design was that it admitted only veterans. For years, this was the only graduate business program that had only veterans as its student composition and was focused on enabling a positive transition from military life and culture to civilian life and culture.

This second volume, centered around the question "Where are they now?" seemed necessary. Understanding what our alumni have gone on to do in their careers, and how they are doing a few years out of the program, may assist in providing insight into the program's impact and allow us to track their activities and growth. Also, one of the book's purposes is to challenge stereotypes and present a different public image of veterans and their lives.

Unlike a more research-focused longitudinal study, which relies heavily on data and statistical analysis to measure outcomes, for this book we employ a qualitative methodology emphasizing personal narratives and storytelling. This strategy not only captures the nuances of individual experiences but also highlights the emotional and psychological journeys of our veterans. By focusing on personal narratives, we intend to provide a deeper understanding of the transformative impact of the MBV program.

With about 700 veterans having graduated from this unique program since its inception, we wanted to learn more about the impact of this experience on the veterans' lives, including the role higher education plays as an intervention strategy for veterans transitioning to civilian life after their military service.

Veterans in Their Own Words

The veteran co-authors were invited to participate in this project, and they were presented with a prompt that focused on reflecting on their transition from military to civilian life. They were asked to write stories that captured the challenges and the personal growth they experienced. The goal was to create a narrative that not only highlighted their struggles but also showcased their resilience and triumphs, including the influence of military values on their civilian careers. The authors were encouraged to explore their personal history, address the challenges of transitioning, reflect on how they overcame obstacles, and discuss their contributions to their career, family, and community. This storytelling project was designed not only as a means of personal reflection but also to provide insight into how higher education and peer support can serve as critical intervention strategies for successful reintegration after military service. Also, writing and sharing one's own personal story can be a powerful experience in itself for the author.

For more than a year, the veterans submitted drafts of their personal stories to us. During the week, we read and discussed newly submitted drafts and provided written feedback to each individual, including suggestions for additional drafts until we felt their stories were ready to move forward in the process.

This part of the writing was more challenging than we had anticipated, as well as more challenging for the authors themselves. The goal for each author was to write an honest and candid personal story of their life's critical events and outcomes, one which they could personally own and share. Many of the authors submitted multiple drafts as they got deeper into their storytelling, often uncovering painful memories they had avoided in the past.

About halfway through the writing process, we began holding Saturday discussion meetings for those who wanted to check in and discuss their progress with the other authors, and we continued to hold the meetings every other week. We avoided unconstructive criticism and provided both our support for the veterans and the space for mutual support from their peers in their common journey. Confronting their experiences in the process of writing their stories, which were intended for sharing with other veterans and interested readers, proved more challenging than most anticipated when they volunteered to collaborate on this book.

The Saturday groups were often small, but the veterans who participated received individual time to discuss their progress, with the group acting as a support mechanism, and total group discussion time focused on both the project and their personal lives. Before long, it became apparent that the group process was taking on a life of its own. In addition to writing discussions, the group became a peer-to-peer mutual support network that provided additional learning opportunities where members could interact, give and receive feedback, and begin to appreciate how this process could produce learning and insight about oneself outside the actual writing, and in addition to the sharing. Listening to others share about themselves and their writing amplified the awareness of the unique identity of each person and the similarities in the challenges they faced. It also became a meeting space for many of the authors to engage with other veterans outside their normal daily routines.

Veteran Stereotypes, Veteran Stories

There are many reasons to catch up on the lives of the MBV alumni. One is a curiosity for knowing how the veterans are doing with their transitions, careers, family, and community, by asking them to share these details with us. A closely related reason is documentation of their personal successes and achievements.

Another reason is to change the public image of veterans and their lives. At one end of the spectrum of stereotypes about veterans, they are described as "heroes," and, at the other end, as broken, homeless, suicidal, or mentally or physically crippled individuals.

A more subtle stereotype comes from a pervasive anti-military bias. This bias tends to blame the soldiers for our wars and combat.

The primary example is how Vietnam veterans were treated returning home from the war. They were spat upon and had rocks thrown at them. There is a sense that because the mission and operations of the military may involve killing the enemy, those who served our country are seen as responsible and complicit in the act of war itself.

Another example of bias was evident during the withdrawal from Afghanistan in 2021. Veterans of the war in Afghanistan faced a mix of emotions as they watched the rapid collapse of the Afghan government and the chaotic evacuation of civilians and military personnel. The public discourse often failed to distinguish between the policymakers and the soldiers on the ground, leading to misplaced blame and guilt among many service members. Some veterans felt a profound sense of betrayal and abandonment, reminiscent of the sentiments expressed by Vietnam veterans. This kind of bias not only overlooks the sacrifices made by these individuals but also perpetuates a misunderstanding of the true nature of military service and the dedication required to serve under such challenging circumstances.

"Thank you for your service" intends to address this attitude of negativity. However, it may have the unintended consequence of distancing civilians from the person on the receiving end of the gratitude and distancing the civilian even further from learning anything substantial about a veteran's experience. This also reflects the fact that, for many civilians, unless they have been part of a military family or have a close relative in the military, there is little direct knowledge of service and military life. Returning veterans often mention this lack of understanding when reintegrating into civilian cultures, including within their own families. Many veterans are left feeling isolated and often long to be with others who have had similar experiences and feelings.

Our purpose here is to challenge various stereotypes and sources of information by adding a different, more personal approach by having veterans themselves tell their own stories rather than depending on others to give their opinions and summary interpretations.

Finally, through establishing different, contrasting narratives about veterans who have successfully transitioned from military to civilian lives and careers, we want to demonstrate the power of personal storytelling as a more accurate and in-depth reporting of veteran experiences and the potential for some healing and growth through writing, understanding, and sharing. In this context, healing refers to the process of deconstructing and reconstructing one's own story, allowing individuals to make sense of their experiences, come to terms with past traumas, and build a coherent narrative that integrates their military and civilian identities.

The Shared Bond of Military Service

From 2008 to 2010, California's Assistant Secretary of Veterans Affairs toured the state and met with various colleges and universities to test the interest in providing a graduate program for veterans that would address veterans' transitions from military to civilian life.

At USC, Karla Wiseman, an executive director of executive education in the Marshall School of Business, put together a team of veterans, some of whom had graduated from USC's Executive MBA program, along with Professor Arvind Bhambri, the academic director of executive education, to draft such a program. They designed a one-year, cohort-based integrated business program with a heavy emphasis on entrepreneurship.

Listening to input from veterans, the new program design would focus on the specialized needs of veterans transitioning into civilian business and other civilian organizations. Being sensitive to time and cost, the team settled on a ten-month program with an executive format of Fridays and Saturdays every other week for two semesters. This would become the Master of Business for Veterans (MBV) degree, which was unique at the time and was offered only to veterans with a bachelor's degree and considerable leadership experience.

Most universities integrate veterans into regular MBA programs. The new MBV program grouped them into one special lock-step cohort design. This did not stop veterans from entering other MBA formats, or other majors, but offered the ability and focus to specifically address the needs and concerns of transitioning veterans in a program full of veteran students.

Essential to its purpose, the MBV program creates a safe space where veterans from all ranks and services work and grow together. By eliminating rank within the program, veterans are encouraged to interact as equals, fostering a supportive environment that promotes mutual respect and understanding. This unique setup helps bridge gaps and build a strong sense of community and camaraderie among participants, especially for veterans' post-military service.

In the military, rank often defines relationships and interactions, creating a hierarchy that can be difficult to navigate. The MBV program intentionally removes this barrier, allowing veterans to engage with one another without the constraints of rank. This approach encourages open dialogue, collaboration, and mutual support, which are essential for personal and professional growth.

In the MBV program, the shared bond of military service—regardless of branch or rank—creates a unified cohort. Veterans from different branches of the military, who might have never interacted closely during their service, find common ground and shared experiences in the MBV classroom. The intentional mix helps veterans appreciate the diverse perspectives within the military community, fostering a deeper understanding and respect for each other's experiences.

A safe space is particularly important when discussing sensitive topics such as transition challenges, mental health, and personal vulnerabilities. Veterans feel more comfortable sharing their struggles and successes with peers who understand the unique challenges of military service and reintegration into civilian life. The program's structure and culture promote an environment where veterans can deconstruct their military experiences, reconstruct their civilian identities, and find paths to personal and professional success.

The twenty-five personal stories included in this book are written by military veterans, all of whom graduated from USC's MBV program. These graduates were members of the first six cohorts of the program. The stories take a look back at the veterans' developmental history, their military experience, and the challenges of transitioning from military life to civilian environments.

Robert B. Turrill, PhD
Heriberto Arámbula, PhD
January 2025

Chapter 1

Storytelling and Veteran Reintegration

Since September 11, 2001, more than 4.4 million individuals have served in the US military, each embarking on a journey that extended far beyond their time in service. The transition from a disciplined, regimented military life to the unpredictable world of civilian society is fraught with complexities and challenges that are often overlooked in mainstream discussions on veteran reintegration. With this chapter, we wanted to delve into what makes veterans' transitions from military to civilian life such multifaceted experiences. As we have found in putting together this book, storytelling can play a role in the transition process that is not only restorative but also an essential mechanism for identity reconstruction and understanding, which each story highlights differently.

Storytelling serves as more than just a recounting of experiences; it forms a bridge that links the structured environment of military service with the fluid nature of civilian life. It enables veterans to articulate their experiences, process transitions, and forge new identities in the post-service world. The goal is to shift from a deficit-based view, portraying veterans as individuals needing repair, to an empowering narrative that recognizes their resilience, diversity, and potential. By illuminating the paths veterans navigate in returning to civilian life, we hope to highlight their inherent strength and adaptability.

A Veteran's Experience

Returning from combat service in Iraq, Dr. Heriberto Arámbula found himself in an environment that was both familiar and foreign. His profound sense of disconnection made clear to him the loss of the military's strict regimen and camaraderie. It was a feeling of being unmoored. In the military, time was a constant, measured, precise companion. Back home, it transformed into an abstract entity, with the days blending into each other and world events becoming distant.

This sense of detachment, common among returning veterans, underscores the challenges in transitioning from military to civilian life—a fundamental alteration of daily life and identity. Life in the military is defined by order, discipline, and clear purpose. In contrast, civilian life often feels unstructured and fluid, lacking clear-cut objectives and community.

Navigating the complexities of transition, Arámbula discovered the profound value and power of education and storytelling. While engaging in dynamic-critical pedagogy, emphasizing critical thinking, dialogue, and reflection, he began to understand storytelling's transformative power as a pathway to personal and transformational growth. It became a means to deconstruct his experiences in the service, allowing him to analyze and understand them in a broader context.

Through storytelling, he pieced together the fragments of his military experience, viewing them through a different lens and weaving them into the fabric of a new civilian identity. The narratives became more than stories; they were a bridge connecting Arámbula's past and present, and they led to deep, personal transformation. Narrative reconstruction was not only cathartic but enlightening, providing a sense of meaning and context to his service and its impact on his post-military life.

In the transformative journey from military to civilian life, a crucial but often overlooked aspect is the path of self-discovery and acceptance. Veterans, through introspection and various self-exploration methods, embark on a quest to reconnect with their inner selves, a process pivotal to their overall well-being and integration into civilian life. The journey is not just about healing but also about uncovering and embracing the layers of identity that may have been overshadowed by service. Tools like mindfulness, art, equine therapy, gardening, and journaling also serve as conduits for this exploration, offering veterans a means to process their experiences, understand their emotions, and reconcile their past with the present.

Challenging Deficit Ideology

Deficit ideology tends to focus on the limitations, barriers, and problems associated with veterans' reintegration. This perspective often paints veterans as inherently damaged or lacking, emphasizing their struggles and their inability to adapt to civilian life. It suggests that veterans are fundamentally broken and in need of fixing, which can lead to a narrative that sees them more as burdens than as individuals with potential. This viewpoint risks reinforcing negative stereotypes and may overlook the strengths and resilience veterans possess. Storytelling directly counters deficit ideology by showcasing the strengths, resilience, and potential of veterans.

On the other hand, a growth perspective adopts a more empowering and optimistic view. It recognizes the challenges faced by veterans but emphasizes their potential for growth, learning, and development. This perspective views the transition to civilian life not just as a series of hurdles but as an opportunity for personal and professional development. It acknowledges the unique skills, experiences, and perspectives that veterans bring to civilian life, seeing these as assets rather than liabilities. A growth mindset focuses on possibilities, resilience, and the capacity to learn and adapt, fostering a more positive and supportive narrative around veteran reintegration.

Combat trauma, particularly moral injury, adds complexity to these narratives. Moral injury is often described as deep psychological wounds caused by actions that transgress an individual's moral or ethical code. Unlike post-traumatic stress (PTS), moral injury involves spiritual and existential crises and demands a nuanced healing approach.

The relentless focus on veterans as broken and in need of fixing caused Arámbula emotional and psychological pain and triggered nightmares and flashbacks. During his doctoral program, he faced significant distress upon learning how veterans were portrayed in research journals. This distress was compounded when he failed to meet the 24-hour notice requirement for an assignment extension on the literature review he was conducting on veteran issues of transition. The academic culture's focus on procedural formalities over individual well-being, as evidenced by the professor's response to a request for an extension, exacerbated his anguish.

Such experiences highlight the critical need for a paradigm shift in how we view and support the veteran community and other marginalized groups. Adopting an asset-based perspective, which represents a distinct departure from the traditional deficit-based mindset, aligns closely with a growth-based standpoint. It focuses on recognizing and nurturing the inherent strengths, skills, and potential contributions of individuals. This shift is vital if we are to change the narrative around veterans' reintegration and promote inclusivity in community building. Voicing our unique stories, languages, and perspectives is a powerful tool in this process. It not only fosters greater awareness and consciousness but also challenges dominant paradigms, promotes deeper understanding, and paves the way for more inclusive and supportive strategies. In this context, the concept of living in "in-between" spaces, akin to the idea of Nepantla, becomes particularly relevant.

Nepantla, popularized and expanded in meaning by Chicana scholar Gloria E. Anzaldúa, can be understood as navigating between different worlds or realities. It resonates with the transitional experiences of veterans. The term encapsulates the essence of moving between the structured life of military service and the often-unstructured civilian world, highlighting the complexities and challenges inherent in this liminal space. Anzaldúa's interpretation of Nepantla helps us appreciate the multifaceted identities of veterans and their unique experiences post-service, encouraging an approach that acknowledges and leverages these diverse backgrounds and skills for holistic growth and integration.

The Alarming Reality of Veteran Suicide

The rise in veteran deaths from suicide and self-injury, now reported to have reached forty deaths per day, presents a crisis for our veterans' mental health.[1] It's a grim reminder of the feeble and inadequate support of current reintegration strategies. These figures combine the Department of Veterans Affairs (VA)—which previously reported

[1] "The State of Veteran Suicide: Mission Roll Call Report." America's Warrior Partnership. 2024. Retrieved from https://www.missionrollcall.org.

seventeen suicide deaths per day[2]—with a 2022 interim report published by America's Warrior Partnership and Duke University that calculates an additional twenty deaths per day resulting from self-injury, including drug overdose. These statistics underscore an urgent need for more effective mental health care and support systems and raises questions about existing support structures' efficacy. Despite available benefits and programs, why do so many veterans find themselves engulfed in profound despair? Exploring this phenomenon requires considering both systemic and individual factors. Systemically, there's often a disconnect between offered services and veterans' actual needs. Support programs may not fully address the complex nature of transitioning from military to civilian life, leaving veterans feeling misunderstood and isolated. Addressing veteran suicide requires not only systemic changes but also a shift in narrative. Storytelling can play a crucial role in this shift, by giving voice to the struggles and successes of veterans and fostering a culture of understanding and support.

Mental health issues, combined with societal reintegration challenges, make a formidable obstacle. Veterans return with invisible wounds, manifesting as PTS, depression, anxiety, and moral injury. These challenges are compounded by readjusting to civilian life, finding employment, reconnecting with family and community, and reconciling military experiences with current reality. Culturally, the military ethos of strength and self-reliance can hinder veterans from seeking mental health help. The stigma around mental health in both military and civilian spheres interferes with access to necessary care. The barrier, coupled with a general lack of understanding about veteran-specific mental health issues among health care providers, can lead to inadequate treatment. Addressing veteran suicide requires a multifaceted approach that enhances mental health services' accessibility and quality and works toward destigmatizing mental health care among veterans. It's about creating a supportive environment where veterans feel understood and genuinely cared for.

---

[2] "VA National Suicide Data Report: 2023." U.S. Department of Veterans Affairs. 2022. Retrieved from https://www.mentalhealth.va.gov/docs/data-sheets/2023/2023-National-Veteran-Suicide-Prevention-Annual-Report-FINAL-508.pdf

The alarming reality of veteran suicide and self-injury underscores the urgent need to destigmatize mental health care. It's imperative to create a supportive environment where veterans can engage in self-exploration without fear or shame. Overcoming this stigma is a critical step in enabling veterans to reconnect with themselves, understand their mental health needs, and seek appropriate support. As they navigate this journey, the role of families and communities becomes integral, providing a safe and understanding space for veterans to rediscover and accept their evolving selves. Through storytelling, they can articulate their struggles and triumphs, making their internal battles visible and relatable, which in turn fosters empathy and support from their communities. A collective narrative helps to dismantle the barriers of silence and isolation, paving the way for a more open and compassionate dialogue about mental health.

The Role of Family and Community

The journey of reintegration for veterans intertwines with family and community life. It's a process extending beyond the veterans' experiences, enveloping their families and communities in a shared experience of transition and adjustment. Insights from conversations with Gold Star families reveal the universal nature of these challenges. Their stories highlight the collective experience of adjustment and transition, touching every family member. Gold Star families often find their experiences mirroring the reintegration struggles of surviving veterans. They navigate a complex landscape of loss, change, and adaptation, supporting the reintegration of other family members or friends returning from service. This dual journey underscores the need for support systems that address the entire family unit's holistic needs. The support required during this time transcends traditional veteran-focused services. It calls for a broader network of community and familial support, providing resilience, reducing stress, and promoting well-being for veterans and their families. Leveraging these networks effectively means recognizing the unique challenges faced by military families, especially those living in remote military installations or dealing with cultural and geographical isolation.

A holistic approach to veteran reintegration support acknowledges the interconnectedness of the veteran within their family and community networks. It's not enough to provide support in silos; there must be an integrated effort that considers the emotional, psychological, and social dimensions of reintegration. Families and communities play a pivotal role in this process, offering a sense of belonging, understanding, and shared experience that can be profoundly healing. Critical questions about the nature and sources of support for military families guide the development of support systems and services that truly meet their needs. By embracing a holistic, family- and community-centered approach, we can create an environment where veterans and their families thrive as they embark on their new journeys. Its purpose is to build resilient communities that recognize sacrifices, honor experiences, and support all their members' well-being. Therefore, by telling their stories, veterans not only find a means of healing but also help to educate and empower those around them, building a community where everyone's experiences are valued and supported.

Our Service Continues

Moving forward, we must continue shifting from a deficit-based view of veterans to recognizing their resilience, adaptability, and potential. Building supportive, empathetic communities that understand and value veterans' experiences involves enhancing the accessibility and quality of support services, fostering a culture that destigmatizes mental health care, and encouraging open dialogue. We must also acknowledge the importance of addressing specific issues like moral injury, which adds another layer of complexity to the transition experience.

The reintegration journey for veterans is a testament to the strength and adaptability of the human spirit. It highlights the importance of community, empathy, and understanding. As a society, we share the responsibility to support this journey and listen to and learn from those who have served. By doing so, we not only aid veterans' healing processes but also enrich our communities with their diverse experiences and perspectives. The narrative of veteran reintegration is one of complexity, resilience, and hope. It continues to evolve with each veteran's journey, requiring our collective attention, understanding, and action. Storytelling is central to this evolution; it is through sharing their stories that veterans can convey the depth of their experiences and the challenges they face during reintegration. By embracing these narratives, we support veterans to reclaim their voices, fostering a deeper connection between themselves and the broader community. Storytelling transforms individual experiences into collective knowledge, empowering both veterans and those around them to work together towards a more supportive and inclusive society. Recognizing the transition to civilian life as an opportunity for growth and contribution is essential. Veterans, through self-discovery and acceptance, are uniquely positioned to offer insights, wisdom, and resilience to their communities. By focusing on their strengths and potential, we shift the narrative from merely overcoming challenges to thriving and enriching the community fabric.

Storytelling has emerged as a pivotal element in this process, serving as a powerful tool for healing, understanding, and bridging divides. More importantly, it acts as a medium for veterans to own their narratives. This sense of ownership is not just about accepting their past but also about acknowledging and valuing their personal growth and evolution. It enables veterans to articulate their experiences, fosters empathy within the community, and aids in constructing new identities.

By embracing these stories, we build a richer, more inclusive understanding of the veteran experience. We honor their contributions and sacrifices, and in doing so, we create a society that values resilience, growth, and the transformative power of shared human experiences. Through their narratives, veterans continue to serve—enriching their communities, inspiring future generations, and fostering a more empathetic and connected world. The positive effect of storytelling cannot be overstated. It not only heals and empowers those who tell their stories but also educates and connects those who listen. In sharing their experiences, veterans offer invaluable lessons in resilience, adaptability, and the human capacity for growth, inviting all of us to participate in a collective journey towards understanding, compassion, and community.

Chapter 2

Ser Un Hombre
by
Heriberto Arámbula

Dr. Heriberto Arámbula, a Pat Tillman Scholar and public pedagogue, served for eleven years in the US Army and US Coast Guard, achieving the rank of sergeant. His service included a combat deployment to Sadr City, Iraq, in 2004.

Heriberto holds a BA in history and political science from Arizona State University, where he graduated summa cum laude. He received an MA in teaching and a Master of Business for Veterans (MBV) degree from the University of Southern California. He earned his PhD in educational and community leadership from Texas State University. His dissertation, "Story of Re-Creating Home: Learning as a Tool for Healing Post-Military Service," focused on veteran trauma and social integration post-military service.

He is the founder of the Journal of Interactive Veteran Experiences (JIVE), a vital platform for veterans to share their stories.

A Dream of Baseball

On Opening Day at Dodger Stadium, in 2016, I was honored as the military hero of the game. As I stood on the field, the cheers of the crowd enveloped me—a tangible expression of gratitude for my service. The atmosphere was electric, charged with the energy of

55,000 fans. It was an overwhelming experience. But an even bigger personal dream was about to unfold, one that connected my past, present, and future in a way I had never anticipated.

In the spring of 1997, I had tried out for my high school's baseball team after playing only intermittently during my childhood. I didn't have much time for extracurriculars because my parents worked a lot. Needless to say, I was not very prepared for high school baseball. The coach told me, "You don't have what it takes." Out of all my friends, I was the only one cut. At fifteen years old, I considered it my first failure.

I had a choice to make: give up on my dream of playing baseball, or work on my skills and try again next year. That summer, I worked on my baseball skills with my buddies. I was determined to make the team. I went to the park daily and worked diligently on fundamentals. My friends hit countless ground and fly balls for me to catch. They threw batting practice to me until the sun went down. It made for a memorable summer.

I worked harder than I had ever worked on a skill. And it paid off. I made the team the following year and played every game until I graduated. As the starting catcher, I batted .441 and stole twenty-three bases.

Growing up, I had always admired Sandy Koufax, not just for his prowess on the baseball field but for the person he was—his integrity, his decision not to play on Yom Kippur, and his overall demeanor. He was more than a sports hero to me; he was a symbol for standing up for your beliefs and excelling despite challenges. Later, as an undergraduate, I delved into his biography and wrote a paper on him that explored his life's journey and impact. His story resonated with me deeply, but I never imagined our paths would cross.

On Opening Day at Dodger Stadium, I found myself standing next to none other than Sandy Koufax himself. The moment was magical, a convergence of my admiration for a baseball legend and the public recognition of my military service.

People asked if I got his autograph. The encounter meant so much more to me than a signature on a piece of memorabilia. I got a chance to hug Sandy, exchange a few words with him, and connect, however briefly, with the man who had inspired a huge portion of my life.

The Essence of Ser Un Hombre

This is my story—a tale of a Mexican kid who grew up in Los Angeles learning to navigate life's complexities and drawing strength from the spirit within us all. My journey spans the battlegrounds of Sadr City, Baghdad, to the academic halls of the University of Southern California, culminating in a doctoral degree at Texas State University. It interweaves the raw and sometimes turbulent realities of urban life with the introspective world of academia. I want to illustrate the quest for meaning, understanding, and reintegration that intertwines personal growth with collective experiences.

In my early years, adulthood was a distant, almost abstract concept. In my mind, reaching adulthood was never defined by age or milestones; rather, it was about embodying un hombre—a man—in the full cultural sense of my Mexican heritage. The notion of ser un hombre was deeply ingrained in me, steeped in the values of strength, honor, and resilience. This concept was not just about personal growth; it was about becoming a provider and a protector who exemplified the virtues that defined manhood in my community. These values, taught from a young age, laid the foundation for my understanding of maturity and responsibility. The essence of un hombre was a guiding force, shaping my perceptions and aspirations as I navigated the various stages of my life. It was a narrative of growth, not just in age but in character, influenced by the cultural and familial ideals that surrounded me. As I matured, this deep-seated understanding of manhood continued to influence my choices and paths, particularly as I faced pivotal moments in my life, such as joining the military or stepping into the roles of husband and father. Each life stage brought me a step closer to fulfilling the essence of un hombre through a journey that was both personal and deeply rooted in my cultural heritage.

At the Crossroads

As a young man stepping out of the halls of Gardena High School in 2001, I didn't just want a future, I wanted to fill a role that was deeply rooted in my cultural fabric. That year, the events of 9/11 steered my journey in a direction I had not envisioned. I enlisted in the US Army in the summer of 2002 and found myself transitioning from the familiar streets of my youth to the unpredictable terrain of war-torn Iraq. This shift was not just a change of location; it was a metamorphosis of my very being. In the military, ser un hombre took on a new dimension. It became about brotherhood, about facing the horrors of war alongside those who became my family. It was about shared sacrifices and the unspoken bond formed in combat. These experiences, both harrowing and enlightening, laid the groundwork for a path that would eventually lead me to academia and self-discovery.

A young man stood at the crossroads of youth and adulthood. Graduating from high school at seventeen, I was filled with a blend of anticipation and uncertainty. Before I decided to enlist, college loomed on the horizon, a beacon of hope and opportunity. My choice to attend California State University, Fullerton, was influenced not just by convenience but also by familial responsibility. It was where my journey into higher education began, albeit under the shadow of a national tragedy that would redefine my path.

My transition to college life was far from seamless. The cultural shift from the diverse environment of high school to a predominantly white university campus was jarring. I grappled with feelings of isolation and disconnection, finding myself a stranger in a world that seemed so different from what I had known. The struggle was also internal, and my academic performance reflected my turmoil. A 1.67 GPA marked the end of my first academic year—a stark contrast to the effortless success of my earlier schooling.

In this period of self-doubt and confusion, the events of September 11, 2001, unfolded. Something profound stirred within me. It was a clarion call that resonated with my desire to serve and protect. The sense of duty, instilled in me from a young age, found a new outlet. It was then that the path of military service emerged as a beacon in the fog of uncertainty that surrounded my early adult life.

The decision to enlist in the US Army reflected a confluence of personal conviction and circumstantial drive. It was a step toward embracing the essence of ser un hombre in its most primal form—to protect, serve, and sacrifice. In the summer of 2002, I stepped into a new world, one defined by discipline, rigor, and the rising shadow of conflict. The transition from civilian to soldier was a transformative experience, reshaping my identity and fortifying my resolve.

My journey in the army began with basic training at Fort Benning (now Fort Moore), GA. Here, in a crucible of physical and mental conditioning, I was stripped of individuality and remolded as a part of a larger entity—the military. The training was grueling, a relentless test of endurance and character. But beyond the physical demands, it was a rite of passage that instilled in me the values of courage, loyalty, and unwavering commitment to the person next to me. As I adapted to military life, I developed a profound sense of belonging. The bonds formed in the barracks and training grounds were unlike any I had ever known. These were bonds forged in the shared anticipation of impending deployment, in the knowledge that we were preparing to face the unknown together. As an infantryman, the reality of my role became increasingly clear—I was training for war, for a conflict that was as much about defending ideals as it was about combat.

Amid this transformation, a sense of purpose crystallized within me. The decision to serve was no longer just about fulfilling a cultural role or responding to a national tragedy. It became a personal crusade, a commitment to stand in defense of values I held dear, to be part of something greater than myself. This period marked not just a transition into adulthood but a redefinition of my very essence—from a young man seeking his place in the world to a soldier prepared to face the ultimate test of his convictions. In 2004, we received orders to deploy to Sadr City, Baghdad, as part of Operation Iraqi Freedom II.

The gravity of the mission was not lost on us. We were entering a conflict mired in complexity and controversy, a battleground far removed from the life we had known. This was a profound chapter in my life. The streets of Baghdad were a labyrinth of uncertainty, where danger lurked at every turn. The bond of our unit became our anchor, providing not just tactical support but emotional strength.

The Ambush

April 4, 2004, is etched in my memory with a clarity that time cannot diminish. It was the day the reality of war exposed itself to me in its most horrific form. While patrolling the streets of Sadr City, our unit was ambushed. The attack erupted into a relentless firefight. The chaos that ensued was surreal and devastating. The battle raged for what seemed like an eternity, claiming the lives of eight brothers-in-arms and leaving more than seventy wounded. Each loss was a grievous blow, a stark reminder of the fragility of life and the brutality of war. One battle-buddy we lost that day, Specialist Ahmed Cason, was more than just a fellow soldier; he was a friend, a brother. His sacrifice and bravery were a testament to the selflessness that defines the very essence of military service and brotherhood. Surviving the ambush produced mixed feelings of relief and profound guilt. The question of why I was spared, when others had not, haunted me. The images of that day, the sounds of gunfire, and the cries of the wounded, became a recurring nightmare, a constant reminder of the price of war. It was a turning point in my life, a moment that shattered my innocence and forever altered my view of the world.

During my time in Baghdad, I witnessed both the horrors and the humanity of war. The loss of comrades, the strain of constant danger, and the moral complexities of combat left an indelible mark on my soul. Yet, amidst the turmoil, there were moments of profound connection—between fellow soldiers, with civilians caught in the crossfire, and with a world far removed from my own. These experiences shaped me in ways I could never have anticipated, forging a perspective that guided me in the years to come.

The Silence of Peace

Returning home in 2005 was another transition unlike any other. I came back physically unscathed, but I carried deep, complex wounds in my heart and spirit. The memories of combat, the loss of my comrades, and the violence I had witnessed lingered in my mind, casting a shadow over my return. The transition was challenging and never-ending, marked by a struggle to find my place in a world that seemed simultaneously familiar and foreign. The sense of purpose and brotherhood I had experienced in the military was replaced with a feeling of isolation.

Coming home meant not only readjusting to a different pace and lifestyle but also grappling with the emotional and psychological aftermath of my experiences. In the silence of peace, I longed for the days of my youth, untainted by war. Yet I knew that longing for the past would not ease the burden. Eventually, I turned to reflection and introspection, trying to make sense of my experiences and find a path forward. The journey of healing was gradual and often painful. Ten years after my deployment, I finally asked for help. It involved confronting the traumas of war, acknowledging the impact they had on me, and learning to live with the memories. This process was not one I could undertake alone. Support from family, friends, and fellow veterans played a crucial role in my journey towards first acknowledging that I had a problem and then asking for help.

Amid this personal turmoil, I found solace in education. My return to academia was more than a pursuit of knowledge; it was a quest for understanding and a means of processing my experiences. As I delved into my studies, I discovered a newfound purpose. Education became a tool for healing, a way to rebuild my identity and redefine my path.

Redefining Ser Un Hombre

In the aftermath of my military service, the cultural concept of ser un hombre which had guided my youth underwent a profound transformation. No longer did it solely encompass the traditional virtues of strength and stoicism; it now included vulnerability, introspection, and emotional resilience. Working on a re-definition was not a departure from my roots but an expansion, incorporating the lessons and experiences I had garnered through my service and the trials of war. The evolution was challenging and liberating—it embraced a more holistic view of what it means to be a man. My new understanding included the courage to face my vulnerabilities, the strength to seek help when needed, and the wisdom to understand that true resilience is often found in the ability to adapt and grow.

The journey of redefining masculinity also meant reevaluating my role within my family and community. It involved learning to balance the traditional expectations of being a provider and protector with a deeper emotional presence and engagement. Balancing is not always easy to strike, but it becomes an essential part of my personal growth and healing process.

A Thirst for Knowledge

My return to higher education was slow and nonlinear. While I was in active duty, I had been enrolled in the local community college and took one or two classes at a time. Unknowingly, at the time, I was on a long quest to understand and find a means of processing my experiences. My undergraduate studies began to provide context to my experiences, helping me piece together a broader understanding of the events that had shaped my service and time in war. Higher learning became another pivotal chapter in my journey. Driven by a thirst for knowledge and a need to make sense of my experiences, I immersed myself in the world of books and ideas. This pursuit was not just academic; it was deeply personal and healing. My studies became a sanctuary, a place where I could explore, reflect, and rebuild.

The focus of my academic journey was not arbitrary. Drawn to subjects that resonated with my experiences, at Arizona State University I pursued history and political science, subjects that offered insight into the complexities of the world.

Throughout my academic journey, I faced significant challenges. One major obstacle was the struggle to adapt to civilian academic life after years of military service, which involved overcoming a lack of understanding and support from academic institutions regarding veterans' unique needs. Additionally, balancing the demands of rigorous coursework with the responsibilities of family life while dealing with the physical and psychological effects of military service added layers of complexity to my educational pursuits.

New Perspectives

My journey continued through 2015, as I pursued further education in USC's Master of Arts in Teaching (MAT) program. The experience was transformative, allowing me to bridge my military background with my passion for education. The program emphasized critical thinking, reflective practice, and innovative teaching strategies, all of which resonated deeply with my desire to make a meaningful impact in the classroom. The MAT program provided a supportive and intellectually stimulating environment where I could explore the intersections of my military experiences and educational aspirations. Engaging with fellow educators and mentors, I discovered new perspectives and pedagogical approaches that enriched my understanding of teaching and learning. The rigorous curriculum and hands-on teaching experiences equipped me with the skills and confidence to create inclusive, engaging, and effective learning environments for my students. I realized the profound potential for education to be a tool for personal and societal transformation.

My time at USC reinforced the value of storytelling in an educational context, allowing me to draw upon my own personal narrative to inspire and connect with students. The MAT program was not just an academic pursuit but a journey of growth and empowerment that further shaped my identity as an educator.

Building on my experiences at USC, I enrolled in the Master of Business for Veterans (MBV) program, joining Cohort IV (2016-2017). Although I didn't create a business, the knowledge and skills I gained were instrumental in developing innovative ways to address veteran issues of transition and reintegration. The program's focus on entrepreneurship and business strategies helped me conceptualize new approaches in order to create supportive and inclusive spaces for veterans. The MBV program equipped me with tools to think creatively about solving complex problems. Under the mentorship of Dr. Robert Turrill, who inspired and supported my aspiration to pursue a PhD, I honed my critical and strategic thinking abilities. His guidance was invaluable, enhancing both my professional and personal growth. While I continued my educational endeavors and completed my doctoral studies, the lessons from entrepreneurship stayed with me.

The MBV program's rigorous curriculum, which blended academic excellence with practical business skills, is tailored for veterans who are leaders. The faculty's mentorship and the program's support for continued growth and contribution have been instrumental in my journey. The MBV program is an exceptional pathway for leaders transitioning into various fields, demanding a high level of professional and academic achievement. My experience in the program was outstanding, and I continue to be connected to my cohort and others.

The Redwood Paradigm

My academic journey included the development of a unique conceptual framework I have come to name "The Redwood Paradigm: Interconnected Growth and Resilience." This paradigm, inspired by the majestic California coastal redwoods, symbolizes the resilience, interconnectedness, and communal growth that have been central themes in my life and the lives of many veterans. The coastal redwood, known for its towering presence and longevity, thrives in a mutually supportive ecosystem. This became a metaphor for my understanding of human resilience and community. In this paradigm, individual growth is inextricably linked to the health and support of the wider community, reflecting the experiences of veterans who, like the redwoods, flourish best with mutual support.

This framework challenges the traditional notions of individualism, advocating for a collectivist perspective where personal achievements are intertwined with communal support. The Redwood Paradigm underscores the importance of an environment where each individual can contribute to and draw strength from the collective, a philosophy I found resonant in both military and civilian life.

My conceptual framework was integral to my doctoral study. My research delved into the complexities of veteran reintegration and emphasized the transformative power of storytelling to bridge the gap between military and civilian experiences. Researching my dissertation allowed me to explore the multifaceted challenges faced by veterans as they transition back to civilian life. I highlighted the need for inclusive spaces in higher education and society at large where veterans can share their unique personal narratives and insights.

My work underscored the need to shift the narrative from a deficit ideology to one that recognizes the strength, resilience, and potential of the veteran community. My research also advocated for policy changes that embrace the power of storytelling and community in supporting veterans. By sharing our stories, we facilitate a process of healing, understanding, and growth, not just for ourselves but for the community that receives them. This process aligns with the principles of The Redwood Paradigm, where each story adds to the richness and vitality of the larger narrative, much like each redwood contributes to the health and resilience of the forest.

Defending my dissertation and developing The Redwood Paradigm were more than academic achievements. They reflected the synthesis of my personal journey and professional aspirations, and they represented my commitment to use the lessons learned through my experiences—as a soldier, a student, a veteran, and a family man—

to foster understanding, resilience, and growth within the veteran community.

The Journal of Interactive Veteran Experiences

The culmination of my academic and personal journey led to a new and exciting venture: the creation of the Journal of Interactive Veteran Experiences at Texas State University. This initiative was not merely another academic project but a mission born from a desire to give voice to the untold stories of veterans, stories that echoed my own and many others.

The inspiration for the journal stemmed from recognizing the vast, untapped reservoir of experiences and wisdom within the veteran community. These narratives, rich in lessons of resilience, sacrifice, and transformation, needed a platform to be shared, understood, and appreciated. I envisioned this journal as a bridge between the military and civilian worlds, a space where the gap of understanding and empathy could be narrowed. The Journal of Interactive Veteran Experiences was conceived as a response to this need. It was designed to be an interactive platform, encompassing varied forms of media in order to capture the full spectrum of the veteran experience, from triumphant moments to profound challenges, and from pride in service to the vulnerabilities of returning to civilian life. We aspired to present a holistic view of what it meant to be a veteran.

Establishing the journal was a significant step toward healing, education, and inspiration, not only for veterans but for the broader community. It provided an avenue for veterans to share their stories on their terms, and it facilitated the process of reflection and understanding. Additionally, it served as an educational resource, offering insights into the veteran experience for researchers, educators, and students.

The project was also a testament to the transformative power of storytelling. In sharing our narratives, we not only get to process our own experiences, but we contribute to a collective understanding and empathy. The journal embodies my belief in the power of sharing stories.

As I introduced the Journal of Interactive Veteran Experiences to the academic and veteran communities, I felt a sense of fulfillment and purpose. This was more than a publication. It was a community, a testament to the enduring spirit of those who have served, and a contribution to the ongoing dialogue about the veteran experience.

Family and Community

Recently, I attended the twenty-year reunion of my unit, the 2/5 Cavalry, commemorating the battle of April 4, 2004. The battle was vividly depicted in Martha Raddatz's 2008 book, *The Long Road Home*, and it marked a significant moment in our unit's military service and in our personal lives. The reunion was more than just a gathering of former comrades; it was a poignant occasion for storytelling, healing, and reconnection.

As we shared stories from our service, the dynamic of family and community support became evident. Gold Star families—those who have lost loved ones in the line of duty—attended, bringing with them deep remembrance and resilience. Their presence highlighted the collective nature of our experiences and the importance of community in the healing process. Through storytelling, these families could share their loved ones' legacies, keeping their memories alive and honoring their sacrifices.

The reunion also provided a space for spouses and children to connect. Their perseverance and adaptation underscored the vital role that family plays in supporting veterans. Spouses and children, who often navigate the challenges of military life and reintegration alongside their loved ones, found solace and understanding in their shared experience. Storytelling showed us that healing is a communal process. The reunion, with its reflection and reconnection, emphasized the importance of bridging the past and present with narrative. It served as a powerful reminder to foster empathy, understanding, and resilience within our veteran community and their families.

"My Steadfast Support System"

Heriberto Arámbula reflects on telling his story

As I reflect on my journey from the streets of Los Angeles to the battlegrounds of Iraq and finally to the halls of academia, I see a tapestry woven with threads of resilience, growth, and the support of those around me. The lessons learned from my military service, coupled with the unwavering support of my family, have shaped my path and purpose. In the words of Nelson Mandela, "The greatest glory in living lies not in never falling, but in rising every time we fall."

Throughout the challenges of military service and the rigors of academia, my family has been my steadfast support system. Each member has played a unique and pivotal role in my journey, providing love, inspiration, and invaluable lessons.

More than just my high school sweetheart and wife, Cindy is a fellow combat veteran who has shared the trials and triumphs of military life with me. Her journey continued beyond the battlefield as she pursued her studies to become a structural engineer. Her strength, on the frontlines and in her academic pursuits, has been a constant source of admiration and inspiration. Cindy's resilience and understanding have been my sanctuary through the uncertainties of reintegration and the challenges of academia. Her unwavering support and deep belief in me have been guiding lights in my darkest moments.

Eddie, my son, has been one of my greatest teachers. Diagnosed with autism, he has opened my eyes to new perspectives and ways of understanding the world. His love for chess and video games reflects a mind rich in strategy and creativity. Being Eddie's father is an honor. His unique view of the world and his approach to life's challenges continually teach me about patience, acceptance, and the beauty of seeing the world through different lenses.

Celeste is a committed and dedicated ballet student. Her drive and determination are commendable. Her passion for ballet and her relentless pursuit of excellence in that demanding discipline are a testament to her strong character. Celeste's commitment mirrors the discipline and focus I have strived for in both my military and academic careers. Watching her grow and refine her talent in ballet has been a source of immense pride and joy.

Rosie, the youngest, already aspires to be a pediatrician. Her caring and compassionate nature is evident in her dream to help others, reflecting a maturity beyond her years. Rosie's innate desire to care for the human spirit is inspiring. She has even been a source of comfort to me, coming to my aid during times of illness or distress. Her empathy and kindness are qualities that make her aspiration to be a pediatrician not just a dream but a likely future reality.

Chapter 3

A Lifelong Dream
by
Amber Culotta

Amber Culotta served for more than fourteen years in the US Air Force. Her service included two combat deployments during Operations Iraqi Freedom and Enduring Freedom. Amber spent ten years as an enlisted E-6 aircrew member on the KC-10 as an in-flight refueler and C-17 instructor loadmaster. During that time, she earned three Air Medals. She then commissioned as an officer, attaining the rank of first lieutenant.

Amber joined Boeing as a flight technical data lead in the Commercial Derivative Aircraft division. She currently serves as the treasurer of the Boeing Veteran Engagement Team and is an active member of Boeing Women Inspiring Leadership.

Amber's academic credentials include a BA from California State University, Los Angeles, and a Master of Business for Veterans degree from the University of Southern California.

Amber is a proud mother and wife, balancing dual civilian and military careers with her spouse, who still serves.

Out of My Small Town

When I was thirteen years old, I made the decision to join the military. I have always had a determined spirit. Even then, I knew it was my ticket out of a powerless childhood.

My parents enforced strict rules, harsh punishments, and controlled every aspect of my life. This included if or when me or my siblings could shower. There were higher expectations for our behavior and academic performance than my peers. The environment focused solely on our value according to how well we fulfilled our responsibility to the family. At the time, the structure and stability brought me comfort after my mother lost her custody rights due to multiple domestic violence incidents a few years prior.

We had a neighbor who had served in the US Air Force and when I mentioned my interest, he advised that the Air Force and the Navy were the better options for women—and that aircrew was a great option for the career-minded. To me, the military was more than just a potential job. It meant travel, better accommodations, and a rewarding career. I threw myself into studying for the Armed Services Vocational Aptitude Battery (ASVAB). I was overjoyed to score what I needed to pursue the job of my dreams.

Shortly after, I selected my job, and my parents signed the papers for me to begin the delayed entry program. I was seventeen years old. Like many military members, joining the military was my opportunity to get out of my small town of Linden, California, and to get out of my difficult childhood. It also meant my parents were proud to send me off to basic training. Pride from my parents was uncommon in our home. I watched my older sisters endure their derision whenever they asked to move out. But here I was, leaving home with their approval, guaranteed healthcare, housing, and a paycheck for the entirety of my enlistment.

First Day of the Rest of My Life

I left for basic training just days before my eighteenth birthday, and I will never forget the freedom I felt on my flight over to San Antonio, Texas. As we landed, I smiled. I was here; I had done it. The warm Texas air and the screaming military training instructors (MTIs) welcomed me to the first day of the rest of my life. During basic training, I kept my head down and followed every rule. I found familiarity in the structure. It was easy to find my place. Turns out, a lifetime of being yelled at, ordered to do chores, and having little freedom was the perfect preparation for this type of training.

I went on to aircrew qualification next, which included Fundamentals, Water Survival School, and Combat Survival School. In Water Survival School I realized I was doing things that most people in the civilian world never experienced. I had to push myself to achieve a goal despite how uncomfortable I was.

I was forced to jump into the ocean, and I hated every second of it. I had only two hours to overcome my fear of heights and the ocean. Initially, my nerves were getting the better of me and I kept moving to the back of the line until I was the only person left. When I finished, I was so proud of myself. In the midst of that pride, I noticed that no one around me cared that I had completed a task that terrified me. The instructors pushed out classes every week, and no one else seemed to have as much trouble mustering up the courage to complete the tasks. The experience taught me that I have to be proud of me. I had to achieve my goals without a cheering section and without validation from others. It taught me that even when I am all alone, I can do anything if I push myself. What I accomplished in those few days is something I will carry with me forever.

In Basic Boom Operator School, I learned about air refueling tankers and basic weight and balance principles. During graduation, we were assigned a military base and an airframe. I was assigned the KC-10 aircraft at Travis Air Force Base, California. The aircraft's primary mission was to complete aerial refuels and transport cargo worldwide. With their installed drogue system, the aircraft could also provide fuel to Navy and Marine Corps aircraft.

The KC-10 was only located on the East and West Coasts. I felt relieved that I would get to stay close to my family and the boy from high school I was chasing after. I regret now not requesting an overseas assignment on the KC-135, the other Air Force tanker that had bases all over the world.

Because I was stationed only an hour away from home, there was pressure on me to be around more to help with family problems and smooth over conflicts with my sisters. Things seemed to have fallen apart after I left, and I think my family thought if I was around more, things would go back to the way they were. I did not have a driver's license because my parents wouldn't allow it. My family picked me up on Fridays as soon as my duty day was over and brought me back to the house. I was expected to spend the entire weekend with family and obey the same rules I had grown up with. It was getting harder for me to focus on my training back on base when I was always on edge dealing with family conflict.

My job required me to be away from my duty station for more than 200 days a year. I did not have to use leave to see my family and friends when I was at my home station, and I still got to see the world a few days at a time. I think I would have grown more if I had left the proximity of my hometown and the relationships I had initially tried to distance myself from. They were unhealthy, and I spent a long time trying to figure out how to respectfully get the space I needed, in order to grow, find myself, and experience life. When I was on base, I was one person, and when I went home, I was another.

Refueling Alone

I started my qualifications to become an in-flight refueling specialist. The training was intense and high-pressure. It required a level of perfection I was not accustomed to. We did months of simulator training that led to my first observation flight. They had me sit in the seat and connect the boom to an incoming aircraft on my first flight. I could not believe it! I was realizing my dreams, and it was both terrifying and exhilarating.

I quickly found myself in charge of refueling aircraft in the air, speaking with pilots over radio, connecting with control towers, and coordinating normal and emergency situations with our crew in the front of the aircraft while I was in the back refueling. It had taken roughly one and a half years from the day I left basic training to be a fully qualified in-flight refueler. I spent the next six months flying as many missions as I could to gain experience. While my friends back home went to college, I headed out for my first deployment on September 11, 2005, just two weeks after I turned twenty years old. I called my sister and cried for hours. I was nervous and afraid. I had no idea what to expect.

One night in particular defined my first deployment. An emergency request to refuel a C-130 came through. The aircraft was lower than our normal flight altitude range, so we had to operate in minimum radio and lighting conditions. Before this, I had yet to refuel a C-130, and I remember noticing how close the propellers were to the refueling receptacle. I was terrified, but I remembered my training. I took each minute as it came, and we were able to successfully refuel the aircraft. Some nights I would refuel upwards of fourteen aircraft during a six-hour mission. At times, I would have one fighter jet connected to the boom receptacle receiving gas, two on the wings waiting to be next, and several more spaced behind us in tow. Over my two deployments, I refueled more than 200 aircraft and offloaded 3.8 million pounds of fuel.

On my second deployment, I did not want to come back home. I wanted to focus on flying, getting in shape, reading, and completing college classes. I was confident in my abilities as a boom operator, and I was deployed with friends that made the experience not just tolerable but enjoyable. After that deployment, I was approaching the end of my four-year enlistment contract and had to decide my plans.

While I was getting more comfortable in the job, someone in a higher leadership role relentlessly tried to tarnish my reputation and progression. It began in initial training. He was the head instructor and acted inappropriately with one of the other female students. Their relationship was something other students and instructors noticed and were concerned about. When I was questioned about it, I told the truth.

That man actively spent the next four years trying to sway my performance evaluators into giving me negative feedback and reviews that directly affected my boards and promotions.

My first supervisor had to gather information from him about my performance during the time I spent at training school. The things he wrote about me were not reflected in any of my previous training reports. In many other instances, different senior leadership stepped in and fought for me and my work performance. They made sure to put me in a different squadron away from him. Then, towards the end of my last deployment, he was senior boom operator. I felt trapped with him as my direct supervisor, but I wanted to focus on the things I could control.

Up to that point, I believed fairness and the truth would always prevail. But military culture can be a hard to break through.

I have come to learn that women connect with each other through shared stories and experiences. I grew up with four sisters and we shared everything with one another. In the military, there were few women around me, and the men never spoke about how difficult their experiences were. There was that feeling again: I am alone, and I must be the only one struggling.

I made the decision to leave active duty and attend college in person. This was something I dreamt about when completing assignments by nightlight in my tent during my deployments. I was so young when I joined, and I longed to find myself outside of my military identity.

Creative Outlets

When I first transitioned from active duty, I was welcomed back into civilian life during an economic recession. Despite all I had accomplished, I felt like I was behind my high school friends who were graduating college and entering the workforce. Thankfully, the GI Bill helped cover the cost of my tuition and some of my living expenses.

I became a waitress during college because the hours were flexible. The service industry was staggeringly different from military life. A few years prior, people had gone out of their way to thank me for my service, and now restaurant guests treated me like their personal servant.

I chose to major in art, more specifically fashion merchandising. It was a passion of mine, and I hoped to pursue a career that allowed me to enjoy going to work every day. I was accepted into the Art Institute of California in Hollywood. I had previously attended their online Pittsburg campus while on active duty. It was inspiring being around other creative students, and it was the change of scenery I needed. Although I excelled in my courses, paying the cost of private school without loans became a challenge. Under the GI Bill, only partial tuition was covered. When I decided to continue my education at a junior college to decrease tuition costs, I learned the Art Institute was nationally accredited and not regionally accredited. This meant I would have to complete all the courses I had already completed in design school. Once again, I felt left behind. I longed to join the workforce to start earning income for a more secure future. I spent two years at a junior college before transferring to California State University, Los Angeles, where I went on to graduate with a bachelor's degree.

During college, I lent my time to various internships, but it seemed that my military skills did not translate well into the creative world. I was older than most of my peers, and yet I felt more experienced than other entry-level job applicants. At the time I wished that I had selected a more transferrable specialty code in the military, like X-ray technician.

I spent time at a graphic design company assisting an event planner and working with a celebrity stylist. While working for the event planner, I started a small creative design business called Lyon By Design. As time went on, the more I wanted to be a person that people admired and serve a purpose higher than myself again.

C-17 and Self-Advocacy

I wanted to return to the military as a reservist after graduating college with my degree. However, I was told by my recruiter that unless I had an engineering or medical degree, the Air Force would not be interested in having me as an officer. He suggested I stay enlisted and continue my aviation career as a C-17 aircraft loadmaster. With my previous experience loading cargo on the KC-10, this cargo transport aircraft would be a good fit. The C-17 is the second largest aircraft in the Air Force's inventory and can provide aeromedical support, airdrop, and carry a variety of cargo, vehicles, helicopters, and personnel. My job was to safely load and secure the equipment in the cargo compartment, ensure weight and balance principles were within limits, and provide safety to all passengers onboard. Looking back, I know now that the recruiter impeded my goals of serving as an officer and I should have sought out another opinion.

When I was a child, I learned that it's up to me to advocate for my future. Now I find it helpful to focus on the things I can do in the present to help contribute to my ultimate goal. When opportunities come, I do not want the reason I am not selected for them to be because I failed to advocate for myself. If someone is standing in my way, I will seek another route to help me achieve my goals. If I am dissatisfied with a situation or process that I am passionate about, I will make a case for why it should change. This mindset is one I pass onto others so they feel empowered to change career fields, pursue promotions, and make their goals a reality.

The time I spent as a C-17 loadmaster allowed me to achieve the rank of technical sergeant, become an instructor on the aircraft, and gain experience as a supervisor. One of the most rewarding experiences during this time was teaching a Load Planning and Equipment Preparation course. I built relationships instructing various departments of the government that are tasked to rapidly deploy or respond to situations, including the Department of Homeland Security, the FBI, Border Patrol, FEMA, and every other branch of the military.

A Love of Teaching

Instructing the course meant imparting my years of airlift knowledge and experience to others. They, in turn, shared current trends and experiences they were facing on the airfields. This exchange made it possible for us to tailor our course material so that it was more applicable to the users. The role of instructor helped me grow as a person and develop my love for teaching and helping others.

I had a student who struggled with the course. He stayed later than the other students and still failed the progress checks. I was dedicated to see him succeed, so I wrote out each step that helped me when I first started teaching the course, and I passed the notes on to him. He practiced at home, and when he came back the next morning, he was a different person. He repeatedly thanked me for helping him and he was so proud that it finally "clicked." He said he was going to take my notes on his upcoming deployment. That made my heart feel so full. Nothing makes me feel better than when I help others succeed or feel more empowered to do their job.

When I speak at STEM programs or Career Days, I try to shed light on the military reserves. In high school, I had only heard about active duty and thought that was my only option when joining the military. The reserves allow members to serve during a weekend and then go back to some normalcy, which I think is so important, especially for young women.

Maintaining a sense of who you are and having a support system outside of the military proved vital for me. Reservists also have many options—they can go to college and serve or enter a career field within the reserves that is the same as their civilian role, or take on a completely different career on the civilian side. There are wonderful benefits to being in the military, but you can have all of those while also pursuing a full life on the civilian side. I like to share my stories as an enlisted aircrew member so women can see that there are many aviation career fields to choose from aside from being a pilot. I also want them to have a connection and resource for honest advice if they choose to pursue a career like mine.

Mentorship

During my time as a loadmaster, I met one of my mentors, Jengi Martinez, who went on to play a pivotal role in my life while flying on the C-17. Jengi graduated with Cohort II from USC's Master of Business for Veterans (MBV) program.

Jengi thought I would benefit from the program; she thought I could learn about expanding my design business with their entrepreneurship focus. I was the first person in my family to complete college, and I had never thought about pursuing a graduate degree. At the time, I was doing well in the reserves and was proud to have earned a bachelor's degree. The more I thought about the program, the more I knew it would be a mistake not to pursue it. It would be challenging and time-consuming, but investing in myself and my future was well worth it.

The MBV program changed my life in every good way possible. The program reminded me that we never stop learning, which I had forgotten about somewhere along the way.

I learned a lot about business and personal finances. Growing up, my father was the sole provider for our large family, and we often went without. Money was always discussed in a negative and stressful way. Having now learned about accounting and finance, I was able to educate myself, pay off all my debt, and start working toward building wealth. I started to have a positive relationship with money instead of being afraid of it. Later, during Officer Training School (OTS), I volunteered to speak about finances and share what I had learned. I was the only woman who spoke on the panel. Afterward, exchanging information with other women, I had the opportunity to review their finances and provide them with actionable advice to help them achieve a more stable financial future. Additionally, I now have a daughter to whom I will teach smart money management.

The ropes course in the MBV program made me aware of my preferred independent tendencies. Some obstacles forced me to lean on others in order to succeed. I learned it was okay to not know how to do everything. We all have our individual talents, and some things come easier to others, and when all of those talents converge, you have a strong team. At the ropes course, I realized that others around me wanted me to succeed. They cheered me on, and we all became closer from supporting each other in our weakness and vulnerability. I realized my ego was keeping me safe but small.

When building a team, even though I prefer to put on a professional exterior at all times, I now make a point to do activities where we can connect with one another on a more personal level. It's vital to the health of the team that we get to know one another outside of the work we do. We become more human to each other.

Risks of Entrepreneurship

One of the main areas of the MBV program that interested me was entrepreneurship. As someone who loves research and collecting data, I enjoyed the customer discovery projects. One of the key takeaways from my research of small business owners was that running a small business was challenging. It had always been a dream of mine to open a small business, so I found this advice important. If I could find a job that allowed me the time to create and do things that inspired me, it did not also have to be a full-time job.

During the course of the program, we worked together and then individually on creating a business. I put a lot of groundwork into creating an apparel line for female veterans, and with my background in design and photography, I felt like it was the perfect business for the course. I could design, photograph, and market the line with little to no overhead.

I designed the logo and ten T-shirts and had them printed and styled. Then I photographed the shoot, created an Instagram account for the brand, and set up my online shop. I was extremely nervous to go live with the brand, but once I did, I got nothing but positive feedback from my fellow female veterans and friends.

After navigating marketing in 2017, the political and social media climate contributed to my decision to close the online shop. I remain very proud of myself for taking an idea from conception to creation.

I learned a lot about myself, including new strengths and weaknesses I had. I also learned that it is more than okay to try something and change your mind. It is important to self-reflect, take note of patterns, and change course when you need to. There is a lesson in every experience.

Negotiations

In the MBV program, my classmates came from all stages of military journeys. Some were still on active duty, others were reservists, some were looking to transition out, and others had been separated for some time. It was amazing to watch them explore other careers, while some even made significant changes within their current roles. We were all transparent with one another, and I think this contributed to us capitalizing on marketing ourselves and having honest conversations, including many about salary negotiations.

This, combined with hearing my peers discuss their salaries, allowed me to know where I should expect my salary to be based on my skill set.

The military failed to equip me with the tools I needed to negotiate in the civilian workforce. While the military does a great job of making salaries transparent—and, to their credit, salaries between men and women of the same rank are equal—we do not negotiate our pay. Since entering my civilian role, I have noticed that many of my veteran peers have a difficult time advocating a fair salary for themselves, even after being in a job for a while. They wait for someone to recognize their contributions and are not vocal about their accomplishments—a "silent professional," as a coworker liked to say. While it may be beneficial in the military to wait for your superiors to say when you are ready for more, these "silent professionals" tend to fall behind their peers in compensation on the civilian side.

A helpful tool I gained from the military was the annual performance reporting process. It taught me to keep track of the actions, results, and accomplishments of my team. It let me convey clearly to management the positive impact my team made on the business at the end of each year. I have passed this documenting process onto others so that they, too, can advocate for themselves for promotions and advanced roles.

Full Circle

When I first started at Boeing, I had the opportunity to become the focal point for the newest US Air Force KC-46 tanker program. I lent my experience as an in-flight refueler. My skills were finally translating to something real and beneficial in the civilian world.

I entered the civil sector with more than ten years of operational aviation experience. I held a master's degree in business with proficiencies in emotional intelligence, operations, analytics, and strategy. I felt confident with leveraging my military leadership skills. For the first time since entering the civilian world, I knew my worth. So I was ecstatic to work on a program for the Air Force while advancing in the civilian world. I had earned the confidence to speak about the aircraft and how Boeing's product would directly impact the customer, because I had been the customer.

I went on to become a trainer for new employees at Boeing. I updated the course materials and created step-by-step instructions that assist employees as they take on additional or advanced roles within our department. I implemented after-training surveys to give trainers feedback about trainee experience and areas where we can improve the training program. This has been one of the most rewarding aspects of my job.

Shortly after, I became a team lead. I was honored that management selected me, and I felt ready for the new responsibility. This leadership role has forced me to reflect on and understand my personal journey from following to leading—which is still a work in progress.

In the Air Force, my aircrew training allowed no room for error, and we trained by exhibiting proficiency in one category before being allowed to move on to another. There was a final evaluation to ensure we were capable of doing the job solo. If at any time we showed a lack of proficiency, we would become unqualified to do the job. I have carried this standard of training and expectation into every job since. It served me well. My managers knew they could trust and depend on me because I put integrity first in every situation. My work ethic spoke for itself. It was the only thing I could fully control.

I allowed very little of my personal life to cross over into my work life. I believed that if I let it, it meant I was unable to do my job well, and that I would gain a reputation for being problematic and difficult. Being imperfect was unacceptable. I wanted to contribute fully and not give anyone a reason to find issues with my performance. This type of black-and-white thinking about proficiency was no longer serving me because as a lead, I was now responsible for other people's work as well as my own.

Having grown up in a home where I did not get to have an opinion or speak up for myself, I find having to deal with disrespect to be extra trying. My parents were only proud of me when I was doing well. So, if I feel disrespected, I instantly feel like I am doing something wrong. My body has a physical reaction in those situations, likely calling back to tactics I used to survive my childhood. Mentally, I can also lose confidence in what I am hoping to achieve with my team. I assume I need to change something about my delivery or something else about myself to make their performance better—I must be failing them or disappointing them in some way. I like to think that I create an environment where others can feel free to bring anything to the table, but if there is a lack of respect, it makes me question if I am doing something that shows them they can speak disrespectful. In some cases, after we have had a discussion and decided the best way forward, those who can be sensitive to coaching will think they can defy the plan or flat-out tell me "No."

In those moments I have to be more assertive, which is not in my nature. It's difficult for me to imagine being disrespectful to those who rank above me. After all, being respectful was a tactic that kept me safe growing up. The leadership role at Boeing has pushed me to focus on finding my voice with confidence to be an effective leader that others admire.

Officer Training School

I put my Deserving Airman package together to compete for a slot in Officer Training School (OTS). What I learned in the MBV program was instrumental during my board panel and Deserving Airman selection. I articulated why I wanted to transition to an officer. One of my biggest goals in becoming an officer was to be a leader who could advocate for others. I wanted to help more airmen become the best versions of themselves.

When I left for OTS, I was very fortunate to be employed with a company that supported my commissioning. When I came back to Boeing a couple months later, I was greeted with a "Congratulations" banner, and my leadership offered refresher courses if needed. Having an employer supportive of veterans, reserves, and guardsman has made all the difference in the world. I do not have to worry if my professional progression will suffer in order to complete my duty.

Family Strategy

As a woman with career goals, I had to be strategic about planning for a family. The enlisted-to-officer process took just shy of three years for me. It was a long three years, full of setbacks and disappointments, and after each one I had to wait longer to have a child. I could not imagine being away from my child to go to training for a couple of months. The OTS age deadline was thirty-five and I attended at thirty-four. I also knew that if I had a child at thirty-five years old, it would be considered a high-risk pregnancy with the likelihood of more health issues for the baby and myself.

In 2021, during the pandemic, I gave birth to my beautiful daughter. When I returned to work, I felt like a different employee than when I had left. I could no longer just focus on work without worrying about my tiny infant away at daycare.

Running on little to no sleep, I was utterly exhausted. While some of my managers thanked me for getting back on the buck, it was not until another female manager shared what she remembered about juggling work and children that I finally felt seen and understood. Her words let me know I was not alone. She mentioned that she and her husband did not get through a full five-day workweek during the first years of their children's lives. It comforted me to know that this was just a phase and that it was okay to be a working mother, and more importantly, how vital it is to be a visible working mother in the workforce.

It is now a mission of mine to advocate for other working parents and to support men using all their parental leave so they can be there for their children and partners. I want to create a culture where family is important, one that recognizes that we are stronger when we support our personal relationships. That way, we can come to work and be our best selves.

The transition to motherhood has also forced me to let others in. "It takes a village" is completely true. I worked up until the day I had my daughter because I felt guilty missing eighteen weeks of work to be with her. I wish I had prioritized my mental and physical health before the biggest change in my life happened.

My friendships with women and mothers have helped me tremendously with sharing and talking about what we go through. The female manager, who tossed me a life jacket when I was sinking, has continued to be a mentor to me. She has been a sounding board and has given me a different perspective. I feel extremely fortunate to have her in my corner and guide me. Although this chapter of my life is new for me, I know it is not unique. I can be a better friend to my other mom friends and those who will come after me by letting them know that they, too, are not alone.

One of my favorite things about the company I work for is its stance on global diversity and inclusion. We have local resource groups across all our sites. After being at the company for a while, I feel comfortable with my competency in my job, and I am eager to lend time to making a difference for others. I am an active member of Boeing's Women Inspiring Leadership group, where we discuss topics that impact women. I find the time we spend together helps us collectively develop and share opportunities and gain tools to thrive in our personal and professional lives. We can share resources, navigate challenges, network, and build a safe, meaningful, valuable connection.

I served on the Boeing Veteran Engagement Team (BVET) as the secretary of the Southern California chapter. More recently, I was selected as the Treasurer for Boeing's BVET Enterprise Board. This role will allow me to continue to advocate, support, and honor veterans companywide. We create opportunities to cross-collaborate with other veterans around the globe; grow ourselves as business leaders; and serve our fellow veterans with the resources they need. Every two years we have a veteran summit, and I am excited to be a part of it next year. I am proud to be part of a company that chooses to foster an inclusive culture.

The Balancing Act

My husband and I juggle civilian careers in lead roles as well as hold senior leadership ranks in the military. Continuing reserve duty on the same weekend as my husband has become increasingly difficult now that we have a child. A small but mighty support system has volunteered to help us. They are friends-turned-family, and without them, we could not do what we do. It's almost as if they are serving alongside us.

I have found that as you take on more in the military, sometimes your civilian career takes a back seat, especially during periods when training or upgrading is required, and vice versa in a civilian career. As you take on more at work, you have less time to dedicate to the military, especially to doing more than the obligatory one weekend per month, two weeks per year. And, as you move up in rank, you will likely be doing more than the minimum to prepare your team for the upcoming drill weekend. My husband and I have to take turns with who can push ahead in their career—we cannot both go full steam ahead in both our military and civilian careers.

Being a dual military family lends another set of challenges when it comes to childcare. Being away from our daughter for one weekend a month adds to my anxiety when our daughter is already in daycare five days a week. It makes it hard to push through the next six years to receive a full retirement.

Recently, I decided to go into inactive reserve, which gives me the leeway to take some time and evaluate if I want to completely separate from the military. This time around, the decision-making criteria are different from when I was younger. My priorities have shifted, and I know I will never regret spending precious time with my daughter. I want to create a healthy work-life balance for my family. For me to be present, something has to give. Similar to my experience with moving on from active duty, I want to see who I am and what life is like without having to consider my military obligation first.

My daughter will not likely remember me in uniform, but the experience I gained is something I can share with her. I hope that she will be proud of me. Being a woman in the military is not easy, but our presence is necessary and shapes the military for the better. You can lead at any level, and that is something I have learned and done from my time as an enlisted member.

I often say everything I have today I owe to the military. But also getting an undergraduate degree made me eligible for the MBV program, which led to a USC connection that helped my resume get noticed in the Boeing hiring system. Being a veteran gave me the ability to afford a home with the Veterans Affairs loan. None of those things would have been possible if I had not discovered the drive within myself and acted.

Trauma Patterns

Shortly after getting a civilian job, all the issues I had run from kept rearing their heads in patterns I could no longer ignore. I was so anxious and stressed all the time that I decided it was time to work on myself. I spent every week for two-and-a-half years in therapy working on those areas. I read more books than I can count relating to women serving, narcissistic parents, and parental abandonment. I was trying to make sense of what happened to me so I could have the tools to react and respond differently going forward. By doing this work I helped not only myself, but my relationship with my partner and my daughter. I am beyond grateful that the things I learned in my life may help her navigate her life. I hope she can look at situations as lessons and move on from them, versus dwelling on why they happened in the first place. I hope she learns not to spend thirty years being angry.

When I think of all the things I went through and endured, I feel brave and strong. There is a fire within me, and that is why I am still here. I was meant to go through all of the things I went through in my life to break generational patterns of trauma that afflicted my family, so it ends with me.

We need to take up space and share our voices. It's a basic need. As a child whose many needs went unmet, I have to force myself to take up space. I made myself small, obedient, and easy. It was the only way I was accepted. I continued this cycle with all of my personal and professional relationships. Because my parent's love was conditional, I tried to be the perfect daughter. I was not allowed to show emotions and learned to hide my feelings. I gave away the attachment to myself and my ability to tune into my intuition for my survival. This belief system was reinforced when I joined the military, a place that honors expertise, self-sacrifice, and discipline above all else.

I want to leave behind this belief that I have to be perfect and quiet. I want to stand strong in who I am as an individual and not feel that I will lose acceptance or love for disagreeing with someone or speaking my truth. Particularly, in times of conflict, my flight response is unconsciously triggered. I feel vulnerable and anxious, and I consciously must redirect to respond differently. I listen to my body more and do breathing exercises often. Trying to undo years of habits is difficult and is something I work on every day. I know the way I was raised and trained is not the only way to live life. I get to write my story.

A couple years ago, my best friend passed away tragically and suddenly. It shattered my ideas about the world, about life, and the meaning of just about everything. I went through a very dark time trying to navigate what life looked like without her. The year she passed happened to coincide with milestones that she and I had anticipated attaining together. I got engaged, married, and became pregnant. The sting of going through the first holidays without her that year was difficult and unbearable at times. After her death, I now believe we are here to learn lessons, to grow, and connect. I have been on a journey to discover what I value the most, what gives me the deepest sense of fulfillment, and how I can use my passions and talents to serve others.

Not serving the full twenty years in order to receive full retirement and benefits feels like I am quitting. I should be putting on captain pins, but instead I will leave as a first lieutenant. I called my female military friends who still serve, and others who have retired, to hear their thoughts on my decision. I am grateful for that network of women, many of whom I met through the MBV program.

Sometimes it only feels worthwhile if I complete the full twenty years of service. The reality is any amount of time you serve is enough.

"A Strong Veteran Community"

Amber Culotta reflects on telling her story

I have been ready to write down my story for some time now. I mostly read self-help books, and, as I read, I jot down statements that resonate most with me. As I reflect, I have wanted to put it all together so that I can help others by sharing the best ways forward that I have found, to use our circumstances as insight and self-awareness to grow and develop.

When I started writing my story, I wanted to present the best version of myself. The version where I have it all figured out and learned life lessons fairly quickly. However, through our feedback sessions, I understood why all the messy details were important too. Sharing all parts of my experience helped me recognize themes and patterns in my life and what accountability I am taking to change them. Recounting some of these details was painful at times and putting it on paper meant others might know things about me which I wanted to keep hidden. Sitting with those emotions helped, and writing through them in order to share them, and share the place I am in now, was very therapeutic.

What I want people to know the most is that our stories are powerful. We can learn from them or stay living with them. We do not have to let what has happened to us in our lives define who we are forever. You can change the course of your life and continue rewriting the story. You can pivot and start new at any time. I do believe that starts with healing. Telling your story and sharing your experiences helps others.

We are not in this alone, we are a strong veteran community. I hope everyone knows we want one another to be the best versions of ourselves. To not only be "successful" but to be healed from any trauma we may have brought with us to the military and gained from our time serving. For those whose experiences may be holding them back from healing and finding inner peace, hearing about others' successes and what helped them navigate life can help. It helps us all move closer to being happier, whole, and healthier.

Chapter 4

I Wanted to Help
by
Andrew Vandertoorn

Lieutenant Colonel Andrew Vandertoorn served in the US Air Force Reserve for twenty years. He is a command C-17A evaluator pilot at March Air Reserve Base in California.

Vandertoorn conducted military operations across the Middle East, particularly in Iraq and Afghanistan, specializing in tactical aeromedical evacuation. He helps shape squadron personnel and operations as chief pilot and assistant director of operations while helping enable safe military operations. He flies for a major US airline and works as a flight instructor.

Vandertoorn's academic credentials include a BS in aviation management from Arizona State University, where he graduated summa cum laude. He earned a Master of Business for Veterans (MBV) degree from the University of Southern California.

Service, the Family Business

Growing up, I was the youngest of three. My sister was the oldest with my brother in the middle. It was primarily the three of us and my mom. Unbeknownst to us at the time, we would become quite the service-oriented family: my mom recently retired after forty years of nursing; my brother is a captain in the fire department; and my sister is a supervisor in law enforcement.

My father is from the Netherlands, and he worked in the Dutch navy before immigrating to the United States. He was a radio officer on commercial and military vessels. Over the years everyone in my family has worked really hard, supported each other, and stayed focused, which allowed us to excel in our public service jobs.

Lately it seems to be less attractive to go into public service. Let's be honest—the private sector pays way more than a basic military salary. Everyone has other personal reasons for joining the military at this point. As the youngest in my family, I had the luxury of learning what a career of service would cost you and what you would gain from it. My family's career paths couldn't have been more different, but the end results were the same. They have accomplished some really cool stuff, and I'm thankful I got to watch that growing up.

I saw how important it was to give back, to help, and to make a difference. Once I found my passion of flying, I immediately wanted to use it to help others. I attribute that to my family. They inspired me to achieve the best I could while helping others, and they demonstrated the value of service to the community. Those invaluable lessons built the foundation of who I became later in life.

I Wanted to Go

The military had always been my ultimate goal. As the youngest child, I watched my family give back, work on their goals, and support their teams. Naturally, I wanted to follow suit. Once I got the flying bug I never looked back. I researched the best schools, programs, and instructors. I did everything to be the best candidate I could be.

Going into the military post-9/11 meant something very different than before. In peacetime, you focus on all the training, opportunities, and locations you might experience. After 9/11, you would definitely be involved in combat operations. Some people got scared, but I didn't. I actually wanted to go, so I worked hard and proved it. I finished all of my civilian training at the highest instructor levels. I used that credibility to pledge and apply for the military. It was an extremely competitive environment, but I finally got selected. That was one of the greatest days of my life.

One thing I still struggle with is celebrating myself. I think it's common among military servicemen and women. When we achieve these milestones, we should step back and seize the moment and realize that what we accomplished is not ordinary, it's extraordinary.

Today, I have flown hundreds of sorties for aeromedical evacuation, tactical operations, and humanitarian assistance. And I still do it. People ask me why I still do it, and it's always an easy answer: I want to give back and I like it!

The road to get here has been littered with challenges. The training was hard and continues to be hard. Staying positive isn't always easy, but I tell myself I won't give up and I'm going to work harder than everyone else. When I didn't have a good day, I went back to study, worked with my classmates, and studied more.

Good Change

When you see all the people who have attended USC's MBV program and shared their experiences, like those collected in the first volume of *Transitions in Leadership*, it really shows you that you are not alone. As veterans, whether we have shared a common experience, a life event, a failure or a success, we have something in common. Regardless of our backgrounds, upbringings, or current conditions, we all have something we can learn from each other and apply to our own lives. Particularly when my peers spoke about personal challenges and what they did to overcome them, I took note. They inspired me to keep trying in my own way.

Personally, I have never been a fan of change. I often get lulled into a routine of normalcy, and change takes extra effort and energy. Sometimes it seems like if you change too much you lose direction, but if you don't change at all you become stuck and unable to adapt. It's human to stick with what you know and what's comfortable.

I would spend a great deal of time, effort, and energy to do things right the first time. I thought that if something was done to the highest quality the first time, I wouldn't need to drastically change anything but occasionally tweak my approach. Stories from my classmates proved change can be a necessary good thing. My background instilled in me the habits of always doing my due diligence, trusting but verifying, and making a good plan. I thought doing all of that would avoid potential mistakes or threats, even in my personal life. But rarely can we plan or foresee everything. One thing I learned from USC is to stay resilient, keep a constant stream of effort, and pivot if necessary.

On the one hand you can't be complacent, and on the other you can't settle or accept stagnation.

One of the reasons I went to grad school was to build new relationships. I have been included in other veterans' life events and memories. It feels good to be included and continue to build on those relationships. It was USC that provided the foundation for the natural creation of those relationships, which is a pillar of the program. It's a great feeling to give back in turn and write a recommendation for another veteran who wants to get into the program. It's an even better feeling when I see their enthusiasm and enjoyment while going through the program.

The primary way I have changed in decision making is reducing the significance or weight of making decisions. I used to put a great deal of pressure on a decision and worry about the outcome ahead of the implementation. I have reversed that thinking, and now I like to see what happens. When I took the stress away of trying to predict the outcome, I focused more on refining the decision itself, thus making a better decision.

Unprecedented Times

The last five years have proven to be anything but consistent. COVID, the Afghanistan withdrawal, and political discourse have really shaped the present. You think you're in control of your life, then you realize you really aren't. From 2018-2020 I was firing on all cylinders and had a good rhythm of military and civilian work, fun and family. I got married and had several great USC grads at my wedding. I got accepted into a leadership program in Phoenix, where I met a variety of new people.

Then in 2020, I volunteered for ninety days of orders to support my squadron. What was advertised as a nice quiet time proved to be anything but. I found myself hauling the quick reaction force to various locations after the United States killed an Iranian general, creating insurgencies in Iraq. That was an exciting month-long whirlwind. Then we started hearing about the COVID-19 virus, and 195 Americans traveling from China were quarantined at my base. Even at that point, I figured the situation would be contained.

Within a few weeks, I was flying a C-17 from lockdown city to lockdown city. It was the craziest thing I have ever experienced. The entire process was changed because we couldn't all be together to work, plan, and fly. Communication was a daily challenge. We had daily email updates and Zoom calls. Luckily, we had great leadership. The world was shutting down everywhere we went, right before our eyes.

As COVID progressed, the airlines also shut down. I figured my job was in jeopardy as I flew empty planes across the country and stayed in locked-down hotels. After several months, it became apparent that I had to make a choice. Several full-time positions with my squadron had opened up. Meanwhile the airlines were furloughing employees. I asked myself where I could help the most.

It was a difficult decision to leave my civilian job in a time of such uncertainty, but it was also an easy decision because it felt like the right thing to do. The deciding factor was this: I wanted to help. I was the chief pilot of my squadron, and as we had hired twelve people, I wanted to be there to help them. I did everything USC taught me. I was flexible, I examined every opportunity, I pivoted, and most importantly I didn't overanalyze a decision I knew was right.

When I look back, I'm glad I made the choice. We are always transitioning, and I took a different path than what I expected. I pivoted from the easy, consistent, and convenient route into a far more complex adventure. The journey hasn't been easy, but it has been rewarding to watch the squadron grow and see the impact we have. We have faced unprecedented times in terms of budgetary concerns, manning, and missions. It often feels like we solve one problem and ten more pop up, then more and more....

Afghanistan

The hasty withdrawal from Afghanistan came unexpectedly. The main surge we participated in finished in two weeks. That pales in comparison to the amount of time we put in to rebuild the country. My entire career in the US Air Force was Iraq and Afghanistan. I spent thousands of hours bringing in supplies and personnel and taking out wounded soldiers on emergency aeromedical evacuations. I wanted to see a successful resolution and a permanent change in that country's future. To see all those people left there who had helped us, and all the supplies abandoned, hit me like a gut punch.

Whether they flew a jet or assisted back home, everyone worked together but no one liked what the US was doing. I was most upset about the destruction of the schools. The US was a big proponent of the educational mission, setting the foundations for learning, and teaching skills. I look back at my time in school while young and at USC and I think how fortunate I am to be here and how powerful education really is. Most of the schools in Afghanistan didn't survive. We were speechless. It took time to process.

For more than twenty years, our squadron supported the schooling effort. People made so many sacrifices and gave it everything they had. Then poof, overnight it was all over. The decision to withdraw was so painful because through the deployments, ops tempos, grueling missions, and shifts in strategy, most valuable was the time we had spent away from home. You can't get that time back. We tried to make the country better at our own detriment. Afghanistan and the withdrawal will be studied for decades. For the veterans who were in it, especially for a lengthy amount of time, it was a huge loss.

It was impossible to pivot that fast without a hardcore reality check. We experienced the full spectrum of emotions, especially sadness and confusion. Some people retired over it. It was really tough to watch something we had worked so hard on, and during the best parts of our lives, fall before our eyes and get quashed, all while we were being told that China was the new enemy.

The Military Machine

I see the military and the Department of Defense (DOD) as a massive ship with a tiny rudder. No matter what, that ship is always going forward. As service members we are the smallest piece that helps steer the ship. We can move the rudder a little left or right. But what is equally assured is pressure from the opposite direction that resists a course correction.

I watch younger officers try to make improvements and change the ship's direction only to be told, "This is how we always do it," and the ship goes back on course. It takes an enormous number of resources and effort to make a noticeable change to course, and often you will hear that the ends don't justify the means. This causes an immense amount of frustration and a huge loss of morale.

Given the complexities of the issues, it's impossible expecting to solve them and alter the course of the ship. I found the best answer is to accept when it's a tough problem for which I don't have all the answers or solutions. Do the best you can to guide the crew, mission, squadron, or servicemember. Most people, including myself, respect that.

Lessons in Management

The transition back from flying civilian to flying full-time military wasn't the easiest. The primary reason I went back was the great team we had. They were one of the best in decades. If there was ever a time to be a part of a great team, to do good things, and help the squadron, now was the time. We were on a roll when the operational constraints of COVID made it very hard to continue.

The team had nine months working together before our key player got promoted. It was a blow to the team, but it was well-deserved for him. His replacement didn't come close to his capability and leadership, so it was a huge shift, and a hard one to adjust to. We pressed on and did the best we could. However, it takes a much greater amount of time, resources, and sweat equity to achieve the results you're used to when you go from a high-performing leader to a low-performing leader.

Our great bosses got replaced with bad bosses. Rather than trust us and not ruin a good thing, extreme micromanagement ensued. If there is anything that can kill morale, it's aggressive micromanagement. After a few weeks, everything we worked hard to create was vaporized. People were very upset. But the boss is the boss, and orders are orders. They lacked faith and trust in what we built, in our process, and in us. In the weeks that followed, I tried to use lessons in negotiation, strategy, and leadership from the MBV program to navigate the way forward. Sadly, it didn't work.

The ones who suffered the most were the youngest troops. They became completely frustrated with the events. I empowered them to speak up. There's a bias against us when the old guys speak up. They think we're negative or cranky. So it was important for leadership to hear from the younger troops. It empowered them to take ownership in the organization. But it also helped validate our concerns.

The breaking point came when an aircrew got in trouble on a trip. The boss was super upset. Even though the investigation wasn't completed, it didn't matter, the crew were deemed negligent and the bosses wanted them punished. The boss wanted everyone to come to a briefing to get yelled at. The fallout from this was distrust of senior leadership, a huge reduction in participation, and an erosion of faith in the squadron. Ironically, the rush to judgment and punishment was all for naught, because the crew was cleared when the investigation finished. But the damage was done.

Time has passed since that incident, but the scars remain. As mid-level leadership, my primary goal was rebuilding trust. When I went back to the military full-time, I wanted to do something that would make everyone happy. I have the mindset to leave something better than I found it. Hard to do in that environment. But two things came to mind, one involving a water leak in a roof, and the second being new jackets for the squadron.

Our building was built in 1980, and over the last ten years a small leak had developed above the briefing table. That table was where everyone said hello, briefed missions, studied, and said goodbye. For decades, missions had started and ended over that table. Leaders and commanders had come and gone.

Maintenance slips were filed and lost. Small repairs were made, but nothing worked. We had to put trashcans under the leaks. The carpet got wet, mold and stains spread everywhere, and the air got stale. I did everything to get someone to listen: phone calls, mold reports, emails, surveys. Nobody budged, and the leak got bigger and bigger. Eventually the room became unusable because of the water and mold damage. Naturally everyone was sad, because that was our briefing room. Commanders would say it was the number-one priority on the repair list, and a week later it would be number twenty-one because someone would restack the priorities.

You might think that fixing a hole in the roof and buying some jackets is easy, who cares? Anyone who knows government knows it's not easy at all.

The issue was about helping my people when no one else would. It might have been an easy fix for a civilian company that responded to employees, especially with health and safety involved. My takeaway was that lower- and middle-level leadership can have a much broader and deeper focus on caring for their people, because higher-level leadership is focused on two things—money and data analytics. They want to know how much they are spending, what the productivity is, and what's their return on investment. That's known as putting a price tag on your people. You have to prove capability, but you also have to know when and how to ask for help. If you ask for help, you had better explain why, otherwise it won't go well. A middle-level manager like me knows better at both an acute and chronic scale where the need is, how to fix it, and what will increase productivity as well as return of investment.

But we may only get fifty percent of what we need. That fifty percent comes with a directive to get a hundred percent output despite fewer resources. I'm often left to explain why we didn't get fully funded. I emphasize the win is that we got some money, and I will keep working on getting the rest, but in the meantime, I still need a hundred percent. You can only motivate hundred-percent results with fifty percent resources for so long before a disruption of morale. Most of the time people will perform because they have a desire to serve and they are answering a call higher than them, but you still need to make your people your first priority and show them that you care.

I realized trying to keep everyone happy was impossible. Family, friends, relatives, my spouse, in-laws, bosses, military and civilian responsibilities—it was impossible to keep them all balanced. Someone was always upset. I didn't know what to do. It was a hard lesson for me to learn. I put a lot of pressure on myself to make everyone happy. Eventually I ran out of gas for myself and wondered how I could fill up my tank.

A Small Piece

When I graduated from USC's MBV program, I was full of energy and had a great desire to change things for the better. I was armed with tools to help me start new projects, facilitate growth, and develop programs. I approached my civilian employer and asked where I could help, and the answer was quite underwhelming and disappointing. They weren't interested in the slightest. On the military side it was better because my squadron actually needed the help, and they were grateful for the assistance.

I enjoy teaching so I went to work training the young pilots. It was rewarding to see them get excited about doing something new or mastering a skill, because I contributed a small piece in helping them get to that point. I continued to be the chief pilot, which allowed me to lead the hiring team and essentially shape the future of the squadron. Selecting a new pilot, getting them through training, and giving them their wings is probably the most rewarding thing I do. It's incredibly rewarding to give someone the opportunity to do something amazing, change their life, reach their goal, and make an impact not only in the squadron but in the military as a whole.

It's roughly a two-year process to select and train a pilot. There's a great deal of ups and downs. When it's all over, and they have smiles on their faces, it's worth all the hard work. This is one of the reasons why it was more fulfilling to go back to work for the military full-time. These small things had a big impact on me and helped fuel my desire to stay.

While the military is more rewarding for the opportunities to teach, mentor, and be involved in the hiring process, it comes with some significant negatives. One of the biggest drawbacks to going back full-time has been the number of duties that has piled up on me. The military is undermanned and underfunded in many areas, which creates a train of never-ending tasks. Making it worse is that many of them are unnecessary and inefficient and don't really provide a measurable return on the investment of time and effort.

The worst part has been the leadership drain. Many of the best mentors have retired. They got to the point where they chose family and quality of life over rank. They were successful outside the military, so they didn't need to rely on the military for a job or for their fulfillment. Now the people at the higher levels are disconnected from the troops, out of touch with reality, and unable to achieve anything greater than position-based leadership. There's no foundation or empowerment for transformational leadership, where a positive change can be instilled and create a catalyst for long-term improvements.

This is incredibly frustrating for me, especially after USC, because of what we learned from case studies, projects, and leadership courses. It's a leader's job to listen, take input, and make changes. But, for example, we provide feedback on why everyone is leaving, what we need to change, how to keep recruiting the best talent, and our leaders don't listen. This happens in strong job markets and when the military isn't looked at favorably. Ten years ago our leaders were better and had more resources at their disposal, which is incredibly disheartening considering today's military budget.

Sadly, the biggest frustration since graduating USC is realizing I can't fix, instill, or empower change at a larger level for the betterment of the organization. I can do so in small areas, but overall I can't improve the major problems. Also the amount of busy work I receive takes me away from the teaching, mentoring, and hiring aspects that keep me fulfilled. So, I have struggled to find a balance and the optimum position to make a difference. One thing I have always heard since first going into the military is "do more with less." Unfortunately, we have done more with less for so long that there's nothing left, and we're at the bare-bones levels. The new saying is "get to yes." We get an "ask" and in turn we explain what it would take. They reply with a response of half what we asked for. They say, "Get to yes, give me what you got." I want to give people the best opportunity and tools to excel and often I can't.

Victories

We have put on multiple airshows, participated in exercises and missions, and completed inspections where the squadron has really shined. The old saying that it takes a hundred attaboys to erase one bad apple applies here. Our primary objective at this point is to make the squadron stand out. We are focused on rebuilding morale with squadron celebrations, and we have increased recognition of achievements, and renewed focus on the history and legacy of our squadron, and all the good we have accomplished over the hundred years our base has been around.

Additionally, we refocused our efforts on mentorship from both top-down and bottom-up. We emphasize the importance of individual development, both professionally and personally, to let people learn and grow to their true potential. We do this to re-energize the senior group, the mid-level group, and the young group. All three groups are pivotal to the squadron's success and are the interconnected fibers of the squadron's heart. Whether giving back, solving one little problem at a time, or standing up for our guys, I'm trying to keep our heart beating with everything I got.

A Future of Service

I recently completed twenty years of military service. I could retire if I wanted, but I'm not going to. I like my squadron, our missions, and the camaraderie. While the military is extremely important to me and a large part of who and what I am, I have begun to look outside the military and my normal circles to give back.

In 2018, I was invited to attend a leadership institute at my alma matter in Arizona. It was the first class of that program. I enjoyed the new opportunity to give back to the community. Since then I continue to volunteer to help the program grow through a variety of causes.

One in particular is for a fellow veteran named Pat Tillman. He was an Arizona State University Sun Devil who turned down an NFL career with the Arizona Cardinals to join the Army. He was subsequently killed in a friendly fire incident. Every year there's a marathon to raise money for Pat Tillman Scholars, students who get an opportunity to attend ASU. I have helped plan the race for four years now.

I have also hosted several mentorship days for pilots on both military and civilian career paths. I put myself in their shoes and remember what it was like to do something no one in my family had done before. Several of my own mentors are still my friends today, and I wouldn't be where I am without them. I remember how much they helped me and how important it was to my future, so I pay it forward.

Hindsight is twenty-twenty. Over the years I have been faced with incredible decisions. Would I have changed anything, knowing now what I didn't know then? No. When someone was hurt in Iraq or Afghanistan, crews like mine were airborne within hours to get them. When someone was within an inch of death, we made it happen no matter the barriers. I was part of their journey to safety and recovery. We couldn't let down our wounded, and we couldn't let the bad guys win. Maybe it was an ethos, a motto, camaraderie, or "no man left behind," but I saw it everyday during the war. I was helping, but also, I was using my skills, and everything I had been training for since I was sixteen years old, to actually make a difference.

Have those experiences affected my decision making, life choices, and career decisions since then? Absolutely. After flying those missions for so long, when you really put it in perspective you value life and helping people. We might not be flying the wounded out of combat zones these days, but that doesn't mean my fellow service members don't need help, resources, direction, mentorship, a friend, or just a hello in the hallway.

One of the most important things people need is to be together. There is no greater connection to experience than to look back with others and say, "We were part of something that helped people." One of the greatest events I organized was a large squadron get-together. We had to fly to a base to do training, but I organized time for everyone to get together and reconnect. It was just epic. They were long days, but the happiness and bonding healed some old scars. The good feelings were tangible in the air. To me that's what happiness and leadership are about: making people better.

The past few years have been challenging, but I have achieved more than I ever thought possible. I got promoted to lieutenant colonel, finished more than 5,000 flight hours in the C-17, and I became an evaluator pilot. I work with a great team, I have twenty years of military service, and I got married and bought a house. If you told me when I was sixteen taking my first flight that I would have all this, I would have laughed.

In about a year I will return to my civilian job. My primary goal is still to help and make things better and build a great team. I have hired more than fifty pilots, and I hope to hire even more in the time I have left. Hopefully, the new hires see the same love for the squadron that we do and will carry it forward to the next generation.

While it's impossible to keep everyone happy and fix all the problems, I did find that it was possible to build a team that could safely accomplish a mission, have fun while doing it, and learn and grow together at the same time.

When other people make big changes in their careers, I ask them if they are going to miss it. Everyone answers differently, of course. When I ask myself the same question, the answer is hell yes! Whether it was a good scenario or a bad one, at any moment I was doing what I loved.

I have many great memories from USC, including the classes, the veterans' dinners, the projects, and the bonds I made with friends. One memory stands out. I was part of a crew of USC graduates that flew over the Coliseum during a football game. It was such a great moment that showcased what a bunch of USC vets can do. It was a very unique, humbling and rewarding thing to be able to give back to the school, and it was quite an honor to be part of.

Flyovers take a huge amount of work involving coordination, planning and, most importantly, very accurate timing. We stress about routing, air traffic control, other airplanes, and focus on the five seconds when we are center stage in the show. Then we move on.

As I think about that moment, and the transitions in the past and the ones upcoming, I realize that time passes constantly. It's always moving, and it can be your friend or your enemy. Except for how it makes my hair fall out, I look at time as my friend. It has allowed me to build my career, and it has taught me invaluable lessons.

"Don't Normalize the Incredible"

Andrew Vandertoorn reflects on telling his story

The best part of going on this storytelling journey was reconnecting with MBV alumni, sharing our stories, and learning more about each other. Genuine relationships have always been important to me, and this gave me the opportunity to keep those alive. I started researching the MBV program ten years ago, and I realize now how many ways USC has woven red and gold in the daily fabric of my life. The most impactful part of this mission is the reminder that we've been through so much. We have to stick together, and this experience has been a catalyst for that to continue. Plus, the project is an inspiration. It shows a talented group of people paying it forward and giving back in the hopes of helping and encouraging future alumni.

The biggest realization I had during the process was that it's impossible to solve all the problems or challenges you face. The most important thing you can do is put people first and keep on trying the best you can. Additionally, everyone's stories inspired me to look again at my own goals and re-prioritize. I'm making a big effort to separate my goals from my employer's, and to make mine a higher priority.

Don't normalize the incredible! Service members downplay the impact they have. We normalize what we do because it becomes the daily norm, despite the fact our unique operations are an incredible feat, full of achievements that demonstrate our capabilities. To squadron-mates, our day-to-day might be boring, but to our families, friends, or to the public it is amazing. This journey helped me take a step back and look at what we do and what we have accomplished and say, "We're incredible!" It's been a humbling experience. I learned a great deal and I'm armed with more confidence, humility, dedication, and perseverance because of it.

Chapter 5

In Pursuit of Eudaemonia
by
Antonio Randolph

Major Antonio Randolph is a second-year resident of the US Army's Command General and Staff College at Fort Leavenworth, Kansas, in the School of Advanced Military Studies.

He received a BA from the University of Michigan-Dearborn and a Master of Business for Veterans (MBV) degree from the University of Southern California. He is a graduate of the US Army's Strategic Broadening Seminar at the University of Louisville; and of the Field Artillery and Logistics Support Operations Officer (SPO) courses at Fort Sill, Oklahoma, and Fort Gregg-Adams, Virginia, respectively.

He has served as the battalion operations officer-in-charge for the 1-182 Field Artillery HIMARS battalion of the Michigan Army National Guard (MIARNG); commander of the Joint Force Headquarters, Headquarters and Headquarters Detachment in Michigan; and the 631st Troop Command sustainment and property book officer.

As a civilian, he is the delivery leader for the CF34-10A commercial jet engine platform at General Electric-Aerospace.

Competence and Capability

Since the first volume of *Transitions in Leadership*, I have continued to apply force to mass to create acceleration in four areas. In no particular order, they are my career, my family, my God, and my interests.

With each stimulus, my predictive powers have improved, but I have yet to accurately forecast all outcomes. All in all, my intent is to get better, do better, and be a better version of me, each and every day. In this regard, I am winning.

After graduating from the MBV program in 2017, I went back to the grind on parallel paths of professional commitment to the military and my career with General Electric-Aerospace. Initially, the challenges were thick, especially interpersonally. I intuitively developed a strategy and an effective concept that I referred to as the *"operational t"* as a result of my three-and-a-half years of running research and development test sites in rural Peebles, Ohio.

My hypothesis was that you can still manage an organization if you are misaligned with your peer group, your supervisor, or those who report to you, but only if the misalignment exists in no more than one of these three dimensions.

If one of the relationship links becomes a point of contention, or even failure, you can develop a path to success in spite of it by strengthening the other two interpersonal connections. Any more than one failure point leaves you vulnerable (but it should also cause some serious self-reflection and course correction). Of course, good performance is still a prerequisite and will always be the foundation of one's relationship with any company. Without it, the *"operational t"* is a moot point. Also, like most hypotheses with anecdotal data points, it is subject to unforeseen failures if not adapted based upon changing context.

As the GE-Aerospace jet engine development and test site leader, I oversaw the first and second engine test for the GE9X commercial engine program. This is the most powerful turbofan commercial jet engine in the history of humankind. In 2019, the Guinness Book of World Records presented the team with an acknowledgement for achieving the thrust record of more than 134,500 pounds.

From test operations, which I thoroughly enjoyed, I was able to move on to a non-operational role as an individual contributor with a small staff. Material fulfillment broadened my scope of the business and gave me touch points with some of the most esteemed leaders within our company's supply chain. Protocols and etiquette presented me with a learning curve, but performance remained the top concern. Influence replaced authority as the tool of choice. Suppliers respond better to those who can leverage interpersonal tact.

My professional climb has recently culminated as I interface with customers directly on the commercial side of the business. The jump has bumped my head up against the executive level. Beyond performance, the persistent challenge for me has been displacing bias by convincing people that it is in their best interest to support me for the sake of their operational and professional success. I acknowledge this is a challenge for the common professional, but I also understand that the norms, beliefs, and values of some individuals are latent with misconceptions and irrationalities that affect their judgment towards inclusivity.

The military, specifically the National Guard of Michigan, has pushed me in a new direction. My good fortune has coincided with some sound leaders who have afforded me the chance to learn, the leeway to take risks, and the guidance to make me a better soldier. With the exception of one antagonist, my experiences have been exceptional. I picked up a second knowledge branch, logistics, to add more perspective to my decision making, and I took assignments that exposed me to the civilian-military arena and movers-and-shakers like Kentucky senator Mitch McConnell.

I had a successful second command at Michigan's Joint Forces Headquarters as a major, and I recently completed the requirements for my current rank at the US Army's Command and General Staff College at Fort Leavenworth. This college has a rich history, as virtually every general since Ulysses S. Grant has either been the commandant or a graduate of the school. My thesis allowed me to do research related to my primary branch, field artillery, and I gained an additional master's degree. I was blessed to be selected to stay another year at the highly selective School of Advanced Military Studies (SAMS) where I have another thesis currently in the works.

In 2023, I received the President's Writing Award at Fort-Gregg Adams (formerly Fort Lee), Virginia, and my article, "Evolving Readiness: Train to Support Future Sustainment Operations," was published in the Spring 2023 edition of the US Army's Sustainment Bulletin.

I do not take lightly the privilege of gaining competence and capability towards actualization. I have always had every intention of becoming better in order to share of myself and make others and the organizations that I serve in better.

Family and Faith

In my family, I have found solace and peace of mind. I have experienced the benefit of unconditional love. Good outcomes derive from good applications, more often than not.

I am an optimist posing as a pessimist who is actually a realist. I strive to live cooperatively on the brighter side of life as I reach towards actualization, which means sustaining a life in full bloom for as long as humanly possible. By chance, I have stood in the presence of greatness.

By happenstance I was fortunately born with two fully functioning hands. I have tried to keep them clean and out of trouble. I speak on behalf of them. The right hand has a particular distinction. It engages in formalities, including its official responsibility of conducting greetings. In adulation and respect, I have shaken the hands of many renowned men and women. Crossing paths with such interesting people has been fascinating.

I was nineteen years old when I shook the hand of Coretta Scott King in Atlanta, Georgia. She was delivering a speech on the campus of Morehouse College that day. As I reached into the aisle to touch her hand, I noticed her impeccable posture, the elegance that accompanied her every motion, and her grace. Then she delivered an inspiring speech to a group of extremely impressionable teenagers.

Also in Atlanta, on a separate occasion, I found myself in a meeting hall standing with Maynard Jackson, Andrew Young, and Julian Bond. I shook their hands in sincere awe. I had seen the documentaries highlighting their participation in the civil rights movement and the Second Reconstruction. I have since welcomed abominable threats to my livelihood and person by referencing the determination and fearlessness of their leadership in the movements.

I spotted Lieutenant Colonel James Harvey III at the Detroit airport. I noticed his Tuskegee Airmen hat. As a fellow military officer, it was an honor to shake the hand of the trailblazer and legend. At the time, I wasn't a huge proponent of selfies, but I did make the request.

I treasure the picture. He recently turned 100 years old. I kept his picture in full view at my desk throughout my attendance at the Army's Command and General Staff College.

For fifty-two years, my father, James Randolph Sr., worked for Chrysler Automotive. He was a hard worker, an incredible provider, and the love of my mom's life. His hands were huge, and I wondered one day if I would ever have a grip so strong or be a father as good as he was.

My brother, James Randolph Jr., is a facsimile of my dad, but with an extra seven inches of height. He is also a hard worker with an incredible story that includes playing college basketball at Southeastern Louisiana University. He has had a very successful career as an engineer working on jet airframes, naval ships, space shuttles, and rockets. Like our dad, my brother has been a consistent team player who routinely leads from the front.

My father instilled in me the value of hard work by his example. He advised me to make my own way through a combination of being present and reliable. His guidance of where to apply these actions was critical as well. "Make your way in the spaces where the decision-makers reside," he said. From my mother, I understood that a dogged persistence, applied to the "art of not giving a care" would be of significant utility, and that continuous learning, on the same loop with continuous improvement, would round out my armor for tackling the challenges and problems of the world.

God has always been my first mover, the anchor and basis of my behavior and decisions. With free will, I have the option of choice. With compassion I have the responsibility to consider the considerations of others. His endorsement of understanding and growth written down in the Bible aligns with my adage to create win-win scenarios within any association I choose to make, from friendship to employment. It is the right thing to do. I pray throughout the day, and I have discovered creative ways to give tribute back to the world, which includes connecting good people with their needs and wants in the form of jobs and providing financial assistance to those in dire need.

I would be remiss if I didn't highlight my belief that my ancestors paved the way for me through prayer, and that allies, with good hearts and just souls, have helped me realize the promises of the Lord.

My knees and joints have managed to keep me upright and functional as my hair turns white. In other ways, I have aged like a prized car headed to a Barrett-Jackson showcase: my intellectual motor has revved up. In terms of pursuing personal interests, I have been blessed to have the chance to amass an increasing book collection. Used bookstores and online sellers have allowed me to expand on the cheap, and clutter has piled up in my basement with stacks of boxes of books. It has become a family affair, too, as my wife and kids have often accompanied me with their own literacy needs in mind. I can only smile.

Learning improves awareness. As connections surface between ideas and facts, opportunities for critical and creative thinking become apparent. Eventually, analysis and synthesis lead to publications and acknowledgments. The intellectual boost after graduating from the MBV program was fueled by Dr. Robert Turrill's parting words of wisdom (to paraphrase): "If the goal is to learn, the path is never-ending, as it should be." Unique value equals legacy. My interests in education, and its creative uses, are helping me build a legacy to share.

"A Better Me"

Antonio Randolph reflects on telling his story

Epiphanies rarely come while talking. One must listen to the mind. This writing journey allowed me the chance to stop and reflect during a very busy time in my life. Although the majority of the words I logged never made it past the draft cuts, the comfort of transitioning my thoughts to a document proved to be cathartic. Aligning the ideas with the value-adding premise of this book is important and necessary. Some of us have the "gift for gab" on paper and can effectively communicate what our comrades would have said. In any case, the message needs to get out, and the stories and their themes seep through the cracks in the minds of those in greatest need—those on the verge of despair, those on the verge of success, those dangling in the middle. "We are there with you" is the message. And, in some respects, we have already seen the mountaintop.

Veterans can move briskly onto the next phase of their lives without understanding that they are not alone and that success on the civilian side is achievable. Our unique value is typically not highlighted by higher institutions, and our post-service development is even lesser of a concern.

In the chaos of coursework at the US Army's Command and General Staff College, time was my biggest obstacle. The demands of my academic experience, and my course through the harder matriculation path to a master's degree, made additional responsibilities a significant challenge.

The shadow side of critique is paralysis. Dostoyevsky's protagonist in *Notes from Underground* finds himself on a self-proclaimed higher plane of critical thinking while extolling lesser men of action. I would argue that there must be a balance, a median where the theoretical is manifested through the practical into the actual. Intellect without the force of will produces nothingness. So, when my thoughts have sailed into the ether of romance, my proclivity now is to yank them back into action. At times, I sat stagnate with no impulse to create, so I just started typing; and it has made all the difference.

Never in my life have I felt freer. Slowly but surely, my passions emerge as more active components in my life. I have made some hard pivots to get to this moment. Along the way, I have discovered a better me, a bundle of extraordinary opportunities, and some exceptional people.

Chapter 6

Betting on Myself
by
Benjermen Reyes

Benjermen Reyes served for ten years in the US Marine Corps and attained the rank of gunnery sergeant. After joining as a supply administration specialist, he served as Marine security at US embassies in Jerusalem, Rome, and Havana. In 2011, he deployed to Afghanistan to train a combat logistics battalion of the Afghan National Army.

Benjermen has a BA in business administration and economics from American Military University, and a Master of Business for Veterans (MBV) degree from the University of Southern California. He has a real estate broker's license, a contractor's license, and a career technical education teaching credential.

Benjermen works as a broker associate with Open Door Real Estate in Bakersfield, California, and as the vice president of Westpac Construction. He coaches youth wrestling and sits on the board of the Kern County Wrestling Association, which promotes youth wrestling in Bakersfield.

New Reality

After graduating from the MBV program, many of us are in different chapters of our lives. For me, I would say that I stand at a new beginning. After serving in the military for ten years, it feels like I am living two different lives—and like the past ten years really didn't mean much.

Education was always a goal and a major influence on me for getting out of the military. I wanted to use education to bridge the gap of what seemed like "lost" time. I knew my experiences, leadership, planning, and everything else I learned wasn't actually lost, but finding a way to show others and help them see its value has been hard.

My current journey began with being accepted to USC's MBV program. At the time, I was working for the city of Bakersfield. I had recently passed my real estate license test. I notified my immediate supervisor that I would like to sell real estate and that I had been accepted to USC. The following week, I was called for a meeting with the finance director and treasurer to discuss my job. They weren't too fond of the fact that an employee would be selling real estate and going to school on top of the job I did for the city.

They let me know that I could pursue my education, but I would not be allowed to sell real estate. I felt like my leaders, my supervisors—after everything I had done for them—didn't trust me to perform my duties while simultaneously pursuing education and selling real estate. I felt betrayed, and I wondered if that was my new reality. It meant nothing that I could do my work for the city in my sleep. The very next day I gave a two-weeks notice. I chose to bet on myself. They let me go that same day.

I asked myself what I really wanted. What drives me? What is my *why*? I missed helping others, I missed putting out fires, and I missed being a leader.

I jumped headfirst into real estate and decided to learn as much as I could to better serve people as they made one of the biggest purchases of their lives. I began the MBV program not long after. I felt a strong bond with my cohort, and I started to see the many opportunities that could be captured by pairing our skill sets with civilian education.

When school finished, everyone set off to accomplish their goals. We were armed with knowledge and connections with like-minded individuals who were all engaged in their own pursuits, their own battles. For me, I wanted to bridge the gap of business knowledge for the younger generations in low-income areas in Bakersfield. To accomplish that goal, I needed more schooling in order to learn more and meet people who could eventually help me teach future students.

Shortly after graduation, another MBV grad and I started a general contracting business. But reading about business and doing business are two different things. We not only had to learn the best practices, laws, and regulations, but we had to put what we learned to work. It helps having someone go through the journey with you. We improved many homes in Bakersfield by installing solar panels, and currently we have a subcontract for services with California State University, Fullerton.

Each job, each business interaction and decision, has grown my partner and me as business leaders and decision makers. We often reflect on the guest lectures we attended at USC and on the case studies we read in the program. Combined with our military experiences, we have a guiding light for our business. Whether building vertically or horizontally, including decisions about how subcontracting will be a handled, and the potential for different outcomes, we work as a team.

One Choice

During our last week in the MBV program, we were asked to write a personal value on a rock and to speak on that value, which was meant to seal a chapter of our lives. On my rock I wrote *family*. With the many people in my life who have passed, and the realization that life is not guaranteed, I wanted to spend more time with family as part of achieving my goals.

This means I have choices to make. With each choice something has to give. One choice will change another. Right now I choose to spend time with family and friends and find a career that will let me do that while having an impact on the lives of others in a positive way. To do this I need to transition to teaching, as the schedule works better, so I can spend time with my daughter and wife.

Judge me by what I leave behind. I began coaching my daughter's wrestling team, and I am now more focused on becoming a teacher, specifically to teach business. I have enough information and experience to teach one of the things I love. I want to be the coach for young people that Jesse Toledo was for me—the teacher that uplifts and connects students with the subject. I want to make my city a better place, and I believe the school district, wrestling room, and football field are great places to start. I want to help in an area many people complain about, the educational system.

In the first volume of *Transitions in Leadership*, I gave credit to three men who were like father figures to me. Since then, both Raymond Gallegos and Jesse Toledo have passed away. Charlie Rodriguez is battling a severe case of Alzheimer's disease.

My younger brother Daniel Reyes also passed. And there are other passings that have affected me: Erica Cordova, a childhood friend my age; and Phillip Campas, a high school friend and fellow Marine. Realizing how final death is, I constantly look at where I am, what I am doing, and what impact I am making.

"Permission to Be Proud"

Benjermen Reyes reflects on telling his story

As I look back, this project provided an opportunity to complete an After Actions Report on my time in the MBV program. I finally made time for myself to acknowledge the good and the bad, and to help create a better way forward for myself.

Despite going through some personal trials which I could have easily used as an excuse not to write, I decided to write. My story is not extraordinary. It's one of many stories veterans go through post-military. It is always easy to reflect when everything is great, but reflection during challenging times creates the most growth and change for the better. As I write this reflection, I am in a far better place, mainly because in the past year during the creation of this book, I refocused on the why that originally started me on this journey.

I have two major takeaways from this experience. The first is giving myself permission to be proud of what I have accomplished. I have gone so far and come so far that the previous version of myself would not recognize me or even think I was possible. I have to pause and consider all I have accomplished and give myself credit. A lot of people, especially veterans, don't give themselves enough credit. We have such a mission-oriented mindset that always strives for perfection and preparation for the next mission. There is no time for a pat on the back. I had a hard time writing what I overcame because I did not believe that it was anything to be proud of.

The second biggest takeaway is that we are not alone. As a veteran, a USC alumnus, especially an MBV graduate, there are so many people willing to help each other. Most people rely on themselves to get through tough times, but sometimes that is not enough. There is help for you, and you deserve to be helped. Whether it's help with a career, personal issues, or just insight, help is there. Being the change I wanted to see in the world is how I started this journey, and it's how I will continue.

Chapter 7

Starting at the Top
by
Bridgette Austin

Bridgette Austin is a combat veteran and accomplished entrepreneur. Her service included six combat tours to Israel, Kosovo, Kuwait, Saudi Arabia, the United Arab Emirates, and Afghanistan. She received military decorations that include the Military Outstanding Volunteer Service Medal, the Global War on Terrorism Service Medal, the Joint Commendation Medal, and the Combat Action Ribbon.

Upon promotion to chief petty officer, Bridgette shifted from aviation mechanic to career counselor. She received a BA in administrative management from Excelsior University; a certification in diversity and inclusion from Cornell University; a Master of Business for Veterans (MBV) degree from the University of Southern California; and a proposal and grant writing certification from San Diego State University.

She is currently the CEO of B. Austin Consulting, which collaborates with the non-profit industry to ensure today's foster youth are adequately prepared for a successful transition to independent living.

School Days

I grew up in Southern California far from my birthplace of Gary, Indiana. From the age of two to the moment I raised my hand for the oath of enlistment, and to this day, I have been grateful to my mother for her foresight and willingness to move me and my sister to a place where we could freely express ourselves in ways that were seen, in other places, as unusual. Southern California allowed us to live freely and be whoever we wanted to become.

We moved around quite a bit, starting in Compton, where we lived with my uncle until Mom got settled. Then we moved to Long Beach where I went to preschool and kindergarten. From there we moved to Fullerton where I attended elementary school and my first year of junior high. The next move was to La Habra where I finished junior high and most of high school. In the middle of my senior year, we moved back to our old family home in Compton, where I commuted back and forth to school until graduation.

Growing up in predominantly white schools had its advantages and disadvantages. I put in a lot of hard work to ensure I was able to hold my own among my white classmates. Even when I was expelled from sixth grade for my rebellious behavior and bumped up to junior high ahead of schedule, I held my own.

However, one of the greatest disadvantages of these environments was the lack of cultural experience that could have given me a deeper understanding of the racial and generational traumas that had been unconsciously passed down to me in their multiple masks and denials.

Much of my trauma is linked to having grown up in an incredibly tumultuous and dysfunctional home with an abusive stepfather. While my mother stopped at nothing to ensure my sister and I had everything we wanted and needed, sadly it was at the expense of her own happiness, which resulted in her alcoholism and depression until she found peace and serenity with her higher power.

Where and how I spent my childhood and teenage years played a significant role in how I became the person I am today. Graduating from high school, I had to make a difficult choice between accepting a full scholarship for basketball or a partial scholarship for volleyball, and this was the first of many decisions where I chose my passion over the more financially beneficial choice. I went off to play volleyball for a year at California State University, San Bernardino.

Thirty years later, and after a number of colleges and universities, I finally achieved my bachelor's degree.

"It's Going to Change You"

A friend once expressed to me how "we all have some shit deep down we need to work through," which I vehemently agreed with. He asked if I had ever considered the possibility that many of us didn't "join" the military but instead "we ran away" to the military as a way out of the dark places we called home.

I have always been driven by an insatiable desire to excel in all aspects of my life. While my willingness to misbehave may have diminished over the years, my eagerness to push boundaries and achieve greatness has grown stronger. This drive has manifested itself in various areas of my life, from the classroom to athletics to the workplace. Throughout my career, I have often found myself in situations where I start off in a role that seems beyond my current capabilities. However, with little to no effort, I have managed to create the space I need to learn the necessary skills and excel in any task or job in record time. This ability to quickly adapt and acquire new skills has been a defining aspect of my professional journey. That's what I call starting from the top.

However, at some point I found myself trapped in a seemingly never-ending cycle of mundane and repetitive work environments. I was stuck in a vortex, unable to break free from the monotony that defined my professional life. My perspective shifted dramatically when I heard about my cousin's exciting adventures around the world. While I had been stuck in a mundane routine, my cousin's lifestyle of constant travel and adventure sparked a desire for change within me. It wasn't until then that I considered joining the armed forces.

"It's going to change you," they told me on multiple occasions—they were my friends, my family, and my family friends. I wasn't ready to believe that an institution of any kind could make such an impact on my character. I was certain that at my age I would be able to resist change in any form, in any way, and on any level, especially never altering my unique qualities. My mind was strong enough to resist, my principles would not be persuaded, and I had the maturity to overcome mental challenges.

But I soon came to endure mental challenges and more as a queer woman of color who entered the armed forces during the beginning of "Don't ask, don't tell" (DADT).

DADT was the official policy on the military service of non-heterosexual people, instituted during the Clinton administration. The policy was issued under Department of Defense Directive 1304.26, and was in effect from February 28, 1994, until September 20, 2011. By the time I enlisted, in November, 1993, I was still in the transition phase of my queer experience, so I didn't think the policy would affect me. In other words, I was bisexual, and I figured I could toe the line when necessary or at least appear to. I was wrong, however; the policy did change me. It changed me in ways I am still learning to manage.

Military Sexual Trauma

We established ourselves in the military in a way similar to the first critical months on any new job. Only this time the job was working for Uncle Sam, and there was no easy way out. Unlike the at-will employment of civilian jobs, where an employee is free to leave at any time for any reason, with no adverse legal consequences, the armed forces doesn't make it quite that simple.

Many of us were assigned mentors early on to help us navigate the rigorous eight-week program known as basic training. There were several opportunities to secure support from the more senior and seasoned service members; however, those options weren't quite so readily available to folks that identified as LGBTQIA+.

Since I was older than the average recruit, I was entrusted to lead our class as the recruit master at arms (RMAA), where I was responsible for the configuration and cleanliness of our division spaces. There I learned the importance of positional leadership, as well as the privilege that goes along with it. Upon completion of recruit training, with a brief apprenticeship school, I set off on my first tour of duty to Barbers Point, Hawaii.

My struggle to find community continued amidst the chaos of my deployment to the island installation of Diego Garcia.

That said, I was able to utilize the gap in age and experience between me and the younger service members and provide support to them as an advocate, especially to those who were looking for opportunities to find their true selves. I provided mentorship whenever I could because I was a lost soul myself.

I start my story with what happened at Diego Garcia not only because it was the first of many deployments, but because it was where

I learned about military sexual trauma (MST) and the disturbing psychological damage it caused my very dear friend. She was my shipmate and my sister-in-arms. What I learned on my first deployment has had an everlasting impact on my military career and my life to this day.

While much of our time was spent working the long arduous hours that accompany any deployment schedule, we didn't miss an opportunity to unwind.

After another long day of working on the aircraft, my friend and I decided to spend the evening in togas, and we headed out to what was known as a progressive drinking party, but which was no different than any other party we had attended.

I first met my battle buddy at Barbers Point when she was still a new arrival from Sparta, Tennessee. I mention the location because, up until her enlistment, she was under the impression that a microwave was a television. Sparta was backwoods, tobacco-chewing, spittoon-spitting, family-feuding Tennessee. Coming from that environment, my friend was also under the impression that anyone serving in the armed forces could be trusted to their core.

At the toga party, my battle buddy and I somehow got separated, and by the time we reconnected, it was the following day at morning muster. I was alarmed immediately by the empty expression on her face as she stared off at nowhere in particular.

Upon the close of our morning muster, and after all were accounted for, I walked over to check on her, and to my horrific surprise, she yelped out an eerie screech at the slightest touch.

It was at that moment, without her ever saying a word, that I knew she had been assaulted.

A fellow shipmate and I quickly notified our immediate superior. We were instructed to stay with her until medical support could provide additional comfort and care. We didn't see our friend again until the end of deployment. She was admitted to the inpatient mental health ward at Tripler Army Medical Center for treatment. We never heard from her again. I started to process the pain of her loss in an unhealthy way.

The defensive walls I put up in childhood began to reappear, only now as an adult they were covered in an emotionless shroud. In addition to hiding my sexuality, I began to use the shroud, and the walls, to stop myself from feeling on a deep emotional level. For instance, prior to my enlistment, I was an incredible storyteller. My creative gifts faded as I learned to navigate and understand this new system I had become part of.

Needless to say, trust within our ranks took on an entirely different meaning. While many hours of sexual assault training followed, we remained unclear regarding the fate of our friend, and we were never informed as to whether the attacker(s) were ever caught. Not that it would have been our business, but my thought at the time was that some level of transparency with regard to accountability would have gone a long way.

Not only was "Don't ask, don't tell" an even greater issue than I had initially imagined, but it seemed to extend to incidents of sexual trauma, the facts of which were eradicated expeditiously. That is where I learned the phrase, "Rank has its privileges," and I never forgot it.

Band of Rebels

I joined the Navy as an undesignated airman, so my first task was to find a rating (job) and strike (apply) for it.

Finding a rating and striking meant that you shadowed different sailors as they performed their daily tasks, in order to learn what trade fit you best. Unlike the majority of sailors, who are given a rating and sent to the corresponding school directly out of recruit training, I was unsure about my interests. I joined as an undesignated airman so I could decide what job I wanted at a later time, once I was acclimated. I had spent most of my young adult life in the workforce as an administrator, so filling a similar role on active duty was not appealing whatsoever. My primary interests were in some form of aircraft maintenance.

Here is where trust became essential to my chosen military occupations, first as an aviation structural mechanic, and then as a non-destructive inspector. As a woman, particularly a woman of color, working in a maintenance capacity; then as a non-destructive inspector, in charge of inspecting multiple areas on the aircraft and related support equipment, I endured what most men would not have wanted to endure, particularly at the rank of petty officer. As a non-destructive inspector, I was only one of fewer than ten women in the field, and one of only two African-American women in the entire fleet. Credibility, integrity, and trust in my professional performance, inspection results, and recommendations was a long road ahead, as they say. But once I got there, my reputation remained solid as I climbed through the ranks.

A specific event stands out to me now. A mission-ready jet was due for inspection, and my inspection results rendered the aircraft unable to fly until a necessary component was either repaired or replaced. Ultimately, due to the nature of the component, repair was not an option. The aircraft was unable to complete its required flight hours until the component was ordered, received, installed, and op-checked. Multiple high-ranking officers came to me to discuss my willingness to allow for a quick repair of the component in order for the aircraft to meet the minimum required flight hours. I was unwavering and, of course, unaware of the consequences of standing firm in my integrity.

Then there was conflict around a tan uniform belt. I had earned the tan belt by getting my ass kicked all over the sandbox during Marine martial arts training. Although the tan belt was an even better match for the tan battle dress uniforms, I was instructed not to wear the tan but rather the black belt. Until the aircraft incident related above, the belt wasn't an issue. Rebelliously, I continued to wear the tan belt underneath my blouse with much pride and many feelings of accomplishment, to the point I forgot it was unauthorized.

Wearing the belt definitely didn't help my relationship with leadership. My senior logistics specialist (LS) and right-hand man, also African-American, noticed a change in how members of the squadron communicated and collaborated with us on daily tasks. The environment actually became quite hostile, and when I learned that a group of officers were speaking to other tenant commands in the area about alternate support for non-destructive inspection needs, I knew something nefarious was brewing. What I didn't know at the time was that it included a plan to fire me from my leadership role, bust me down in rank, and fly me out of the area of conflict.

"Leader of a band of rebels" was the title they gave me simply for having integrity and not wavering in my decisions. Even as they continued to chastise me, they couldn't resist referring to me as a leader. That was profound. In spite of being oblivious to the true plot to terminate my role as leading chief, I was intuitive enough to understand that he who holds your personnel file controls your career.

Without hesitation, I directed my senior LS to ship our personnel files to our parent command in Sembach, Germany. That action alone literally saved my career. It wasn't until that experience that I realized, even after all my initial struggles as an aircraft mechanic and inspector, that not only was my gender an issue, but most importantly, so was my race and my unspoken sexual orientation.

Although I was eventually removed from my station in Afghanistan, I was not fully removed from deployment. Instead, I was remanded to Sembach with our parent command. While I wasn't necessarily off the hook for "disrespecting" a senior officer by disobeying an order about the tan belt, neither did I lose rank.

I was required to improve my skills in peacekeeping by attending a few courses in team building and anger management. I was also required to work with wounded service members coming from combat triage units who suffered from various long-term and permanent injuries at Landstuhl Regional Medical Center. I read to them, listened to their stories, and kept them company until their loved ones arrived. I kept a journal about the experience. It was the greatest gift I would have never known to ask for.

That same year, after so much effort was made to remove me from the area of conflict in Afghanistan, I was slated to return to the same combat zone—as the leading chief supporting the same unit. It was the same unit that had previously fired me for doing my job and then lied about why I was fired. Had it not been for my spotless record, and their unmistakably inept effort to mock my speech in a racist manner, I would not be sitting here today sharing my story.

After I fought tooth and nail not to return to that unit, only to be falsely reassured that things would be "fine," I was finally removed from the detachment at the eleventh hour, and only after word of a planned "accidental" friendly fire incident was mentioned to my superior officer in the same breath as my name.

Yes, there was an underhanded scheme to eliminate my "black ass" once and for all while deployed, and to make it look like an accident.

It wasn't until December, 2020, during a Veterans Affairs-sponsored minority-based stress and resiliency workshop, held as a type of check-in with minority veterans, that I finally relived the trauma and allowed myself to feel the absolute fear, hurt, disappointment, and sadness of the attack on me planned by enemies of mine who wore the military uniform of the United States.

The workshop was established after all the racial unrest surrounding the George Floyd murder (and other similar attacks on people of the global majority) in an effort to provide emotional and psychological support. A few weeks later, and during what was to be our last session, the Capitol insurrection unfolded on January 6, 2021. We all agreed there was no point in moving forward and the remainder of our therapy sessions were cancelled.

Upon acknowledging that racial discrimination extended into the US military, and while I considered, on more than one occasion, to part ways with such an oppressive institution, I realized the potential advantage of utilizing my knowledge and experience to become an advocate for young black and brown folks, as well as for members of other communities not yet openly visible among the ranks.

The first inclination came when a young seaman apprentice, an African-American woman, approached me about my hair. I was wearing cornrows, and she quietly asked if anyone had approached me regarding the grooming standards established by the Navy's uniform regulations manual (NAVPERS 15665J Ch. 2).

At the time I was a first-class petty officer, and I led an array of policies and programs, so of course my answer was no, I had not been approached by anyone.

When this young seaman learned that I had not been approached about my cornrows, she broke down in tears and explained that her supervisor, another first-class petty officer, had told her that she was out of regulation by wearing braids, and they had directed her to remove them immediately. As she continued to weep, she explained the pain it had caused, and I could see where her hair was severely damaged and breaking due to the intense and immediate trauma she had done to her scalp to remove the braids.

I sat her down with a copy of the uniform regulations manual, and I showed her exactly where it read that she was, in fact, not out of regulation, and I instructed her to retain a copy of the standards with her at all times.

The emotion I felt from her long embrace of gratitude was all I needed to make the decision to remain in service just a little while longer, which turned into more than twenty-three years.

Life Out Loud

Although DADT had not yet been repealed, it was time to unleash my secret (or rather not-so-secret) lifestyle. The rest of my identity was no mystery, based on my cishet appearance and the color of my skin.

As I began to live my life out loud while still an active military member, prior to the repeal—and in spite—of DADT, I also became weary of living on both sides of the fence. Part of me wanted to remain in service and continue climbing the ranks in order to better support my blended community.

But a larger part of me feared climbing the ladder too high with a shadow creeping behind me, and I began to quietly self-sabotage my career. Initially, I didn't realize I was doing it, from failing to do my best in semiannual physical readiness tests, to not putting my best foot forward during high-level inspections.

Partly it was due to how I was growing weary of the game. I missed my partner, and I resented all the opportunities to support me that she had to miss, whether being by my side during milestones, or simply lying next to me each night after a hard day at the office, just like any other couple.

What was all the more painful was learning that, as I struggled living in the darkness and shadows, there were other queer folks serving and living out loud among the enlisted and officer ranks. You would think this newfound information would be welcomed knowledge! It only worsened the pain of the years I had lost not living out loud. And with this new knowledge, my heart only grew colder, as my partner and I grew farther apart.

I wanted to ask the queer enlisted folks how they were able to get away with it. I even rationalized it by making race a factor, thinking (hoping) that other races received more leniency. But once again I felt diminished when I met a black queer couple just after their marriage following the repeal of DADT. The most painful part was that they had been living as their true selves throughout their entire fourteen-year military careers.

Sea Legs

In the Navy, I became certified as an equal employment opportunity leader. Then, due to the nature of how many of us managed the countless stressors and traumas of service through the use (and abuse) of alcohol, my next step was to become a drug and alcohol program advisor. Both roles proved essential both for myself and in order to support, guide, and mentor the members of my blended military community.

Failure was not an option during that critical time in my career as a newly-promoted chief petty officer aboard my first ship, a nuclear carrier vessel with more than three thousand sailors aboard. I was only one of three counselors accountable for the career development of each servicemember.

I was not only newly promoted, but I was also relatively new to my rating as a Navy counselor, not because I didn't know how to manage the program, but because I didn't gain my knowledge through climbing up the ranks in that role. And as a prior aviation mechanic, I had never been aboard a ship. It was imperative that I not only perform my leadership abilities as a chief but do so while learning to navigate a nuclear carrier vessel and training our career development team of 128 career counselors, all in preparation to support more than two thousand military personnel deployed to the Mediterranean and Arabian Sea during a time of conflict.

Here is where I grew my sea legs and learned the importance of "servant leadership" and humility in the face of the unknown. During that time in my career, I learned that just because you wear the rank, it doesn't necessarily mean you have the knowledge of what it takes to get there. The lessons came frequently and steadily throughout the ten-month cruise. My body took the beating of a lifetime as I climbed narrow ladder wells to inspect 3M maintenance processes, which I performed as damage control locker officer and damage control training team leader during general quarters training drills.

I share all of this because it shaped and prepared me for transition back to civilian life, or so I thought. I thought that pivoting from an aviation career to a career in administration wouldn't be as big of a change as it was. I thought adapting to an entirely new leadership role while providing mentoring and guidance to my team and eventually preparing service members for transition back to civilian life, would have been my golden ticket to self-preservation and preparation. What I learned was that I had forgotten a very critical component of the entire process.

What I had forgotten was that I still had so much more to learn about authenticity, vulnerability, self-awareness, and, most importantly, people-centered connections.

"Starting at the top" became so prevalent while on active duty that I carried over that perspective to civilian life. Of course, I attended all the transition preparation workshops I could, including career fairs, boot camps, internships, and fellowships. All of them perpetuated my certainty that I was more than prepared for my transition. However, transition is about more than being professionally prepared to enter the workforce. It includes managing the process of reintegrating into the civilian community as a whole. This was a concept I hadn't remotely considered until I realized that I had to shed the many masks that I had learned to wear in order to perform in my previous jobs.

Even in the midst of transition, I found a way to deflect my anxiety by projecting my energy onto my fellow transitioning service members, whom I was selected to coach. During the career transition coaching process, I learned about the sizable translation gap between transitioning service members and the civilian workforce, so I decided to establish a small business to work with employers, hiring managers, and other human resources professionals to bridge that gap.

Around that time I was asked to speak to a group of candidates in the MBV program at USC.

It was a welcomed opportunity to speak to my experience as a new small business owner, a recently transitioned veteran, and as the only woman of color on a panel of entrepreneurs. My initial excitement quickly turned to terror as I realized the magnitude of what little experience I could share with such an elite group of veteran scholars. Nonetheless, once I sat before the group, my experiences in leadership, as a facilitator, and in public speaking propelled me through the occasion.

After the panel, members of the MBV cohorts approached me to ask more pointed questions. I was also approached by the program's director to discuss the opportunity to become an MBV candidate. I was beyond thrilled for the opportunity to become a Trojan, and I began mentally preparing for it that very moment. Little did I know how much attending and completing the short, rigorous program would change my trajectory as an entrepreneur, a leader, and a human being.

Among my many concerns about reintegrating to civilian life, I needed to find a new purpose, a new community, and a new identity out of uniform, all while rebuilding a relationship with my partner after more than nine years of being mostly apart.

I didn't throw myself a retirement ceremony. I held a quiet gathering on the beach instead. But I don't regret not standing at a podium pretending that my partner had been there right by my side through all the deployments, advancements, and celebrations of my career.

Another concern I had was finding and connecting with my community. By community, I am referring to the queer community, as well as to the communities of color I rarely found myself surrounded with during my early military career. Growing up in Southern California, I had the freedom to explore things not considered acceptable for young ladies and especially for folks of color. For example, ice blocking in La Mirada Park, skateboarding, lunchtime house parties, motorcycle riding, monster truck events, water skiing (even before I learned how to swim), and similar adolescent activities. These were unheard of when I swapped stories with my black and brown peers.

I wasn't quite ready to take on my new role as a civilian employee. I also wasn't ready to learn how to show up as my true self. And did I even know who my "true self" was? Hell, it took me almost my entire career to show up even a little bit, and I wasn't all that great at it back then either.

One particular incident would forever alter my thoughts on civilian employment. While still on active duty, I worked as an intern through DOD SkillBridge, a program established to connect transitioning service members with industry partners for "real world" career opportunities. Upon completion of the internship, I was hired on the spot. I quickly learned that my skillset far exceeded the job requirements. I submitted my letter of resignation, but the letter was quickly returned to me with a counteroffer for further employment as a consultant working on the same fellowship program. Surprised, I responded with uncertainty, and the HR manager responded by repeating that I had a lot to offer and commented with, "Have you seen your resume?" That practically knocked me out of my chair.

While there are numerous programs available to support transitioning service members, not all civilian employers understand or care to properly place us, even though we offer extensive time served, leadership experience, and transferable skills to their organizations. Once I learned this, I immediately began working with fellow service members to help them properly articulate their skills for better industry placement.

If you have ever been part of any kind of affinity group, you can resonate with my experience of finally not being "the only one" in a room (whether that affinity group is your ethnicity, gender, sexual orientation, or level of military leadership experience).

Environments where you instantly relate to others on multiple levels, like in the MBV program, has been the remedy to the pain of my military transition. I no longer felt alone on my entrepreneurial journey.

My initial focus as a small business owner was to market my services as soon as possible, beginning with fancy business cards, a website, and an emphasis on my military leadership experience. What I learned in the MBV program was the absolute critical importance of customer discovery, authenticity, and always putting the customer first. I also learned the importance of knowing when to pivot, and not to fall in love with my first idea.

These initial lessons literally saved me during the COVID-19 pandemic, resulting in a number of accolades bestowed on my team for the services we provided when literally the entire country was on lockdown.

Among the lessons I learned came that five-letter word—"trust"—and to use it lightly when dealing with like-minded professionals. Not just trust, but trust and verify, has been the magic ingredient for any merger or partnership, and any federal, state, or local contract.

What I also learned during my time in the MBV was how to lead and speak from the heart. That is how I began to remove the masks I had learned to wear, even though the masks had gotten me through the most difficult times of my military career. I will never forget the most profoundly honest feedback I ever received when a professor told me, "You can do better," while allowing me the opportunity to retain my original work. It was as if he knew I was simply completing the assignment. But he also knew there was much more I could share.

Throughout the MBV program, my sister and I tended to the critical healthcare needs of our ailing mother, and it was thanks to my incredibly supportive MBV family (Cohort VI) that I was not only able to support Mom, but I was able to keep up with my studies. There was always a member of the cohort standing by me, including my amazing goals partner, to ensure I didn't fall behind.

Transition of any kind can be challenging, but the physical and mental challenges that we as service members bring with us need healing on entirely different levels. Speaking for myself, I entered the military with scars I didn't know I had, so carrying the additional baggage of combat by enemy, as well as friendly fire, required an amount of support I can honestly say I received as a member of the MBV program.

Although my journey may have started off in a tumultuous, dysfunctional, and abusive environment, the opportunity to navigate so many challenges has brought me to my true self. After more than twenty-five years of military service, I am still so grateful to my mother, may she continue to rest peacefully, for giving me the gift of a free spirit. Although it has taken me a while to fully understand what all that gift would bring, I am grateful to be able to share my gift of knowledge, experience, openness, and authenticity. My journey as a black queer woman continues.

"Out of Uniform"

Bridgette Austin reflects on telling her story

The most impactful elements of this experience came while reading my story back to myself. What truly stood out was sharing my childhood experiences and the dysfunction in my home as I now recognize it. I see how it has impacted me as an adult, and particularly as a service member. This is a journey I never thought I would ever be vulnerable enough to share, yet the experience has been remarkably healing.

In recent years, I have begun to perform a search of self-discovery, and in that search, I have reached out to total strangers, which has taught me that I am not alone in my past struggles. I am hopeful that by sharing my personal narratives, others will resonate with them and understand that they, too, are not alone.

My reflection conveys the challenges I faced as a queer black woman serving in the US Navy during "Don't ask, don't tell" and being constantly forced to hide aspects of myself in order to conform to the expectations of a largely conservative, white, and heteronormative environment. Despite the hardships I faced, I also found moments of strength and resilience within my community of fellow LGBTQIA+ service members and allies. Together, we formed a support network that allowed us to share our experiences and provide each other with the encouragement needed to continue serving proudly, even in the face of adversity.

What I would like to convey to readers, as well as to my fellow veterans, is the importance of finding yourself, finding out who you are out of uniform, and allowing yourself to heal the hidden wounds and remove whatever masks you had to wear to fit the occasion. Allow yourself to take the time needed to find your new purpose and live in it!

What I learned about myself unfolded as an emotional hurricane in my heart and throughout my body. This opportunity to collaborate couldn't have come at a better time in my life, and while sometimes I struggled to put my thoughts on paper, it wasn't due to a lack of things to say but instead finding the capability to express so many thoughts flowing at one time. Since allowing myself to sit still, I have experienced nothing but profound realizations. Now I'm in a place of peace to let them flow freely. The obstacles I encountered throughout this journey were in discovering memories from my military experience and from my childhood that required me to work to fill in gaps in my narrative.

This journey has shaped me in ways I never could have imagined, and I am proud to have overcome the obstacles that were placed in my path throughout my military career. While challenging, it was ultimately the most empowering experience knowing that my resilience and determination would ultimately prevail.

Chapter 8

We Aren't Meant to Go Through Life Alone
by
Darren Denyer

Darren Denyer has served as an active duty and reserve naval officer for nineteen years. A surface warfare officer by trade, he is currently a commander attached to US Indo-Pacific Command in Honolulu, Hawaii, having attained the rank of captain.

Darren's command tours include Maritime Civil Affairs, Beachmaster Unit One, and Maritime Expeditionary Security Squadron One. His military decorations include the Meritorious Service Medal, two Joint Commendation Medals, three Navy Commendation Medals, two Army Commendation Medals, the Navy Achievement Medal, and the Combat Action Ribbon.

Darren has worked as a consultant and operations director, including at Naval Information Warfare Systems Command and Commander, Naval Surface Forces, Pacific Fleet.

His academic credentials include a BA in international affairs from the University of Colorado Boulder, an MA in international security and strategic policy from the Naval War College in Newport, Rhode Island, and a Master of Business for Veterans (MBV) degree from the University of Southern California.

Family Legacy

To understand my journey, you need to understand my foundation. I grew up in an amazing family. I was born in San Diego, California, to parents who were entrepreneurial in business and service. My father,

Paul Denyer, was a pioneer in the field of home infusion care therapy, a field ripe for growth in the 1980s. My mother, Diane Denyer, was a talented and devoted special education teacher and mother to my half-sister and me. In addition to the professional success they achieved, they are also prolific philanthropists. They have extensively supported the United Services Organization (USO) and their alma mater, San Diego State University. My sister Kim is eleven years my senior and a groundbreaking social pioneer in her own right. Her work with marginalized youth in the Bay Area has achieved national notoriety. All three of my immediate family members have been crucial for me becoming the person I am today. I don't think growing up is easy for anyone, no matter how positive or negative the experience may have been. With such a successful family, finding my place and trekking beyond their shadows was difficult. Despite the challenges and the self-imposed pressure to excel, I have had their unyielding support for my choices and direction in life.

If living up to the legacy of my immediate family wasn't hard enough, I knew I wanted to serve my country since I was young, primarily due to the legacy of my grandfathers. Both of my grandfathers were decorated Navy chiefs. I never had the opportunity to know my maternal grandpa Hostetler. He was killed in a plane crash while serving as an air crewman in an S-2 Tracker that went down on San Clemente Island in 1961. Before this, he flew aboard PBY Catalinas in the Aleutian Islands during World War II. He looked like Walt Disney with slicked-back hair and a distinguished pencil mustache. His photo hung proudly in my parents' home my whole life. His death left an unfillable void that stuck with my mother's family and is still felt to this day. I visit his grave at Fort Rosecrans National Cemetery whenever I am in San Diego, as a connection to the man I wish I had had the privilege to know.

On my father's side, Senior Chief Samuel Denyer was my driving influence to pursue what became an amazing career in the US Navy. Grandpa Sam was a navy corpsman attached to various marine divisions during the island-hopping campaign in WWII. He found his way to the Pacific theater by jumping on a transport ship headed west through the Panama Canal. From there he found himself at Guadalcanal, Iwo Jima, New Britain, and on various ship tours aboard small destroyers in the South Pacific. On Guadalcanal, he was shot five times by a Japanese light machine gun. He remained on station, removed the brass from his body, and stayed in the fight. He was also witness to the first atomic bomb test on Bikini Atoll. If that wasn't enough, he remained in service and served in the Korean and Vietnam Wars, only to retire and join the Merchant Marines. We lost Sam shortly after the attacks of September 11, 2001, and I wish every day I could share with him my own journey in the Navy.

The men in my life were bonified heroes, but the women behind them made their service and sacrifice possible. As a child, my grandmother Diane Hostetler was my happy place. Before she married my grandfather, she was a showgirl performing with her seven sisters. By night she performed and by day she worked as a "Rosie the Riveter" building aircraft in WWII.

Life was not easy for these women. They had to provide for themselves and survive while raising three kids mostly alone. Their strength was inspiring. My grandmother spent her life in constant emotional and mental pain following my grandfather's untimely passing. She was a woman full of life and had a positive attitude that always made me feel welcome and wanted. She took me to Navy and Marine Corps bases when I was young, experiences that stuck with me.

I was also blessed to have my grandmother, Betty Denyer, who is a continued presence in my life and still thriving at the age of ninety-eight. She has always been supportive of my choices regarding the military despite her own experience living through decades during which my grandfather was largely absent. Her sharp wit and determination to persist are a testament to how tough she is. Five feet tall, she looks like a sweet lady, but cross her or challenge her at Rummikub, and you will realize your error. I am beyond blessed to have these women as core influences in my life.

A bonus third family also blessed me. My sister has always been close to me, even though we have different biological fathers. Her paternal family, the Aceves, added such a positive and exciting dynamic to my childhood. Growing up with the additional influence of a fantastic Mexican family was pivotal. I was always welcome at their home and included. My abuelo Lauro served in the army in WWII. He and his family were rightfully proud of his service despite the discrimination that came with being Mexican during that time. His sons and nephews served in the military as well. Lauro and all the Aceves men had their military portraits proudly displayed on the wall, and my first US Navy photo was added to the mix as soon as I was commissioned.

My abuela Tilly was masterfully kind and endearing. She made food that would make any chef weep tears of joy, and she was kind and loving. She fought hard to come to America even though the US made it difficult. She and Lauro raised a beautiful family and showed my sister so much love. They always emphasized taking care of family and community. Nobody was a stranger, and anyone in need would get whatever help they could provide. I'm genuinely grateful I had them in my life for as long as I did.

My sister Kimberly is nothing less than a force to be reckoned with. Growing up, she was always there to take care of me, and she has always been a hero of mine. She left for UC Berkeley when I was in elementary school, and she never looked back. It was hard to let her go, but what she became is incredible. My sister has devoted her life to social equity in the Bay Area. Her path was long, but as a staunch advocate for marginalized youth and the LGBTQIA+ communities, she became a trailblazer in her field. While my sister and I operate on opposite ends of the service spectrum, we both know the importance of helping those in need. Her wife and our nephews are some of the kindest and most loving people on this earth. They will give you the shirt off their back.

My sister is the founder and executive director of RYSE in Richmond, CA. Her goal is to lift up, empower, and provide opportunities to the youth of the Bay Area. Her center is based around social justice, where young people can have a voice and be respected. She and her team have been so successful that her model is utilized in every major urban center outside Richmond. She was one the first recipients of funding from the Obama Foundation and the Mackenzie Scott Foundation. She is using that funding to address health and housing inequity. While we are separated by years and life experiences, she remains a cornerstone of who I am and what I value in life. She showed me that everyone can make a difference, and we must do what we can to be a positive force whenever possible.

Life at Sea

I had it in my head that I wanted to be a Cobra pilot or a zoologist. In high school, I was part of the Civil Air Patrol while maintaining my school and sports requirements. I would never be a professional athlete, and I had difficulty seeing myself entering the business world out of college. Being a tad rebellious and feeling as if I knew everything about anything, I made the call to pursue an education as far east as I could stomach. This led me to an international affairs degree, Naval ROTC, and an incredible study abroad experience in Geneva, Switzerland, through the University of Colorado Boulder.

I couldn't shake the desire to serve in the military, living out in the world doing the "cool stuff" I had heard so much about from my grandpa Sam. My eyesight prevented me from being a Cobra pilot, so I received my Navy commission as a surface warfare officer instead, and I set out on the most exciting and unintended version of a naval career I could have ever conceived. I had my choice of first duty assignments and chose to be part of the pre-commissioning crew of the USS *Halsey* as its electrical officer and, eventually, navigator.

With my degree and gold ensign bars, I stepped out into the world and landed in exotic Pascagoula, Mississippi. It was a challenging and exciting first assignment. I quickly learned that I had a lot to learn. We left the shipyard a few weeks before Hurricane Katrina decimated the Gulf Coast in 2005.

Life at sea is a conundrum. It's simultaneously calm and stressful, a ticket to the world, an office always on the move, and thankfully without cubicles. Much to my chagrin, I came to realize that no matter who you are or how hard you work, the surface warfare community will find a way to kill your spirit. While this, in theory, ensures that the most committed members advance, it can be rough. So, after my first deployment on the Navy's newest warship as the ship's navigator, when the opportunity arose to try something different, I took it, volunteering to serve with Joint Special Operations Task Force–Philippines (JSOTF-P).

I could have never imagined how this experience would alter my trajectory in ways I didn't think possible. Working in that environment was life-changing. I was on the periphery of the War on Terror, not in Iraq or Afghanistan, but working on problems outside of public awareness. I influenced operations in an archipelago off Mindanao in the Southern Philippines. I worked closely with all military branches and the Armed Forces of the Philippines. My commander was an accomplished professional who was credited with helping capture Saddam Hussein. All in all, the experience was transformative and pointed me toward the most extraordinary experience of my military career.

A Whole New Ballgame

Before that deployment, I approached the prospective commanding officer about a new capability for the Maritime Civil Affairs, part of the Navy Expeditionary Combat Command. I sought them out and committed to making the idea a reality.

Pre-commissioning a navy ship is a tried-and-true process refined over decades of ship construction and delivery. But pre-commissioning a new capability is a whole new ballgame. It becomes a defining moment for an organization and its people. The best analogy for the experience I can make is the Navy UDTs in the Korean and Vietnam Wars. I had the opportunity to define the capability, how it looked, and how we recruited, trained, and equipped for missions. Maritime Civil Affairs was an incredible capability and staffed with some of the finest sailors I ever worked alongside.

We were not SEALs, but we trained as small units, deploying worldwide in five-person teams. Our primary mission was overt access, influence, and information operations. We trained across the spectrum, with one end resembling the US Agency for International Development (USAID) and, on the other end, conducting tactical training at facilities such as Blackwater. Going out into the world alone and unafraid was the name of the game. From supporting military task forces to combatant commands on diplomatic missions, my team worked in places such as Columbia, Suriname, Guyana, Djibouti, Uganda, and Rwanda in support of Combined Joint Task Force–Horn of Africa, and we supported the response to the Haiti earthquake in 2010.

During the War on Terror, the military branches experimented with new capabilities as they could access what seemed like an unending tap of funds. The funding stream began to slow in 2010. Although we could conduct our mission for pennies on the dollar, the big grey navy took precedence because maintaining ships is incredibly resource-intensive. The decision was made to transition our unit from the active-duty force into the Reserves. To say I had enjoyed the work would be an understatement. I finally had the adventures Grandpa Sam had told me so much about, while making a difference for people and communities in dire need. I knew, more than anything, that I wanted to continue supporting the mission, so I decided to leave active-duty service and transfer to the Reserves.

I stood up the first detachment on North Island Naval Air Station in Coronado. There we trained, manned, and equipped in ways that made us unique from most military organizations of our size and capability. We trained with NATO civil affairs forces in the Netherlands and expanded our tactical training stateside. We were literally on top of the world, and everyone we worked with performed at a higher level than most, and we all genuinely enjoyed the mission. We were there because we wanted to be there, not because we had to be.

Clocking In

Before leaving active duty, I met my wife-to-be, Meaghann. We met right after Christmas in 2009.

I had returned from Africa to a hometown without many of my friends. San Diego is a great place to grow up, but for many people, it's hard to stay and start lives and families. My wife and I joke that it's better we didn't know each other when we were teenagers, as we would have been like oil and water. But as it turned out, we were magnets that couldn't be kept apart.

Right after meeting and hitting it off with her, I was called to deploy to be the civil affairs advisor to Fourth Fleet in response to the earthquake in Haiti. I got the call early in the morning and boarded a plane within hours. It was an incredibly stressful period—responding to such a massive disaster took its toll on time, energy, and stress. Fortunately, Meaghann joined me in Mayport, FL, towards the end of the response, and we hit it off again. It was pretty evident that she wasn't going anywhere at that point. We moved in with each other upon my return and were happily married on the Fourth of July at her parent's homestead in the eastern Sierra Nevada mountains of California.

In terms of my civilian career, I spent a few months rudderless trying to figure out where to go. In September of 2010, I took a job as a defense consultant working for Booz Allen Hamilton. I started as a support to foreign military sales at what was then called the Space and Naval Warfare Systems Command. It was an interesting foray into the civilian sector. I was initially left in a holding pattern for months as I figured out what I was doing. Once firmly into the position, it became evident that the grass was not always greener on the other side.

Transitioning to the world of clock-in and clock-out individualism was tough. This was a period of loss as I continued to figure out who I was and what I was doing. The job was a job, but the purpose and joy I experienced while part of the Civil Affairs missions was simply not a part of my new world.

I felt alone and anxious about what to do and how to do it. I had no real idea about how to find the right job outside of the whole approach of "write up your resume and reach out to companies that take junior officers," which you are told when you leave active duty. This ultimately led me to accept my first job with Booz Allen Hamilton. I knew it wasn't what I wanted to do, but what did I actually want to do? Answering that question took ten more years and become a central mission in my life. If I have learned anything since then, it's to leave your ego at the door. Ego is not the best running mate in times that require deep introspection and hard choices.

Fortunately, I remained heavily involved in the Reserves and my civil affairs detachment. The happiness, camaraderie, and purpose I needed was still there, and that compensated for the cold, corporate reality of my desk job.

In 2013, we received the call to conduct the first detachment-wide mobilization. This involved six months of workups in Dam Neck, Virginia, to prepare our detachment to support special operations units in US Central Command (CENTCOM) and US Africa Command (AFRICOM). We were prepared to deploy, the most prepared we had ever been as an organization. Deployment wouldn't be the case, however.

The US government went through a process of sequestration, which fundamentally altered the wartime funding of the Department of Defense. The Navy decided there would be no more extended funding to support a mission at the fringe of its primary warfighting purpose. We were two weeks away from getting on a plane and deploying. Instead, we were asked to go home, and on top of that, the capability we worked so hard to develop was disbanded. To be clear, we were very good at what we did, but in the grand scheme of things, we were too small to be of significant concern at the strategic and operational levels.

If I didn't know what I was doing when I got off active duty, you can only imagine what was going on in my head at this point. My whole motivation for leaving active duty was gone. Rudderlessness took on a whole new meaning. I returned to my job at Booz Allen, working as support for the Commander's Action Group at Naval Surface Forces, Pacific Fleet, in Coronado, CA. It was a way to pay the bills but didn't scratch any particular itch or spark any joy.

There was a silver lining, however: our first child was on the way. Our son was born, and a new life journey was ahead of us. Balancing all the wants, needs, and requirements of professional life when you have a child to support completely changes your focus. Devoting the time to work on my transition was made even more difficult, and taking risks was even riskier.

In 2015, we found out that a second child, our daughter, was on the way. I was also going to be recalled for mobilization. To have some control over where and how I would be deployed, I took local orders to support Naval Special Warfare Command. It was a significant stopgap, and I got to return to the world of special operations, where my efforts and actions had a clear and tangible impact.

I was asked to mobilize in support of a Navy special mission unit. It was an opportunity I couldn't turn down, but only with my wife's support did I accept the orders. I headed out three weeks after the birth of our daughter. This meant that my wife would be raising two children under the age of two, acting as project manager for our home remodel, and teaching high school full time for at-risk teens. It was quite literally a life-changing experience for both of us.

I got to see and do things that very few people get to experience, let alone know anything about. I saw combat and learned a great deal about myself. I most certainly did not walk away unscathed from the experience, nor did my wife, but I'm proud of the team and what we accomplished.

I returned home to a house I had never seen, a one-year-old I didn't know, and an extremely active toddler. I was a little broken, but I knew now that I was far more capable than I had previously given myself credit for. This period of transition was one of the hardest I had to face.

Both entering and exiting a world-class team of professionals was hard. The organization itself is a masterclass when it comes to supporting all the members associated with the mission. You prove yourself very quickly when you arrive. When you leave, you are on your own. You are no longer part of the organization. You don't have anyone to talk to about what you did and what you experienced, bad and good; you don't have access to the resources and professionals that kept you above water; and you are left with nothing and no direction on how to navigate.

It was comforting to know that my wife held our family together and kept our kids safe and healthy while I was away. She has always been the type of person who can tell when someone is struggling; it's what made her so successful in working with at-risk youth. When I came home, she knew I needed a change professionally. Booz Allen was highly supportive of all my Reserve deployments, but professionally, I wasn't making the progress I needed.

Big Moves

I found the MBV program and got accepted in 2018. I knew the program would be transformative, but I didn't realize how transformative. Despite working in the civilian world as a defense consultant, it was evident that I had never effectively transitioned to my non-active-duty life. I didn't have the context or knowledge to adequately live effectively in both my reserve and civilian lives. It was as if I needed the wider world to conform to my notion of how it should work as opposed to knowing how to effectively navigate the world as it was.

I drove north from San Diego to Los Angeles every two weeks and stayed at a hotel nearby to get through the program. To say I couldn't do this alone would be an understatement. Without the support of my wife, I could not have been able to make this a reality. Yet again I asked her to take on all duties with our home, our children, and her career while I left fortnightly.

The first day of meeting Cohort VII was a great day. It was the first real glimpse into what the program offered. Based on the cast of characters and their motivations for joining the program, I knew this was where I needed to be. I enjoyed the fact that there was a cross section of experience that cut across services, pay grades, and experiences. The energy on the first day was palpable. Everyone wanted to be there to make something of the experience for themselves. I had never been in a program at any level that had that level of enthusiasm. It was like having my team back.

Change is hard until you realize that by embracing it and being open to it, you go from swimming against the current to riding it into new opportunities. I commiserated with my goals partner from the program about the age-old question of what we wanted to be "when we grew up." I'm still working on that aspect to this day. Opportunity, like change, can be challenging to see and harder to embrace, unless you are ready and open to the experience. For the first time in my adult life, I was truly prepared.

Then COVID made it clear that things could be better in San Diego. We knew we needed a change, and that change came quickly.

I got a call out of the blue in November, 2020. A good friend at US Army Pacific Command (USARPAC) reached out to me. A specialized DOD program needed liaison between USARPAC and US Indo-Pacific Command (INDOPACOM). I was needed in Hawaii in a little over two weeks' time. Never in our lives had we seriously considered leaving San Diego, let alone moving to the middle of the Pacific Ocean. Still, change was calling, and a reason to move to Hawaii was presented wrapped up with a bow on top. It was a leap of faith, but our family embraced the opportunity. I would be working with a commercial company, and I needed to negotiate a stellar compensation package to make a move of this magnitude make sense. It wasn't just a change for me—my wife had to leave a position as department chair in special education in a highly specialized program she had built from the ground up.

The MBV program prepared me to negotiate a package that not only covered the (expensive) moving expenses but bridged the pay cut my wife took.

I landed in Honolulu to start a new chapter. There were caveats, though. My wife Meaghann stayed in San Diego through the end of the school year to provide stability for our children and her students. I was alone in Hawaii until my family joined me. While this was not optimal, it did allow me to pour myself into my new career. The sense of adventure and confidence I felt on deployments began manifesting again. The MBV program had reinvigorated for me what it meant to lean forward and embrace the change. More importantly, it helped me make the experience my own.

My primary duties were slow to materialize due to the classification of the program. I had to be value-added, so I volunteered to manage the development and execution of an ambitious project supporting USARPAC. The project was to build a multi-use, multi-program secret compartmentalized information facility, commonly called a SCIF.

My goal in volunteering to assist on this project was to make myself invaluable to the company and the DOD clients. While I did not have a background for managing programs such as these, I quickly pulled from my leadership experience, knowledge, flexibility, and MBV toolkit to adapt to the needs of the project. The project went better than I could have hoped for, but it was unnecessarily complicated by the government client. As a part of this process, I was given the opportunity to rapidly build my contact list of major stakeholders, key decision makers, and most notably, to build relationships with the subcontractors brought in to make the project come to life. I looked at it as an entrepreneurial endeavor. This mindset fundamentally impacted how I approached the project, managed my time and resources, and subsequently how I continued growing my personal brand.

As my reputation blossomed in Honolulu, it was finally time to welcome my family to the island. All the credit goes to my wife. While I was focused on my professional pursuits, she was working, packing out, organizing, taking care of our kids, and getting everyone, including our dogs, to the great state of Hawaii. They arrived with eight suitcases, all weighing exactly fifty pounds, with three carry-ons and three backpacks, which contained all the essentials we needed while we waited for the arrival of the shipping container that carried the contents of our home.

I could not have progressed so quickly without my wife's support and herculean efforts. In our first month together on the island, we explored and reconnected. It was a beautiful time. My wife was immediately offered a job at the school our children attended. The stars seemed to be aligning perfectly. That is where the story took a turn for us all.

Out of the Blue

About a month after getting to Oahu, living out of an unfurnished condo while we attempted to purchase our Hawaiian forever home, we received some of the worst news imaginable. My wife's brother James and his wife Shauna were killed in a small plane crash outside of Napa,

CA. In one moment, all our lives changed forever. Their son, who had just turned one year old, was not on the plane.

My wife rushed back to California to care for her infant nephew alongside her parents and mother-in-law. Shortly after that, we were named his legal guardians, and on the day of the funeral, I met our family's newest member for the first time. We brought him home as our son, and our kids welcomed their unexpected baby brother.

We had three weeks to process the terrible loss and prepare for the infant we never expected. Change and opportunity blew in our faces with the force of a super typhoon. The emotions surrounding this event are still as complicated as ever. We are so blessed to raise James and Shauna's child, but we wish it had not come at the cost it did. Like all things in life, the pain hasn't receded, but our new normal continues to evolve for the better.

A massive family change coupled with my employer's unprofessionalism necessitated a career change. My Rolodex of contacts came into play. The work I did on the Honolulu project had not gone unnoticed. The primary subcontractor that built our structure reached out looking for someone to join their team in Hawaii with an eye towards the greater Pacific Rim. I had enjoyed working with them on the project, and I was looking for a company with a startup mindset and marketplace position. Such companies are not ordinary commodities in the defense contractor world. It didn't take long to recognize what they could do for me and what I could offer them. My scope and responsibility needed to outpace the growth of the previous ten years of my professional life. As director of Pacific operations, I managed a portfolio of projects from the California coast to the Ryukyu Islands of Japan. Where there was a need, that is where I went.

Nothing Worked Out, In a Good Way

Since completing the MBV program, life has offered growth, discovery, adventure, and hardship. I have had to lean into the changes by viewing them as opportunities not setbacks. Before the MBV program, my post-active-duty career was generally rudderless in terms of purpose, passion, and direction. Most of my issues, as I would learn, were not due to a lack of opportunity but were self-induced and mainly in my head.

Despite the success and adventure of my time on active duty, that sense of purpose and confidence didn't translate well into civilian life. Life looks very different when you take off your uniform. It's not that people are different; rather, in the military, a collective, team-oriented mindset based around being part of something bigger compliments the individualistic goals and aspirations of each person. This symbiosis was challenging to find as a civilian. The MBV program showed me the power of embracing my goals and aspirations and, in doing so, identifying and unlocking the limiters I had unwittingly placed on myself. The MBV program became the first chapter in a new book on life that I continue to write.

Although I make a good living as a defense contractor, my ultimate goals extend beyond a standard nine-to-five job. The MBV program and its many professional advisors emphasized that sacrifice is required to go big and be transformative. I find myself wrestling with the trade-offs needed to set my family up financially while not missing my children growing up. It has been a more difficult task to reconcile than I ever thought possible. I see why many people talk about their dreams and why so few follow through. But my children are not my excuse. They represent a fundamentally important aspect of my life. Taking away from my time with them and limiting my presence weighs heavily upon me. Despite the daunting nature of what it takes, I drive forward. Much of my time in this pursuit resides in conducting due diligence on prospective business acquisitions. I refuse to stop, and I refuse to be stagnant.

Continuous learning seems to be a cliché these days, but it is a constant in my life. Without a desire to adapt and keep my eyes forward, I wouldn't be where I am today. I believe the past is best used as a learning tool, but not at the expense of being an anchor that prevents you from progressing towards a better version of yourself and the life you seek.

Despite my best efforts, nothing worked out how I thought it would. That isn't a negative. Frankly, I am blessed that it didn't work out how I wanted. If I am anything, it is a man who enjoys the journey more than the destination itself. I have learned to let go of many of my preconceived notions about what my place in the world is, what it means to be successful, and how to make the most of the time I have.

Ultimately, it comes down to this: take risks, fail, and get up. Above all, know that asking for help is a strength and not a weakness.

We aren't meant to go through life alone. Everyone knows something you don't. People can help in ways you don't expect and provide perspective from the vantage point unique to their experience. Finally, embrace change, be a good human, and remember that everyone is figuring it out, just like you.

The entrepreneurial mindset and toolbox provided by the MBV program were instrumental to my current success. It was never enough to have the leadership and global experience from my time in the military. I needed other resources and a supportive professional network to find success. I know better now what to look for and how to look for it, all with more knowledge about how to support myself and my fellow MBV graduates. To this end, I have had the opportunity to invest and support two MBV alums: Ty Smith of CommSafe AI, and Eric Johnson of Trident Coffee.

There are no shortcuts, and as Ty demonstrates through his background and daily efforts, the only easy day was yesterday.

"I Am Far More Than This Job"

Darren Denyer reflects on telling his story

There have been times when I worked at not remembering elements of my past, due to trauma and experiences that taught me lessons I likely didn't need to learn. However, when it came to adding my voice to this project, many of those concerns faded to the periphery. The storytelling process and structured recollection, and the content we focused on, allowed me to reflect on and highlight my positive impact, hard-fought lessons, and the people who shaped who I am. Recognizing this has made imparting some of my lessons and advice more accessible and more salient to the book's audience.

This process taught me, more than anything, that I am better served by speaking up and leading than being quiet and reserved. We all have something to offer. Especially in this negative social media era, we have much to offer those we meet and interact with daily. It's hard to remold your identity or look at it from a different viewpoint. The military has been central to my life, but it isn't who I am, because I am far more than my job. This process and the people behind it have reminded me that I have something to offer and a place beyond my uniform.

Getting to this point was difficult for me. My desire to avoid putting my thoughts on paper made this a long and arduous process.

Life has thrown me and my family some curve balls. I had to persevere and push forward and tell a story I thought might be worth telling. I may have never completed this task without the positive reinforcement, patience, understanding, and multiple Zoom calls. It takes a village, and I am grateful to have so many incredible people in my corner to reinforce that I didn't have to go through this alone.

Chapter 9

Better Late Than Never
by
Dave Foster

Dave Foster served for seven years in the Marine Corps. His service included a combat deployment in 1990-91 for Operations Desert Shield and Desert Storm, and deployments to the Western Pacific. He left the service at the rank of captain.

Dave worked on the civilian side of the Department of Defense for twenty years as an aerospace engineer and operations research analyst on projects that included the design and performance analysis of air vehicles, guidance and navigation systems, data links, and multispectral imagery.

Dave's academic credentials include a BS in mechanical engineering from the University of Massachusetts Amherst, an MA in engineering management from George Washington University, an MA in history from West Texas A&M University, and a Master of Business for Veterans (MBV) degree from the University of Southern California. He is currently working on a PhD in history from Texas Tech University, where his research focuses on the adoption of computers by the Department of Defense in the mid-20th century.

A Different Time

This is a story about learning important career lessons very late in life.

I grew up in the Boston suburbs and went to high school and college during the 1980s. I was the eldest child and the only male. In my town, there was still a small-scale agricultural economy oriented to local customers. There were farm stands in every other town, and pop-up booths littered the back roads selling seasonal corn and pumpkins. There were still many apple orchards devoted to cider-making. Indeed, it was the apple crop that brought the railroad to my hometown in the nineteenth century, and my first part time job in junior high was working at the town's commercial orchard business. A year-round dairy business operated in the next town, delivering milk along a few routes and selling various products from a store front along one of the main commuter roads.

But the general trend in our town and the surrounding region across the second half of the twentieth century was suburbanization. The region's industrial economy waned. Most of my friends' fathers worked in the tech or finance sectors and most of our mothers stayed at home. The auto plant and brewery in the next town over closed in the 1980s. The rising computer and electronics industries in the area would gain national attention and be named for the inner beltway encircling Boston: state highway 128, or simply "Route 128."

In high school I spent my time indoors building and tinkering with machines and electronics in the basement and garage, building plastic models of military ships and aircraft, and reading about current affairs and history. Outdoors, I did a lot of yard work, held various part-time jobs, and played team sports at school. My parents were from the West Coast, so we did not have family in the area. But most of my classmates were in similar circumstances—their parents were originally from somewhere else. Virtually all of us lived in two-parent, single-income households. Over ninety percent of my graduating class went on to four-year college after high school and entered some kind of white-collar profession.

Tinkering

I studied mechanical engineering in college because it seemed to be the most reasonable academic field for someone who grew up tinkering with tools, electricity, and various scientific and maker kits. I don't remember the origin of my interest in the Marines, and for whatever reason, I was interested in only the Marines. Presumably this came from pop culture and perhaps from a visit a Marine recruiter made to our high school.

Events of the 1960s and '70s hung heavily over the culture. Revisionist modern images suggest the Reagan era was a time of unbridled patriotic unity that sought to counterbalance the chaotic Vietnam years. Rather, the Greater Boston of my teen years was mild and largely apolitical—perhaps stereotypically suburban. And although many of my friends' fathers had been in the service during the mid-century draft years, joining the military was an uncommon decision.

My parents and my friends certainly thought I was making an odd choice. I did not come from a military family—you would have to go back to the eighteenth and nineteenth centuries to find ancestors in my direct line who had been in the service. Neither my father nor either of my grandfathers had served in the military. But I was interested in joining the Marines after college for the action and adventure.

My university had only Army and Air Force ROTC, so my route into the Marines went through their officer candidate program (OCS). Only about two percent of my graduating high school class had gone from college into the military. It was not a common career option. I had three plausible post-college career options: I could go directly into graduate school, I could enter the job market, or I could join the Marines. Only the Marines had a practical constraint that limited the time horizon for eligibility. If I recall correctly, a newly joining Marine officer could not be older than twenty-seven or twenty-eight to first start. The other two options could—and would—be waiting for me after the Marines. So the choice was pretty clear. I had had enough of school for the time being. And I wasn't eager to jump into the corporate world right away.

At the time I joined in the late 1980s, graduates of OCS came in with a reserve commission. This meant that if I had wanted to stay beyond my initial obligation, I had to "augment" and be selected for a regular commission sometime early in my tenure. But in the context of my initially scant knowledge about the Marines, in retrospect my goals for military service were exceedingly modest: I thought it would be great if I could get a thousand hours of flight time and become a captain by the end of my obligation, which would be just prior to my thirtieth birthday. Not to diminish the dedication and effort required to get there, but it worked out that several adequately successful years in the fleet were all it took to reach these minor achievements, and I'm happy with my experiences and the ultimate outcome of my short time in the Marines.

Early Networking

My plan following active duty was to go to grad school and study structural engineering, then go into some civil construction field—buildings, roadways, bridges, etc. After separating, I enrolled in an engineering program and took classes for two semesters. But as I did some preliminary planning by looking at potential job opportunities, I discovered something I had not anticipated. If I completed my graduate degree, in the plus column I would have engineer education and training. But, as I came to find, the minus was substantial—I had no practical engineering experience.

My military experience did not seem to be valuable to potential employers, or at least I did not know how to explain my military experience and how it could be usefully applied to practical engineering problems. The focus of my conversations about employment opportunities was on my practical technical skills and not on the broader problem solving, project organization, or leadership challenges. Had I been savvier or sought some placement advice, I could have emphasized the latter and presented myself as a more compelling candidate, rather than letting the interviews bog down on the limited issue of my technical skills. But, as I look back, I did not successfully move the conversations into these areas of personal strength and instead remained mired on the topic of technical inexperience.

So now what? I started talking to some of my former squadron mates who were also now off active duty. One of my old buddies had a similar background—a technical degree and a half a dozen years flying. He was originally from the Washington, D.C., area and had found work with an engineering company on a contract with the Defense Department. He said he would pass my resume to his HR guy and see if the company could find a fit, the idea being that I knew something of the DOD's operational side and had a technical degree. This turned out to be true and I was offered a job as a systems engineer working on a development contract for the US Navy.

Failing to Take Advantage of My Advantages

Looking back, I am still rather stunned about my career cluelessness. I knew how to work hard and do a good job, but I was blind to the need to curate a professional network.

It had never occurred to me, and no one explained to me the idea of networking. I had plenty of advantages and potential opportunities, but I just didn't know how the professional world worked. There were plenty of people whom I could have talked to and learned from, and most would have likely been more than glad to give me a hand. I already had a huge advantage, but I didn't see it for whatever reason.

I had grown up within a large professional network: my high school and college friends' parents were predominately business professionals. My father had worked in the tech industry since graduating from college. But the only career advice that I recall from my college years was from one of my father's coworkers who advised that I stop studying engineering, forget about the technical stuff, and instead go into finance. I'm sure the dearth of career guidance was not as extreme as I recall, but other than reaching out to my buddy in D.C., I don't remember undertaking any career networking after I left the Marines.

The people I worked with in my new job as a defense contractor were either civilian DOD employees or former service members with technical or business degrees. I had administrative "collateral duties" as a Marine officer, so I was not unfamiliar with paperwork. The late 1990s saw a wave of computerized democratization, and everyone worked behind a desktop computer. Indeed, my command-line computing skills and ability to modify the local mail server, update the office web page, and troubleshoot the local network were seen as big positives in the decision to hire me. Most of the other former military members were older guys from the Vietnam era for whom desktop computing was comparatively new.

When I was on active duty, I was wholly ignorant of the vast military "acquisition system." I hadn't really thought about how the military supplied itself. I simply assumed that the Marines called Boeing, General Dynamics, Raytheon, etc., and ordered aircraft, radars, weapons, support equipment, what have you. That my flight manual was published by the Naval Air Systems Command never registered nor inspired my curiosity. But now I was working within the system that developed, built, and supported the diverse tools and equipment needed across the military.

I started in the acquisitions system before 9/11 at a time when there was still ongoing and contentious debate about a "peace dividend" and the necessary size and optimal role of the US military. The Asian financial crisis was the main issue in the world of foreign affairs on the news. China was not yet a member of the World Trade Organization, and Russia was ambiguously moving into some semblance of market economics. The United States was the sole superpower, and its leaders were generally sanguine about its apparent global primacy.

After relocating cities and gaining some tentative knowledge about my new industry, I wondered what I should do about graduate school. Would structural or mechanical engineering be useful, or should I look into a new discipline? A coworker was studying engineering management at George Washington University, and after looking into the program, I decided its focus on the business aspects of technical projects would be a good fit. And it was. Classmates came from diverse academic and employment backgrounds, and many worked in various government departments and agencies in addition to the DOD. Through discussions and collaborations on projects, I learned a lot about the ins and outs, ups and downs, methods, and cultures of a wide variety of organizations.

In my job I worked on many types of aviation systems, most involving computer hardware and software. I also increasingly became involved with operations research and systems analysis projects and found that I greatly enjoyed developing and analyzing alternative courses of action. I was generally interested in the technical and operational rationale behind the designs of these many systems and put a great deal of effort into studying the history and interconnectedness of systems and the organizations that developed and used them. Within several years I had earned a reputation as the go-to guy for questions encompassing the backstories, current situation, and future possibilities of the broad ecosystem of DOD guided weapons.

I continued to naively if not myopically presume that my ongoing excellent performance would automatically be observed, appreciated, and rewarded with promotion and greater responsibility, and while my performance was apparently observed and appreciated, it did not lead to bigger and better things. My aloof and fundamentally ignorant approach to my own career development kept me treading water as a strong performer and great resource.

Treading Water for Decades

We can fast-forward twenty years from my time as a contractor to my final years as a civilian employee. I worked on many interesting projects, had positions such as managing a thirty-employee analysis shop, and became the chief engineer for the largest technical project office in the weapons division at the Naval Air Warfare Center. But I had barely moved up the hierarchy.

In 2016, my wife, a career naval officer, attended the MBV program. She had great things to say about it, and I got a look at some of the academic topics and coursework, which looked very interesting. She encouraged me to apply, explaining that the program was awesome and that I would both love and get a lot out of it. However, I was initially reticent. I had two graduate degrees already and wasn't sure of the value another would bring to my career. She persisted, and I applied. I earned my previous graduate degrees simultaneous with full-time work, so I had strong study habits and the ability to manage the demands of work, school, and family.

The MBV program was an entirely new experience, however. The MBV explicitly and overtly infused the academic elements with career preparation. My previous schools had career center resources. But these were external and separate, and I did not make productive use of them. This may very well have been a leading-a-horse-to-water-type situation, with me being the clueless, stubborn, recalcitrant horse.

Finally Figuring It Out

What now seem to be straightforward and obvious career resources— such as the alumni network, resume counselling, and LinkedIn—were fresh opportunities that I committed to taking advantage of while in the program.

Knowing that our family would be moving out of state soon after I graduated, I made a career plan and followed it energetically. I broke up the academic year into four career phases. For the first half of the first semester, I gathered information. I had worked for the federal government for three decades in a variety of roles—active duty, contractor, and federal employee—but I wanted to look into opportunities in the private sector as well, and I needed to learn the terminology used across a range of potential industries. The MBV program brought representatives from different industries for interactive presentations, sponsored career-based events to meet with USC's strong alumni network and set each student up with an established mentor.

With a newly supplemented career vocabulary and dozens of conversations with representatives from many career fields and industries, I used the next phase to focus on creating and refining my resume and LinkedIn profile. Beginning in the new year at the start of our second semester in the program, I sought informational interviews to test the traction of my now substantially improved career presentation.

In the fourth and final phase, I took several actual job interviews. Two included traveling to corporate offices for final interviews followed by job offers. I accepted one several weeks after we graduated.

Flight school was a seminal transformation in my academic seriousness and outlook and, similarly, the MBV program was a very clear transition for my professional mentality. I credit the MBV program with my success in being able to make a big change after thirty years in the DOD umbrella.

The main adjustment was not in the practicalities of technical analytics or project management but my outlook and relationship with my employer. The MBV program illuminated the need to communicate my career achievements and objectives effectively.

My previous mentality had been appropriate for the operating environment of military aviation. I had perfected the ability to compartmentalize, and I carried this mentality into my work. This manifested as a suppression, if not an outright rejection, of anything I determined was extraneous to the task at hand. My somewhat introverted engineering mentality exacerbated this tendency towards insularity, and over the subsequent decades of work I discounted the importance of most elements of the professional milieu beyond my own activities. I had a strong reputation not just inside my organization but also across the small universe of technical and operational organizations in my field as a go-to subject matter expert—what else mattered? The problem was, I did not see the need for nor the value in packaging this reputation in a way that my superiors were interested in promoting. I was doing a great job in a little box but not really going anywhere according to my leadership potential.

Continuing to Learn

A few years after finishing the MBV program, I decided—yet again—to return to graduate school. I have returned to the field of history and am currently working on a PhD with a dissertation closely aligned to my professional experiences: the adoption of computers by the Marines for administrative organization and automated data processing in the 1960s.

Apart from pure academic interest, one of my reasons for studying history is that in my decades of work in technical fields, I have found a widespread disinterest and often blatant disregard for the value of history and the humanities more generally. As tech writer Doug Rushkoff has explained, modern American society has been captured by a mindset that elevates the academic STEM (science, technology, engineering, and math) fields over the arts and humanities. More insidiously, in my view (because STEM is worshipped in a misguided way), STEM has become the avatar of what is considered correct and most valuable. Yet the tools of math and science, as much as I love them, are just building blocks of knowledge. They are not in themselves arbiters, and certainly not determinants, of value. We have all seen or experienced the pitfalls of this mentality, of the exuberant promotion of the seemingly new, which turns out to be superficially different or halfway conceived yet pronounced with certitude as a new correct path forward.

In addition to the intellectual challenge and the benefits of the rigor of the effort, I have a practical outcome in mind. While I do not expect that the nation's universities are clamoring for another PhD graduate—indeed, the higher education production model creates an oversupply—I have found that there is a niche within the non-tenure career track.

Nontraditional, part-time instructors are being used by a variety of institutions, and I have begun to look into possibilities for work as a remote-learning or community college instructor. It may be an interesting side hustle to begin with, but I believe it can evolve into my next occupation. I have found that there are not too many hybrid academics who credibly straddle the technology and humanities fields. I have spent several decades working in large, technically oriented organizations. These have an increasing degree of influence in society where many if not most features of our daily lives are integrated into technological systems and networks. I believe it is important to think about technology in terms beyond the limited scope of STEM, and I am hoping to usefully contribute to the conversation about it.

"Attending Business School in My Fifties"

Dave Foster reflects on telling his story

After participating in the first volume of *Transitions in Leadership*, I thought it was important to assist with the second.

Growing up, I was athletic. I played both varsity and pickup sports year-round. I had the mentality that doing well and feeling good required personal care in physical fitness and nutrition. This was a great mentality for an individual, but over subsequent decades, I discovered that I had completely overlooked the need to take care of myself as a professional. I had been focused on the direct outcome of my efforts—doing well in sports, making teams, becoming a Marine officer. But I had failed to comprehend the greater process that takes place in all organizations. And I paid for this inattention for most of my working life, until I clued in while attending business school in my fifties.

I have spent the nearly thirty years since I left active duty attempting to understand the operating environment and germane variables in technical and organizational systems, in order to improve and refine system design and effectiveness. To me this is a natural way of thinking that goes back to my childhood tinkering with tools and machines.

And yet, into my fifties, I lacked a sense of awareness for the need to apply some of this optimization mentality to myself. I had generally gone in the professional direction that interested me and did a generally poor job engaging with professional mentors to craft a plan. My goal in sharing this story is to encourage others—hopefully a great minority—who may not understand the importance of managing one's own career.

Get mentors, create and take advantage of your career network, hit them up with questions over the years, and make and refine a career plan.

I do not know the majority of my fellow collaborators in this project, and it has been enjoyable and illuminating to learn about the diversities of background and experience amid our community of veterans and classmates that extends across the MBV cohorts. We share many temperamental similarities—adaptability, directness, and problem-solving orientation. Yet, these characteristics were not created ex nihilo by our experiences in the military. They were perhaps at first nascent then refined by our experiences in the military. Learning about my others' paths and seeing their unique approaches to telling their stories and imparting their lessons offered me many unexpected insights, and the collaboration has been rewarding.

Chapter 10

Perfectly Imperfect
by
Jessica Felix-Bradshaw

Lieutenant Colonel Jessica Felix-Bradshaw has served for twenty years in the California Army National Guard and US Army Reserve. Her assignments included the 40th Infantry Division, where she deployed to Kosovo and managed human resources support and morale, welfare, and recreation. In Kaiserslautern, Germany, she served with the 202nd Military Police Group (CID) and Fifth Military Police Battalion (CID). She commanded the headquarters units of both the battalion and brigade, while simultaneously serving as the unit administrative and logistics officer.

Moving with her husband's assignments, Jessica served with the 275th Combat Sustainment Support Battalion at Fort Lee, Virginia, then with the 2-290th in Oklahoma City, Oklahoma, before transferring to the 63rd Readiness Division in Mountain View, California. Currently, she serves with the 79th Theater Sustainment Command in Los Alamitos, CA.

Jessica is a manager at Deloitte and Touché, where she specializes in strategic planning and provides support for executive leadership. She lives with her husband and two children in Seaside, CA.

Sweet and Smart
Growing up, I was known as the sweet or smart one. One time my lola (grandmother) Luming was talking about my sister and me, and I

overheard her say that my sister, Melanie, was the "pretty" one and I was the "smart" one. While some may have found that offensive, I did not, because after being characterized as sweet or smart all my life, that is who I became and what I am known for today. I did not care to be considered the prettiest because, unless I was getting plastic surgery, I was not going to change how my lola or anyone else in my family saw me. Those traits stuck with me. People I meet comment on how sweet I am. Maybe it's something to say to be polite, but it's interesting that those two personality traits are the most visible, even to strangers, and I have carried them my entire life as part of my public persona.

I am a first-generation American citizen born and raised in the United States. My parents migrated from the Philippines in their early teens. My grandparents, aunts, and uncles were all blue-collar workers. My inang, my great-grandmother on my dad's side, worked as a seamstress at one of the clothing factories in downtown Los Angeles. Lolo Eddie, Dad's father, worked in the oil industry. Lola Nedy, Dad's mother, was a hairdresser who owned her own salon in Carson, which was filmed in the background in *Friday* (1995). And Lola Luming, Mom's mother, worked in home health care. My titos and titas worked as nurses, in home health care, or at AT&T. Sometimes, instead of buying new clothes, Lola Nedy and Inang made dresses for me and my sister. I grew up with modest means, but I never felt like I was deprived. We had the basics, and that was good enough.

My sister and I got to travel the world because of our parents. My dad enlisted in the Air Force, then later in the Oregon Army National Guard and the California Army National Guard, where he retired as sergeant first class. My mom, after twenty-seven years, is still a flight attendant for United Airlines.

But at one point, my mom, my sister, and I lived in Las Vegas, and because of my mom's lower income, we were on welfare, and she worked two jobs. One of her jobs was at Blockbuster. My mom took us to work, and my sister and I sat in the kids' corner watching movies like *The Last Unicorn* (1982). That was our babysitter. It was such a different world. My mom's second job was working the leasing office at an apartment complex. Since my mom was working, my sister and I walked to school by ourselves in the Vegas heat. We were seven- and eight-years-old. Whenever my mom and I reminisce, we talk about how crazy that sounds. Knowing it was just the three of us, my sister and I made sure to take care of ourselves, to avoid adding to our mom's struggles.

I remember going to the grocery store at night because my mom was embarrassed to use her food stamps during the day. Fewer people shopped at night. I did not really understand at the time what being on welfare meant. It was our norm, and my mom did a great job shielding us from the struggle to provide for us. I imagine it was not easy raising two girls as a single mom, since during this period she and my dad were on and off.

My sister and I did not ask for much, because truthfully, we did not need much, but seeing how difficult it was for my mom to raise us, it made me promise that I would work hard and not have to depend on anyone to provide for me. I learned a lot from my mom, but the two attributes that stick out most for me are independence and strength. Despite the strife, we were content with our daily routine. Before bed, the three of us recited our mantra together. I cannot remember what the mantra was, but I recall it being on a poster in our home.

During the summer between third and fourth grade, my mom packed up all our things and we moved to Oregon to be with Tita Nenette, Mom's older sister, who took us in with open arms. Just when we thought we were going to start a new life and settle into a new home in Oregon, my dad got into a bad car accident in Las Vegas. My mom left us with Tita Nenette to go back and get him, and that is how my dad reentered our lives.

Living in Oregon was great. We got to grow up with our cousins, and our parents were together, unlike many kids our age whose parents were divorced. I remember them as simpler, happier times. In 1997, my lola Nedy passed away, and Mom left her job as a bank teller, became a flight attendant for United Airlines, and was based out of San Francisco International Airport. She commuted across state lines from Medford to San Francisco whenever she had work. Sometimes my dad, my sister, and I would go to work with her, and we flew to places like Hawaii, New York, and Australia.

Eventually, we moved from Medford to Vancouver, Washington, when I was in middle school, because my dad was in the Active Guard Reserve and had a permanent change of station to Portland Air National Guard Base. We lived in Vancouver for about two years, then during the summer between my freshman and sophomore years, we moved to Torrance, California, because my mom transferred bases to Los Angeles International Airport. My dad was a geographical bachelor while he commuted from Vancouver to Los Angeles every so often.

By the time I went to North Torrance High School, I had attended twelve different schools in California, Oregon, Washington, Nevada, and Japan.

Before I started my sophomore year in high school, my sister and I spent the summer with Lola Luming at her home health care facility in Gardena, until my mom bought our triplex in Torrance. Between my mom's and dad's trips we fixed up the triplex one unit at a time. We did not have a working bathroom or kitchen for a couple years. We ate out a lot and lived in a construction zone while attending high school. It was not until I was a senior in high school that my parents separated for good. I took them being on-and-off for granted because I thought they would eventually get back together. That was not the case.

College and Recruiting

Typically, children follow in their parents' footsteps, which is exactly what my sister did. Straight out of high school, my sister enlisted in the California Army National Guard. She also became a flight attendant, just like Mom. My sister was the face of the California Army National Guard when they switched to digital print uniforms, and she also competed in Miss Philippines Earth, and won the title of Miss PAGCOR International in 2008.

Since I was not the pretty sister, I leveraged the fact that I was smart. When I started high school, my lolo made a deal with me. If I got straight As on my report card, he would give me $400. That was enough incentive to do my best every year in school. I was a Goody Two-shoes and did very well. I was a cheerleader and part of the Associated Student Body Council; I dabbled in basketball and took Advance Placement classes; and I graduated in the top two percent, out of more than 400 students, as a valedictorian of the class of 2003.

I hadn't realized that the University of Southern California was down the street from Lolo and Inang's house in Gardena until I arrived on campus. In my dad's opinion, enlisting in the armed services was not an option for me because of my acceptance to a private university.

He did see me taking the officer path, however, if I wanted to go the military route.

I was accepted to USC as a spring admittance, so I began my college experience at El Camino Community College in the fall. It worked out well because my sister attended El Camino, too, before she was activated for a border mission in San Diego. She showed me the basics of being a student. Although I would have liked to start at USC as a true freshman, it was not my path. Regardless, I thought it was really cool that my older sister was there for me during the transition from high school to adult life. I worked part time at American Eagle, joined a sorority in my junior year, and was on USC's competition cheer team. Instead of focusing on my studies, I partied, went to Las Vegas a lot, traveled, and spent time with friends.

My dad ended up recruiting me into the California Army National Guard as a Simultaneous Membership Program cadet with the intention of commissioning upon graduation. While I kept busy during college, I did not meet all of my military requirements. I did not listen to my dad and show up to drill with my unit as much as I was supposed to. Unlike in high school, I was not a good student, and I barely squeaked by. I was distracted by all the parties and cool happenings on campus, not to mention the football games.

Since my parents could not afford to pay for private school tuition, my dad did what he did best and used his resources to get me an ROTC scholarship. My mom, my lola Luming, and my sister sent extra money when they could, to help offset the costs of college, cheerleading, and sorority life. My dad again helped me make extra money by enrolling me into the Guard Recruiting Assistance Program, which rewarded guard members $2,000 for each person they recruited successfully into the California Army National Guard.

Making extra money by recruiting fit perfectly in my schedule. I kept quite busy and active. For Army ROTC, I participated in field training exercises in the Camp Pendleton area, around USC and downtown Los Angeles, and around the California Science Museum at Exposition Park. We held exercises and pre-mobilization training at Camp Roberts and drill weekends in Azusa, CA. I often visited the San Luis Obispo area to see my sister.

By the end of the Guard Recruiting Assistance Program, I had signed up nine people and earned a total of $12,000. (One of the people I signed up was my cousin, EJ Felix, who is now a sergeant first class just like my dad and attending Officer Candidate School to become the second officer in the Felix family.) Little did I know at the time that the Guard Recruiting Assistance Program would eventually become my worst nightmare.

While it was great that I was accepted to USC, it was also a difficult time because my parents finally separated for good. It caused a rift in my relationship with my dad, with whom I was very close. I was his Mini-Me and he often referred to me as "Li'l Felix." Instead of dealing with my issues regarding my parents' separation, I threw myself into many activities just to keep busy.

I always made time to check on my mom at home. During one of my unannounced visits, I found her cleaning behind the washing machine and dryer. She was crying. I hugged her and told her that everything was going to be okay. It was not easy going home knowing how difficult things were for my mom. Because I was the youngest and going to school so close to home, my perception of this time is much different than my sister's. I saw firsthand how painful that time was for Mom. My dad, on the other hand, seemed fine with his new wife and baby.

As one of the first in my family to graduate from a private university, my mom threw a huge party to celebrate the achievement. It should have been a joyous occasion because their youngest daughter had graduated from college, but it was awkward having my mom and dad in the same room together. It was not easy for my mom to extend the invitation to my dad and his new family but knowing that this moment was not about either of them, she did.

Family Matters

Watching my parents' marriage fall apart made me promise myself that I would not get into any serious relationships. I shielded myself from dating because I did not want to have a failed relationship, especially if there were children involved. I saw how hard it was for my mom knowing that my dad left her for the typical "younger woman," and I hated that narrative. I did not want to feel the pain my mom felt. I saw early on that it was not easy supporting a family, let alone having a happy marriage.

To this day, whenever old things are brought up within my family, I feel helpless and fall apart. I like to think I am a strong person, but whenever I am around my mom, dad, or sister, I have difficulty maintaining my composure, and I become that young girl again who saw my parents' relationship deteriorate. I often take my mom's side and blame my dad for what happened to our family. Not saying that what I believe is right, but for me, my truth is I always felt like my dad was looking for a way out, which he eventually found. My dad got to start over, and my mom was left with my sister and me.

If someone is truly happy with their current situation, they do not not need to bring up the past. It has been more than twenty years, and my parents are still not okay being around one another, which makes it extremely difficult to celebrate birthdays, holidays, or any special occasions. Holidays are a difficult time for me, because while my husband's family is great at celebrating special moments, I find myself not wanting to partake in the festivities. They are a constant reminder of my broken family. If I had the option of fast-forwarding through the holidays I would. I would love to have normal family get-togethers, but it is not possible with my family's dynamics.

Often, I feel guilty for my kids that things are the way they are, because it is not their fault that they were born into my family. I try my hardest to give them a normal childhood, whatever that means, but it's hard to see that they do not have close relationships with their cousins, titos and titas, lolo and lola like I did. Instead of these circumstances breaking me, they push me to do more and be better for myself and my family.

Kaiserslautern

After graduating from college, and upon my commissioning, I attended military schools in Oklahoma and South Carolina; went on a joint assignment to South Korea; and mobilized to Kosovo as the deputy G-1 and lead for the Morale, Welfare, and Recreation program—pretty cool for my first gig. I hosted many events with professional athletes, bands, comedians, cheerleaders, etc., which made being away from home that much easier.

From Kosovo I had the opportunity of a lifetime to be on active-duty orders as a National Guard member in Kaiserslautern, Germany. I was in my twenties, living on my own in a foreign country, with my own apartment (with its own lake), not one but two hoopties, and I was traveling the world! It was amazing, and I loved every minute of it. First Sergeant Patrick Taylor so lovingly referred to my group of friends as the "Lieutenant Mafia," later known as the "Captain Mafia." We traveled all over Europe together, and you could describe our adventures as epic. Eventually, I met my husband Scott during seder dinner with a mutual friend, and the rest is history.

As a woman in the military, it was easy to be jaded with relationships. I never wanted to get married, not only because of my parents, but also because when you are one of a handful of women on a military installation, shenanigans ensue.

Many service members understand what I mean by the term "deployment goggles," which are like beer goggles. Members of the opposite sex you would never touch with a ten-foot pole all of a sudden start looking pretty good and become viable dating options on deployment. Also, as a woman you are automatically sought after because there are so few options in the military. The male-to-female ratio is heavily skewed.

When my future husband Scott and I started dating in Germany, he deployed to Afghanistan as an individual augmentee. Because of how I felt about serious relationships, I was already over him before our relationship began. Having previously made my own mistakes while suffering from "deployment goggles," I didn't think we had a foundation that could survive his deployment. Boy, did that backfire.

We talked in my downtown Kaiserslautern apartment, and as I tried to break it off with him, he agreed with my points and said, "Or we can try" to make it work. His response was different from other guys I dated, and I decided to give him a shot. During his deployment, I looked after his apartment, took care of his car (not a hooptie like mine), and traveled all over Europe with our Captain Mafia friends.

I traveled to the Dominican Republic, and during that trip I asked him, in a drunken stupor via text message, "Are you going to marry me or what?" He responded, "That's the plan." We laugh about it to this day, because the way he tells the story, he got the text in Afghanistan while freezing inside his sleeping bag. His phone buzzed with a text I had sent while living the life in the Dominican Republic at an all-inclusive resort with my bestie and my mom. We started planning our wedding at the Ritz-Carlton Bacara in Santa Barbara.

Because I was in Germany, I could not try on my wedding dress, so I had my sister do the fitting. Because Scott was in Afghanistan, we did not have an official engagement until he redeployed, and that was when he proposed to me in front of our friends. Scott called my dad from Afghanistan to ask for my hand in marriage, and my dad's response was, "Are you sure you're talking about Jessica?" The first day Scott met my dad was at the rehearsal dinner. I almost didn't have a groom because Scott's redeployment date kept shifting, and he barely got back in time from Afghanistan. We walked down the aisle on a beautiful bluff in Santa Barbara, celebrating with our closest friends and family. Twelve years, two beautiful children, and three station changes later, here we are.

Indirect Flights

For the first decade of my military career, everything was picture-perfect. By the time I was headquarters and headquarters detachment commander of the Fifth Military Police Battalion Criminal Investigation Division, I was married and six months pregnant. We left Germany for my husband's Captains Career Course. I joined the US Army Reserve in Fort Lee, Virginia. Then we moved for Scott's Active Component to Reserve Component Program in Mustang, Oklahoma. We found out I was pregnant then with our second child. I had two toddlers and an Army Reserve job, we had just bought our first home, and I attended the Adjutant General Captain Career Course while I was still breastfeeding my daughter.

What else could I add to my plate? A masters! With the encouragement of a fellow Trojan ROTC member and MBV Cohort II graduate, I applied to the program and was accepted.

For ten months, I commuted from Oklahoma City to Los Angeles every other weekend. Every. Other. Weekend. Sometimes I flew solo and sometimes the babies came with me to visit their lolo and lola. Luckily, there was a direct flight from Oklahoma City to Los Angeles, and thanks to Mom I never had to pay for my flight. The timing of the program was perfect because shortly after graduating, United stopped servicing the direct flight between OKC and LAX. Redeeming myself for my sad undergraduate GPA, I graduated at the top of my cohort.

Unforeseen Consequences

Shortly after graduating from the Adjutant General Captains Career Course, I received a call that changed the trajectory of my military career. The call was from a Criminal Investigation Division (CID) agent asking if they could talk to me about the Guard Recruiting Assistance Program I participated in during college more than a decade ago. It seemed like a reasonable ask, and since I had previously worked with CID and was the headquarters and headquarters detachment commander for two CID units, I agreed. What was the worst that could happen?

After an hour of speaking to the agents, I realized I was being interrogated—and in my own home, with my children, husband, and a handyman present. After the agents left, I broke down into tears in the kitchen. I was embarrassed, scared, and angry to be accused of serious crimes. The agents informed me that my saving grace was the statute of limitations, and I was not the big fish they wanted to fry.

I did not realize the severity of what was happening. Not knowing what to do, I called my dad, and he helped me hire a lawyer to rebut the claims made against me. I submitted a rebuttal to my chain of command and waited for a response. I thought everything was behind me. Later, I received an email that said that because of my participation in Guard Recruiting Assistance Program and the findings of the investigation, I was receiving a general officer memorandum of reprimand (GOMOR). I was shattered. Most service members know that receiving a GOMOR is the kiss of death, a career ender.

It was not until I received the reprimand that I recalled some of the details of the recruitment program. From its inception, it lacked rules and guidelines. Before I wrote my rebuttals, I read the current training module information and had a difficult time recalling any of the stated policies or guidelines, which were published by the National Guard Bureau on May 4, 2010, five years after the program launched and when I was no longer participating in the program.

According to the guidelines, all recruiting assistants were independent contractors employed by Document and Packaging Brokers Inc., or Docupak, and were not allowed to share payments or login information with anyone. The only reason I took part in the program was because of my dad. He was one of the top recruiters in the California Army National Guard. He recruited both his daughters, a few nephews, cousins, and countless others. While he was working, why not help his daughter make money by providing leads for the Guard Recruiting Assistance Program? After all, he could not afford to pay alimony or child support. While I know this was my dad's way of helping, it almost ended my military career.

My investigation report included witness statements from soldiers interviewed by telephone who claimed they did not know me. I never saw the witness statements. By the time the investigation took place and I was questioned, there were five-year statutes of limitation that had run out on four potential offenses: identity theft, grand theft, false claim, and false statement, two of which are felonies. I received heavily redacted files that showed a fluctuating number of potential soldiers I was accused of improperly signing up. Considering I had exited the program more than a decade ago, my memory of who, when, why, and how I met any of the individuals failed me.

While I did recall inputting information for some of the potential soldiers, I did not remember all of them. Shady and sneaky are not words that describe my character. Any wrongdoing on my part was unintentional. My integrity has always been paramount. When I received payments from the program, I had no knowledge that anything was wrong with my conduct, nor that receiving the payments was also wrong. The agents informed me that recruiters had received kickbacks from the recruiting assistants. Due to the statutes of limitation, the government could not charge me criminally as a civilian, but they could prosecute me administratively as military.

Later, I learned that problems with the recruiting program and subsequent fraud investigation had been published in places like the New York *Times* and *Army Times*, and high-ranking military officials had testified before Congress.

Problems with the Army's own investigation included guilt by association, compulsory interrogations, investigators who had their own personal financial incentives, violations of the Posse Comitatus Act, and ignoring statutes of limitations, among others.

Receiving a general officer reprimand was the worst experience in my military career, and it continued to plague me for another five years.

The summer after graduating from the MBV program, I was selected for promotion. I was overjoyed, but the joy was short lived. My former executive officer and my battalion commander informed me that my promotion was withheld. I had to go through another administrative process to rebut and defend myself from the allegations of grand theft and identity theft. I went through the whole rollercoaster of emotions again.

Unexpected Strength

I honestly thought my military career was over. Despite all of this, I had to keep going, especially for my family. I took a job at Macy's that paid $14 an hour. Also, I needed more civilian human resources work experience, since the only experience I had was from the military.

Then I accepted a sales position at Effectv, formerly Comcast Spotlight, in media and marketing, as an associate account executive. It was a fun job because I worked with local business owners writing and producing short ads for their businesses which aired on local cable networks. The work pushed me out of my comfort zone because I had zero experience. Sales is tough, but in the first three months after graduating from their associate program, I won an annual sales competition.

After twenty months, I realized I couldn't do sales forever. An opportunity came up when I attended the first annual MBV reunion. Thanks again to the Trojan network, one of my fellow Cohort IV graduates, Josh Lagana, told me about a Deloitte contract in Monterey. Since I was local to the area, he submitted a referral for me. I interviewed that summer, and I have worked at Deloitte ever since.

Meanwhile, I transferred to the 63rd Readiness Division, and for two years, and then two more years, my military career stagnated. I couldn't do anything about it. I did not want to tell people my secret, nor have anyone ask questions about the recruitment program and my GOMOR. Mostly I did not want to call out my dad, but I also didn't want to call myself out, especially because people had a perception of me as a Goody Two-shoes.

I was still hurt and ashamed that I had participated in a fraudulent program because of my own father. When I brought it up with him, he didn't think it was a big deal. He put me in a bad situation, but I also put myself there. I was distracted and did not pay enough attention to the recruiting program.

I interviewed for an aide de camp position for the division's deputy commanding general. But I could not accept the position, even if it was offered to me. I had to tell the brigadier general my secret: I was flagged and could not receive any favorable actions. Only after telling her about my situation did things start to change. I started a new job as a senior consultant with Deloitte, and my husband started a new career in real estate. When I finally received a long-awaited answer in the form of a memo from the secretary of the Army, with the word "retain," I cried ugly, happy tears.

The ordeal lasted four years. When I was promoted to major, I went back and forth on whether I wanted a ceremony. I could have pinned myself and called it a day. But not only did I owe it to my family to celebrate that moment, I owed it to myself.

I gave a speech at my promotion, and I shared my story with my unit. I barely made it through the first sentence before my eyes welled up with tears. I choked them back to make sure I got through it in front of my family, friends, and the soldiers. I wanted to show them that scars are not always visible. My picture-perfect life was anything but. The experience taught me that it's okay to make mistakes, as long as we learn from them. Trust me, I have. Even as I write this now, I get emotional thinking about it. But each time I share my story, I get stronger. I feel stronger. I am not ashamed of my childhood or how I grew up. In fact, I feel blessed to have lived through the experiences I did, because they gave me a source of strength.

"Strength in Acknowledging My Failures"

Jessica Felix-Bradshaw reflects on telling her story

When I started the process of writing, I thought that telling my story would be simple, but as life continues to move forward and my story continues to evolve, I am finding many complicated parts of my persona. As I write this, I want you to keep in mind that this is only a reflection on a story that is a snippet in time and one part of a personal journey.

I like to think that what you see is what you get with me, but I also know that I share things about myself based on my comfort level and my relationship with whomever I'm talking to. Most people I interact with will never realize if or when I'm having a bad day, because I normally keep a smile on my face regardless. I don't wear my hardships on my sleeve, and I don't give off the perception of a person hardened by life.

Reading everyone's story in the first volume of *Transitions in Leadership* was enlightening, because as much as it reminds me of our differences, it showed me how similar we all are. It highlighted our perceptions—how you think you know someone but really you have no idea who they are or what they have gone through. I think about how I see other people and how learning about them helps me understand who they are and where they come from. It gives me the perspective not to judge anyone. While their stories help me learn from their experiences, they also help me understand that the way they see themselves shapes how they see me. I know that people judge me before they get to know me. Not just for the typical demographic and physical characteristics reasons of being female, Filipina, 5'1", but also because most people are often surprised to learn my age, that I am college-educated (from the most sought-after university in Los Angeles, I might add); that I'm in the military, an officer; and that I work for a Big Four company. Am I bragging? Maybe a little, but what I really wanted to point out is that most people do not expect me to be who I am or to have accomplished what I have.

Sharing my personal story compliments what I wrote for the first book and completes the picture of my journey before, during, and after the MBV program. I liked being honest about the stumbles I experienced in my military career and my journey to this point in my life. As I wrote my story, it helped me realize just how messy life is. Perfection does not exist. We are human and make mistakes, and just because someone is a leader, it does not mean they are free of character flaws.

Being vulnerable helped me confront many emotions—not just happiness, but also sadness, anger, and fear. Reflecting with my goals partner, we were amazed with how well Drs. Robert Turrill and Eddie Arámbula know us after this experience. We met every two weeks for more than a year, and we shared things with one another that we would not have shared if this platform had not existed. We gained an intimate view of each other through our stories. We had a strong support system that was invested in each of us and the overall purpose of the book.

The most healing aspect was the thoughtful feedback that showed strong and deep connections from the support of individuals who truly care about us. The feedback helped us dig deeper and make our own conclusions. We peeled back our emotions layer by layer until we got to the root cause of our feelings. Helping us confront the things we all want to forget or ignore was extremely powerful. When I read through my story, although the pain was there, it was not as raw as it once was, and I found strength in acknowledging my failures.

I hope this book helps others realize that they, too, have unique and interesting stories to share. This form of storytelling is healing, because by listening we gain understanding and empathy. Reading my story again, I realized that a lot of cool things happened in my life. This experience made me grateful for all of those opportunities. With hardship comes resiliency, which is not just a buzzword but something each of us must master. These personal connections improve our resiliency and continued success. I know in my heart I will always have a strong support group because of this book and with other veterans. While my story is a single chapter and unique to me, it's part of something bigger.

Chapter 11

Unconventional Paths
by
Josh Lagana

Josh Lagana is the proud father of two boys with a decade of military service. He served as a military police officer, including for two tours in Iraq, and served in the honor guard, the US Army's presidential escort. He graduated as the distinguished leader in the Army Advance Leadership Course (ALC).

Josh became a senior manager at Deloitte in the company's higher education practice. He advises universities and academic institutions on their technological transformations.

Josh is deeply committed to veteran causes. He serves as an executive partner for the Master of Business for Veterans (MBV) program at the University of Southern California; mentors fellow veterans in Deloitte; and coaches Hire Our Heroes programs sponsored by the firm.

Josh's academic credentials include a BA in business administration with a focus on entrepreneurship and small business management; an MBA with a concentration in finance from American Military University (AMU); and an MBV degree from USC.

A Turbulent Beginning

I have embarked on a remarkable journey that took me from high school dropout to senior leader at Deloitte, one of the world's largest professional services firms. For years I hesitated to share my story, feeling embarrassed and insecure about my educational background. But now I believe my story can inspire others by demonstrating that the past should not define the future. My journey has been marked by trials and challenges after military service while I have continued to pursue personal and professional growth.

I grew up in a turbulent environment with parents who separated when I was three years old. My life was marked by my family's constant moving to different cities. The lack of stability made me feel lost and confused. My father, also a high school dropout, worked as a janitor, but he changed careers to driving trucks after being injured on the job. My mother was an elementary school teacher's assistant and became a special education teacher. She had to relocate often for new jobs. Our frequent moves and the constant change in environments made it feel like I didn't belong anywhere.

I grew up with diverse talents, excelling in sports, music, and math. My love of sports, including baseball, basketball, and tennis, was a constant. Music became a refuge for me. My guitar offered comfort from the chaos at home. But my restless upbringing and uncertain future left me without a clear direction. I was a jack of all trades, dabbling in various pursuits—from juggling in my room to night classes at the local community college. All the while I was searching for an identity.

As my family situation was unstable, I experienced the challenges of a broken home. I had physical altercations with my mother's ex-husband, which made me leave home for days at a time when I was only thirteen and fourteen years old. My high school education was disrupted due to a severe illness that caused me to fall behind my peers, leading to my withdrawal from traditional public school, and I had to attend an alternative studies program. At sixteen, I took on full-time work delivering lost luggage at the local airport. I managed my expenses independently and had to confront adult responsibilities due to my family's circumstances.

Childhood Illness

As I navigated the tumultuous waters of adolescence, my high school responsibilities, relationships, and the pursuit of my dreams all competed for my attention. Then life took an unexpected turn.

I fell ill, and the symptoms were severe and perplexing. I experienced extreme fatigue, swollen lymph nodes, and weight loss. My immune system was under attack. My body showed no mercy. Frightened and unsure of the cause, my family and I sought answers from medical professionals.

Weeks turned into months, with numerous doctor appointments, tests, and consultations. The initial concern was that I might be dealing with a severe disease, even cancer. The dread of such a diagnosis was a heavy burden to carry. My mind grappled with existential questions that no teenager should have to contemplate.

After what felt like an eternity, a breakthrough occurred. Doctors discovered that I had a severe case of mononucleosis, or mono. This diagnosis, while a relief, came after an extended period of suffering and uncertainty. I lost significant weight, and the toll on my body was evident.

As my body began to heal, the reality of the situation became clear. The school days I missed had accumulated to an extent that made it impossible for me to return to traditional public school and complete my junior year. The prospect of falling behind my peers was very real. The daunting task of catching up overwhelmed me.

My health had improved but the educational setback was a hard reality. I was at a crossroads and faced a critical decision. I had to withdraw from public school and enroll in an alternative studies program, which meant a significant deviation from the typical high school experience.

I began attending alternative studies four hours a week. The flexibility of this program allowed me to seek employment to support myself. I found a full-time job delivering lost luggage at the local airport. I was now living an independent life, taking on responsibilities that many teenagers can't fathom, such as buying my own car, paying for insurance, and covering personal expenses.

The experience of working full-time made returning to a traditional public-school environment a challenge. I tried to reintegrate for a few months at the beginning of my senior year, but the radical shift from an independent life back to the structured setting of public school was difficult. It felt as if I were caught between worlds. My academic progress had allowed me to catch up with my high school peers, but adjusting to the traditional classroom was a struggle. Eventually, I chose to return to alternative studies.

While facing these personal and academic hurdles, my family situation remained in turmoil. My mother, a young single parent, had to navigate the complexities of raising my sister and me, and her marriages brought additional difficulties. My stepfather was a physically aggressive alcoholic. This environment fostered anger and frustration in me, making home an unpleasant place to be.

These challenging family dynamics, coupled with my severe illness, marked a dark period in my life. I felt a sense of confusion, uncertainty, and a profound lack of motivation to complete high school. While my peers were graduating, I was still struggling to catch up academically. My high school years had become a battle, a stark contrast to the carefree days of sports and music I once cherished. It was during these trying times that I decided to make a significant life-altering choice.

The Surge

As I sought an escape from the chaos at home, I happened to encounter an Army recruiter. The meeting was a turning point, and it led to an unexpected decision. I had always been fascinated with the military, but I had never seriously considered it as a career path.

In 2005, as the Iraq War intensified, the US Army was rapidly increasing its troop levels. This surge in recruitment led to certain loopholes that allowed for enlistment without a high school diploma or GED. One of these exceptions was based on the number of college credits one had earned. As fate would have it, my accumulated college credits were deemed equivalent to a high school diploma (according to the Army's criteria at the time). At seventeen years old, I enlisted, setting off on a journey that would define my early adulthood.

My enlistment in the army marked a moment of profound transformation. I had a newfound sense of purpose and commitment. It was my way of escaping the turmoil at home and pursuing a more structured path. I enlisted as a military police officer, a role that gave me pride and purpose and aligned with my interest in law enforcement.

The initial training was rigorous. I underwent a twenty-two-week one-station unit training program, which was meant to prepare me for my first duty station. However, fate had other plans. Upon completing my training, I was swiftly assigned to a unit that was deployed to Iraq.

As a result, I found myself on a military flight to the Middle East, barely eighteen years old at the time. A mixture of fear, confusion, and doubt swirled within me. It was overwhelming. I questioned my decision but remained resolute in my commitment to serve the country. The experiences I had in Iraq went beyond anything I could have imagined.

Home Trials

Then returning from Iraq brought a new set of challenges. I faced severe anxiety and struggled to adapt to normal life. The experience of combat left me with more questions than answers. To cope, I turned to competitions within the military. I participated in the All-Army Military Police Warrior Challenge and earned third place. I even pursued joining the special forces, although I was ultimately sent back to my unit after an injury. These extracurricular activities provided a temporary escape from the difficulties of reintegration.

Then my unit was deployed again after a year being home. I found myself back in the combat zone. I celebrated my twenty-first birthday in Iraq. While my experiences brought a set of challenges, they also forged my character and instilled in me a profound sense of maturity, grit, and resilience that would serve me well in the years to come.

My time in the Army was characterized by my constant drive to excel. I earned the distinctions of becoming a paratrooper, a pathfinder, and an air assault graduate. I was also a distinguished graduate in military leadership academies. Despite not having a high school diploma, my records labeled me as "college equivalent," which allowed me to pursue an associate's degree and enroll in courses towards a bachelor's. However, the ambitious spirit within me remained unsatisfied. I harbored dreams of becoming a pilot, but medical issues stemming from my time in Iraq thwarted those aspirations. Despite these setbacks, the dream of achieving something more significant than my current military role continued to simmer within me.

To chase my ambition of becoming a helicopter pilot, I embarked on a journey that led me back to California. I was accepted into a flight program at the College of the Sequoias in Paso Robles. I eagerly packed my belongings, signed my exit documents, and prepared to pursue my lifelong dream.

What seemed like a dream come true at first soon turned into a nightmare.

During my first days in Paso Robles, I attended an orientation that left me brimming with excitement. The prospect of flying helicopters and pursuing my passion was electrifying. Then the unimaginable happened. On the morning of the second day of my flight program, we were shocked to receive a visit from the school's president. His somber announcement was shattering. The Veterans Affairs office had notified the school that it was no longer accredited for veterans' benefits. This meant that we, the students, would have to pay out of pocket until the school resolved the situation with the VA. Our dreams had been ripped from our grasp.

In the wake of this devastating news, some students and I questioned the president about the abrupt notice and the circumstances surrounding the loss of accreditation. Our probing questions were met with anger, as the president accused us of overreacting and not understanding the situation. Frustration boiled within me, and I stormed out of the classroom, joined by others who felt the same way. I couldn't help but dwell on the fact that I had invested thousands of dollars in moving across the country only to be met with an uncertain future.

About an hour later, I received a call from the president of the College of the Sequoias. He asked me why I was so upset. I laid bare the financial hardship that the school's loss of accreditation had put on me. I mentioned my recent exit from the Army and how the program's cancellation had disrupted my post-military career. In response, the president made an offer. He proposed a $6,000 check in exchange for my agreement not to pursue legal action against him or the school. It was a perplexing and difficult choice, but I accepted the offer as it felt like my only recourse given my financial predicament.

Returning to my mother's house was not an easy decision, but it was a necessary one. I was at a crossroads, unsure of where life would take me. I applied for unemployment, a stark shift from the life I had envisioned. The turbulence of my past began to resurface. Returning home left me with a lingering sense of defeat.

My post-military experience was one of tremendous difficulty. I was twenty-seven years old, living at my mother's house, unemployed, and facing the challenge of reintegrating into civilian life. My military skills, training, and experiences didn't always translate directly into the civilian job market. My aspirations to become a helicopter pilot seemed unattainable, given my financial situation and the turmoil I experienced. I had to confront the stark truth that reintegrating to civilian life entailed more than merely securing employment; it involved navigating unfamiliar social customs, reshaping my perception of my skills and identity, practicing the art of self-promotion, and immersing myself in an entirely new vernacular of terminology and expectations.

Moreover, I was grappling with the psychological and emotional toll that my military service had taken on me. The transition from a structured, high-pressure environment to the unpredictable civilian world was jarring. My experiences in Iraq left deep psychological scars that began to manifest as PTSD symptoms. Recurring nightmares, hyperarousal, and intrusive thoughts made it challenging to adjust to life.

Degrees of Success

The job market posed a challenge, and I felt ill-equipped to navigate it. I applied for positions in the corporate world, but without a bachelor's degree, I faced rejection.

I decided to revisit the dream of pursuing college education. A degree would open doors and show the path to a more stable future. It was during this period of reintegration that I embarked on my journey towards a formal education and a career that ultimately led me to Deloitte.

My pursuit of a college education was driven by my desire to move beyond the limitations that had hindered me since high school. The path was not without its challenges. With no high school diploma, I faced several obstacles in the enrollment process. Despite the hurdles, I was determined to secure my place in academia. I explored community colleges, seeking opportunities to demonstrate my academic potential. Eventually I was admitted to a community college, and it was there that my journey towards advancing my higher education truly began.

As I continued to excel in my courses, my confidence grew. I transitioned to a four-year university, where I embarked on a major in business administration. My academic performance remained strong, and I sought opportunities to challenge myself. I engaged in various extracurricular activities where I could apply my military discipline, such as student organizations and nonprofits.

Upon graduating with a BA in business administration, I was confronted with a multitude of career options. Deloitte was not on my radar at that time, but life had an unexpected plan in store for me.

During this crucial phase, I chose to pursue an MBA at American Military University (AMU). Motivated by my military background and a growing interest in business, I viewed it as the natural progression of my academic journey. Though I didn't finish the degree then, after I graduated from USC with an MBV degree, I returned to AMU and completed the MBA.

USC's MBV program emerged as an ideal opportunity to immerse myself among fellow veterans, providing invaluable support during my transition. Getting accepted to USC filled me with overwhelming excitement, as I had once believed it to be beyond my reach as a high school dropout. Despite the rigorous curriculum of the MBV program pushing me to my limits, I consistently ranked among its top achievers.

As I neared the end of the MBV program, I began to explore career opportunities that would harness my skills and allow me to continue making a meaningful impact. Deloitte, a global leader in consulting and professional services, emerged as an attractive choice. The firm's commitment to excellence, the opportunity to work on diverse projects, and the potential for professional growth were all elements that resonated with me.

Partnerships

My path to Deloitte was characterized by dedication and resilience. The rigorous interview process, including case interviews and extensive assessments, pushed me to draw upon the skills and knowledge I had cultivated over the years. I was determined to leverage my diverse background and experiences to demonstrate my value to the firm.

In 2017, I joined Deloitte as a consultant. My role primarily involved advising aerospace and defense industry clients on their finance operations and applying my analytical skills and business acumen to address complex challenges. Transition to the corporate world was marked by a steep learning curve, but I embraced the opportunity to grow and adapt.

My rapid growth at Deloitte, combined with my passion for supporting the veteran community, guided me on a purposeful journey of service and personal development. Within just a year and a half at Deloitte, I achieved a significant milestone: I was promoted to manager, an elevation that came with a profound shift in expectations. As a manager, I was tasked with not only delivering and managing services but also driving business growth and managing the scope of work and major deliverables. The job filled me with a heightened sense of confidence, bolstering my belief in my ability to thrive in the corporate world and especially at Deloitte.

Nevertheless, I remained acutely aware of both my strengths and the areas that needed improvement. Having spent only a few years in the corporate world, I fell behind my peers in terms of industry knowledge. I recognized the need to further develop my communication skills, my writing, and my analytical and critical thinking. This recognition led to a conscious decision to challenge myself, both professionally and personally.

In my pursuit of professional development, I sought opportunities that took me out of my comfort zone. At this stage, I had primarily served federal clients at Deloitte, a setting that sometimes brought me back into military environments. Yearning to prove to myself that I could perform in different contexts, I embraced the opportunity to serve higher education clients, a decision that later became a turning point.

The second important aspect of my journey was giving back to the veteran community. Deloitte offered two main avenues of service: supporting veterans internally through mentorship programs, coaching, and training; and externally through initiatives such as veteran recruiting and development programs. I eagerly engaged in all of these endeavors, assuming a lead role in veteran recruiting, mentoring newly hired veterans, and actively participating in coaching and facilitating training programs for veterans both within Deloitte and outside the company.

However, the initiative that held the most personal significance for me was giving back to the MBV program at USC. I was profoundly grateful for the opportunities the program had afforded me and was determined to pay it forward. Becoming a consulting resource to the program, I helped veterans interested in consulting or in Deloitte specifically. Owing to my commitment and advancement at Deloitte, I was invited to be an executive partner for the program. I embraced the role of a mentor to help other veterans find their path.

Balancing these commitments, I encountered another critical turning point in my career. In 2020, shortly after becoming a manager, I was approached by a Deloitte partner with an exciting proposition. The opportunity to join Deloitte's higher education practice, focusing on delivering services to higher education institutions, was presented to me. What made this opportunity truly special was that my first client in this new role was USC itself, the very institution that had been instrumental in my military-to-civilian transition and my career at Deloitte.

Leading USC through a major finance transformation and various student service transformations was deeply rewarding and professionally fulfilling. It was a significant moment in my career, presenting me with incredible opportunities to demonstrate my leadership capabilities on a larger scale. It allowed me to take on more complex engagements and establish myself further as an up-and-coming leader in Deloitte's higher education practice. This paved a strong path for yet another noteworthy promotion to senior manager, which was a humbling promotion and signified my ability to excel in the corporate world.

My contributions to USC, through Deloitte and my independent efforts, merited me the recognition of being one out of only thirty-one Marshall alumni featured in the "Graduate of the Last Decade" section of the business school's centennial book. This honor left me profoundly humbled and at a loss for words. To be acknowledged as a gold alumnus from the esteemed Marshall School of Business was a testament to my journey. Being recognized in that way underscored for me the transformative power of perseverance and dedication to one's goals.

Priorities

Amid these accomplishments, there were significant challenges to overcome stemming from my military experiences. Following my military service, I was encouraged to seek help for a range of service-related conditions, including psychological challenges such as anxiety, sleep disturbances, irritability, focus and memory issues, and migraines. Through this process I was diagnosed with post-traumatic stress disorder (PTSD) and a traumatic brain injury (TBI). Accepting these conditions was not easy, but I refused to be delusional. I knew the importance of acknowledging my struggles. Seeking help became essential, and I discovered that my support system played a crucial role in my recovery.

As I navigated the challenges of PTSD, my family expanded. My wife and I married and soon welcomed our first son. Our second son followed in 2021. These additions to our family gave me a renewed sense of purpose beyond anything I had experienced before.

However, like any parent, I confronted unique challenges in raising my children. We began to encounter severe behavioral issues with our older son, and the possible developmental issues came under evaluation. Balancing the demands of a fast-paced work environment and my military-related health conditions, I wasn't able to be there for my family as much as I should have been. Eventually, I made the tough decision to leave Deloitte and take a job that afforded me more autonomy and freedom to address the needs of my family.

This led me to join Salesforce, a company known for its excellent culture, flexibility, and family support. However, while the job allowed me to give my family the attention they needed, and to address my own health and wellness, I found myself in a role and with a company that did not fulfill me. I missed the sense of belonging I had at Deloitte. Consequently, I returned to Deloitte, which was warmly welcomed there.

Upon my return, I was committed to pursuing partnership and taking a leadership role in growing the business. I requested a growth leadership role in Deloitte's higher education practice, a request that was granted and further developed for me. I was asked to take a national role in leading a specific service category within the higher education practice, effectively positioning me as one of the eleven growth leaders for the entire practice. While I am still in the early stages of this new position, it represents a significant milestone and a steppingstone toward advancing as a senior leader at Deloitte.

Reflecting on my journey, I can't help but feel emotional. My past, filled with mental roadblocks stemming from my lack of a high school diploma, had limited my perception of myself and my ambition. But today I stand as a leader in the higher education practice of the world's largest professional services provider. My story is a testament to the idea that our pasts should never define who we are or what we can become. The person I once considered myself—a high school dropout and an enlisted soldier—transformed into a senior leader at Deloitte, a business investor, a veteran advocate, a mentor to dozens of veterans, and a proud father.

Not only is it my goal to continue advancing my career, but to give back as well and inspire others to embark on their journeys of self-discovery and growth. My experiences have shown me that it's never too late to change your trajectory, rewrite your narrative, and make a significant impact. I am living proof that with tenacity, courage, and a strong support system, we can overcome adversity and achieve remarkable success. In the face of adversity I found a second wind that propelled me to heights I once deemed unattainable.

"The Profound Impact of Storytelling"

Josh Lagana reflects on telling his story

Since embarking on my transition journey, a process ongoing to this day, I never took the opportunity to comprehensively reflect on and articulate the events that shaped my path. This exercise of documenting my story evoked a spectrum of emotions. There were moments that once filled me with embarrassment. Today, I sit with a deep sense of pride in my accomplishments.

As I delve deeper into this exercise, I find myself revisiting the pivotal moments that brought me where I stand now. It has become evident to me now that my journey was marked by determination, perseverance, and resilience. These qualities have not only enabled me to navigate challenges but also underscored the importance of internal strength and a robust support system for achieving personal growth.

This process has shown me the value of regular self-reflection. Documenting the events leading to my decision to join the military, and the subsequent experiences therein, has allowed me to revisit the emotions I felt, the decisions I made, and their enduring impact on my life.

Through this exercise of documenting and sharing my personal narrative, I have come to appreciate the profound impact of storytelling. Opening up about my experiences not only allows me to connect with others but also facilitates a form of self-healing. It has taught me that sharing your story can resonate deeply with others, potentially influencing them in unanticipated ways.

The journey of reflection has reinforced the importance of introspection and storytelling as vehicles for personal growth and connection. It serves as a reminder to pause and reflect regularly, appreciating how each chapter of our lives contributes to an ongoing narrative of resilience and self-discovery.

Chapter 12

Living the Dream
by
Kathy Takayama

Kathy Takayama is a US Navy veteran and former commissioned officer at the Naval Nuclear Power School. She is an assistant professor of practice at the University of Southern California.

Kathy was nominated for the Women's Excellence Award while working at Hewlett-Packard. She has received a commendation from California state assemblyman Kevin Kylie, and she has been honored by the NBA and Sacramento Kings as a Hometown Hero.

Kathy supports causes that impact veterans and women. She founded Hewlett-Packard's Veterans Employee Resource Network in California; served on boards for Women Veterans Giving and the MBV Foundation; and is the faculty advisor for USC's Military Veterans Association. In honor of her late husband, Mike, she leads fundraising efforts to support the Takayama Memorial Scholarship, which provides $10,000 in annual scholarships to support college-bound women.

Kathy holds a BS in applied mathematics from the University of California, Davis, and a Master of Business for Veterans degree from USC.

Optimistic Heritage
My dad said, "You can't always control your circumstances, but you can control the way you respond to them." That was an interesting

149

perspective coming from a man who lived behind bars twice having never committed a crime.

During World War II, my mom and dad were among the 120,000 Japanese American citizens the US government locked up behind barbed wire and monitored with armed guards.

Thirty years later, my dad's A-6 Intruder bomber was shot down over North Vietnam and he was taken prisoner. He survived the internment camps and months as a POW.

He was the most optimistic person I have ever known. We took it to heart when he taught my brothers and me to make the best of any situation.

I, too, am an optimist, and looking back at my life, I have been very lucky. But maybe that is just the default perspective I take in any situation where I could complain instead.

Nuclear Instruction

Growing up, I was sandwiched between older and younger brothers. It trained me to be resilient and fit in with the guys, a skill I later discovered was a significant advantage in my professional career.

When my older brother neared college age, Dad pressured him to take advantage of an ROTC scholarship like he himself had done when he attended the University of California, Berkeley. The military was not what my older brother wanted, though he did follow in Dad's footsteps and attend UC Berkeley.

When my time came to go to college, Dad didn't even think of talking to me about the military. Was that because of the pushback he saw from my older brother, or because he didn't consider women part of the military? I never thought to ask. I headed off to college at UC Davis.

When it came to selecting a major, I picked something easy for me, applied mathematics, and added an emphasis in statistics and computer science. I had loved tutoring students in math when I was growing up. It came easy to me, and I enjoyed helping others discover that math wasn't as scary as they thought, once they looked at it from a perspective different from what they saw at school. I thought of becoming a math teacher or an actuary, which was supposed to be one of the world's least stressful professions.

Although Dad would say he was a career pilot in the Navy, some of his tours included teaching stints at the United States Naval Academy, and, during his retirement, he taught at the Naval Postgraduate School. So I guess I did follow in Dad's footsteps as I pursued a career in education.

My younger brother also ended up following in Dad's footsteps by earning a Navy ROTC scholarship. It wasn't due to pressure from Dad so much as his financial circumstances. During our childhood, my brothers and I knew from a young age that Mom and Dad would pay for half our college tuitions. We were expected to cover the other half. My older brother and I worked every summer and sometimes through the school year to save up for our education. My youngest brother never worked a summer job—the baby of the family always gets away with so much!—so he had no money to go to college. However, with the NROTC scholarship, he attended a wonderful private school called the University of Southern California. He had to take not only the same mechanical engineering classes as the rest of the undergraduates, but he also had to drill, take military classes, and participate in military extracurricular activities. It was a lot! Plus of course, he had now committed to serving in the military after college.

I continued working part-time jobs to pay my way through school, mostly at a small hometown bank as a teller and loan processor. When a military recruiter approached me at the start of my junior year, I was immediately interested. My grades must have been public record back then. Many mathematicians, physicists, chemists, and engineering students got recruited as instructors for the Naval Nuclear Power School (NNPS). The thought of getting paid to do "nothing" for the military (nothing other than keep my grades up and serve for four years) sounded like a great deal.

Having grown up as a navy brat, I knew Mom and Dad would be thrilled when I shared with them the NNPS opportunity. I had a serious boyfriend at the time, and he had a different perspective. I told him I would decline the offer and stay in California if he wanted me to (i.e., if he put a ring on it). He said that I should not rely on him to make my decisions. I said, "Fine!" and accepted the recruiter's offer.

After graduation, I headed to Orlando, Florida, for four years. My boyfriend and I stayed together during my tour in Orlando where I trained both enlisted and officer students who were going on to operate nuclear powered submarines and aircraft carriers. My boyfriend and I survived a long-distance relationship back when there were no cell phones and we had to pay for long-distance calls. I have photos of him stretching the curly phone cord from the kitchen to an adjacent room where he could close the door for some privacy. Whenever I had a break, I flew to California, or he flew to Florida. Not bad for a guy who had never flown before and was petrified of flying. He always hated to be the passenger.

The miliary taught me leadership skills in a very short time. Since our commitment at NNPS was only four years, identifying potential leaders in the first couple years was critical so that those select individuals could lead (and train future leaders) during their second two years. I don't think many of my undergrad peers were managing a department and mentoring colleagues just two years out of college. Initially, I taught high school-level math to enlisted men, but within six months I started teaching reactor theory. I taught that subject for a year before being selected as the division director for the officer mathematics department, where we taught collegiate math for officers. I led all officer math instructors and oversaw curricula and department operations during my last two years.

I also met some of my best friends in the Navy—friendships which are hard to describe, but when we see each other after thirty years, it's like we were never apart.

Networks and Networks

When I transitioned out of the military, I discovered the value of networking. I returned to my hometown, and when I went to reopen my bank account, it just so happened that my manager from the UC Davis branch was there. She had been promoted to an executive role in the bank, and she told me about an opening at the bank's headquarters for a statistician (what we now call a data analyst). Fortunately, I had a resume in the car, and my college degree included an emphasis in computer science and statistics. Right place. Right time. Right credentials.

My desk sat in a large room with twenty other operations staff. Whenever anybody needed help with their spreadsheets or word processing program, I offered to teach them how to use the tools. And if their computer wasn't working right, they called me over to help.

About six months into my data analyst role, the IT manager stopped by my desk. He said I was doing their job anyways, so would I be interested in working for IT? Working on networked servers, computers, and printers sounded like fun to me, so I started my career in IT. Soon I transitioned into project management, which has turned into a lifelong passion. My military background set the perfect stage for my work in this field: organizational skills, attention to detail, adaptability, and leadership. That small-town bank merged with USBank, and I worked there for nearly a decade.

The IT training company we used for our engineering certifications began recruiting me because they were impressed with my skills in the Novell networking courses. At the time, I was working for a boss I hated, so it was an easy decision to quit my project management job at the bank and get back to teaching. The training company was also a consulting firm in the Sacramento region. I taught networking classes half the time and spent the other half providing IT consulting for small companies. Hundreds of companies around Sacramento needed IT consultants, since it wasn't cost effective for them to keep their own IT departments. Even large venues like Arco Arena, where the Sacramento Kings played, hired us as IT consultants to maintain their networks and every computer in the building. This provided a huge benefit for us—the IT classes at our training center stood out because we included real business situations, making them so much more interesting than the competition.

One of my students was impressed with me and recruited me to join Hewlett-Packard. HP was launching networking on printers for the first time. (Right spot. Right time. Right credentials.) Little did I realize at the time that HP would be my home for two decades. I was the R and D program manager responsible for HP's networked printing products for enterprise customers, and I eventually led all cybersecurity on enterprise printers. This means that if you used a networked printer in a business from 2000 to 2015, most likely my team was responsible!

When HP grew so large that they split into Hewlett-Packard Inc. and Hewlett-Packard Enterprise, each job site had to decide if they would become part of HP or HPE. Since my location became HPE, I transitioned away from printers. I led worldwide telecommunications programs that combined servers, storage, networking, and cloud services. Our first product helped launch the 5G network prevalent on smartphones today. Eventually, I transitioned to a chief of staff role and the operations office for telecom business units at HPE, reporting directly to the vice president of Telecom.

It was during this time that I found Women Veterans Alliance (WVA), or rather they found me. (Right spot. Right time. Right credentials.) The WVA founder Melissa Washington saw my profile on LinkedIn and contacted me to ask if I was interested in meeting with a group of women veterans. I wasn't sure if I really was a veteran—I had never fought in combat, never served overseas, and I wasn't retired like my dad. I had only stood at a chalkboard—did that make me a veteran? Melissa said I helped train the men and women who protected our nation, and that they could not have done it without someone like me. I, too, had signed on the line to say I would give my life for my country.

I joined the small group of ladies at monthly dinners. WVA reminded me how great it felt to be in a room with fellow veterans. Without knowing each other, we bonded the moment we said hello. That group of ladies became the foundational members of what has turned out to be an organization of thousands of women veterans across the world. WVA helps women veterans recognize they are veterans (like they did for me), helps connect them to the veteran services they deserve, and helps connect them to educational and professional opportunities. It was through WVA that I discovered USC. (You can find my story and photograph in the VA pamphlet, "Women are veterans, too!" by James and Mara Morrison.)

Michael

I married my college and long-distance sweetheart, Mike, and we had a son. We had a wonderful life and family with lots of sports. Mike coached at the high school: softball, including a California State Championship; boys' basketball; and then when the school needed a women's basketball coach, his initial three-year commitment turned into twenty-nine years. He became the winningest women's basketball coach in the Sac-Joaquin Section and the second winningest of both men's and women's basketball coaches. Mike was selected Model Coach of the Year and inducted into his high school Hall of Fame. Mike coached many of our son's teams as well.

Mike was a financial consultant, which gave him the freedom to lead practices all afternoon, since he met with clients in the evenings when they were available. And because I could take late-night conference calls with my engineers in India, Hewlett-Packard gave me the flexibility to attend our son's afterschool sports.

At HPE I was nominated for the Women's Excellence Award for my work and volunteering. I founded and chaired the Roseville Veterans Employee Resource Network (VERN), leading our team of volunteers once a month on day-long activities to help veterans in the community. I also led the first annual HPE Military Appreciation Event, which was broadcasted to HPE sites worldwide.

I was recognized by the Sacramento Kings as a Hometown Hero and by California state representative Kevin Kiley for my leadership in volunteering in the community. Some of those activities included leading the local Project Management Institute (PMI) Professional Development Day and sitting on the boards of Women Veterans Giving and Rocklin Little League. In addition to volunteering for team mom on my son's athletic teams, I led a golf tournament as the primary fundraiser for Sober Grad Night when my son graduated high school.

We were upper middle class and lived in a big house on a cul-de-sac with a basketball/pickleball court in the backyard overlooking wetlands and a park. We even had a vacation home in Truckee, CA—which was paradise when the Sacramento Valley got too hot—located just a ninety-minute drive into the Sierra Nevada Mountains. We were incredibly happy.

We had a huge network of friends, colleagues, and family in Northern California, so there was no question that that was where we would live out our lives. After leaving the Navy, I had never lived anywhere else.

Everything changed on June 21, 2016. Mike returned home late after spending the day fishing with his best friend, and that night he had a massive heart attack and passed away.

The following six months were a blur. The only thing I really remember was his celebration of life. Two thousand people attended. The student-athletes he coached shared many incredibly touching stories about how he had been a father figure to those who didn't have a father; how he had literally saved the life of a teen who had been considering suicide; and how the success of his teams became the successes of his alumni, from professional basketball players to collegiate coaches. In his short fifty-six years, he made an incredible impact on so many lives.

It opened my eyes to the fact that you never know your last day. Do what you can to make an impact.

I founded the Michael Takayama Memorial Scholarship, which was a good distraction to avoid feeling sorry for myself. Each year the scholarship raises funds and distributes them to graduating high school students who have overcome challenges to go to college. Mike would love to know that he continues to make an impact in the lives of those who struggle.

As of this writing, we have distributed more than $75,000 in scholarships.

"Don't Be So Sure"

Later that year, while attending a WVA conference, I came across USC's MBV program. It caught my attention since my retirement plan had always been to get back to teaching. I loved teaching in the Navy, and I loved teaching in the corporate world. But I needed at least a master's degree to be eligible to teach in higher education. I had already started an online MBA, which was a waste of time and money—I knew more than the professors, and my classmates were barely college graduates with little to contribute to discussions.

At the WVA conference, James Bogle manned the MBV booth with Tootsie Rolls on the table. I said, "I am happy to hear your story for a Tootsie Roll, but I know what it costs to get an MBA at UC Davis, so I know I can't afford your private USC business degree." He said, "Don't be so sure." He explained that the MBV was a one-year program, which already made it half-price. And there was a network of supportive alumni who loved veterans, and there were scholarships to cut the cost even further. I could continue working my full-time job as I completed the program, since classes met every other Friday and Saturday. It was only two Fridays per month that I needed to be away from work. After more discussions and an information session, and a year of procrastination, I applied for MBV Cohort VI. It was one of the best decisions of my life.

By the time I started the MBV, I had nearly twenty years of R and D experience at HP, plus almost a decade in IT—so how much more was there to learn about the business world? Boy, was I wrong. Every weekend I said, Wow, I wish I would have known that a decade ago. Had I understood that fifteen years ago, I would have accomplished so much more in my career!

And it was work. I wasn't skating by because of my past experiences—it was like having a part-time job on top of my full-time job. Every day after work I read case studies and worked on homework, including Zooming with classmates to work on our projects. I offered mentoring sessions for those struggling with the data analytics class, while other classmates did the same for accounting and finance. We worked as a team toward the goal of everybody graduating. I made some of my very best lifelong friendships in those ten months.

"But What About After?"

My transition wasn't from military life to civilian life but rather from a career in corporate America to my semi-retirement plan of teaching at a university. I wanted to combine the teaching skills I learned in the Navy and in IT consulting with my decades of experience at HP to teach and mentor future business leaders. This was the main reason I wanted an MBA, since it would qualify me to teach part-time as an adjunct professor. During the program, I had lunch with the dean of the business school. We discussed the possibility of me teaching part-time in the evenings as I continued to work full-time at HPE for another five to seven years until retirement.

Then something else came up. As graduation approached, Professor Turrill asked me about my post-graduation plans. I shared that HPE had more strategic roles lined up for me, including working on mergers and acquisitions. He said something like, "But what about after that?" I shared my retirement plan of teaching at a university. He said that I should look through the USC catalog and see what classes I might want to teach, then, while I was on campus, set up informational interviews with each of the department chairs. After looking through the catalog, I found one department that offered project management (my passion), operations (my current HPE role), supply chain (a big part of my telecommunications experience), and statistics (the emphasis of my bachelor's degree). This one department pretty much encapsulated my entire adult life.

I scheduled an informational interview with the Data Sciences and Operations (DSO) Department Chair. It just so happened that the day before my interview, MBV Program Director James Bogle and Professor Turrill had met with the DSO department to plan the next semester of MBV. Additionally, the department chair was married to my managerial accounting professor, with whom I got along very well. When I walked into his office, before I even said hello, I could tell he liked me. We chatted for a while as he shared with me what it was like to be a professor at USC. He asked if I had a resume (you always carry a resume), and then he asked if I might be interested in teaching for him in the fall. My jaw dropped to the floor. I wasn't there asking for a job; I wasn't even considering working for an incredible institution like USC. He told me that the adjunct professor they had lined up had dropped out at the last minute and they had nobody to staff some classes in the fall. How lucky is that? A position for which I was eligible, in the one department I had skills? Right place. Right time. Right credentials.

I said I would have to think about it. I was still planning on working at HPE for another five to seven years before retiring—could I really start my semi-retirement plan early? Could I just pick up and leave Northern California?

Many pieces had to magically fall into place to even consider this incredible job offer. After the MBV, it turned out I no longer needed to be in Northern California. If Mike was still alive today, I am certain I would still be living happily in Northern California. We had a great life there. But as it was, I was on my own. Maybe there was some good in that. Like Dad said—make the best of every situation. My son had just graduated with a BS in mechanical engineering and an MBA and already had a great job working as an engineer at Tesla's Gigafactory Nevada.

At work, I had already been cutting back on my volunteer leadership activities because the MBV took all my spare time. Then I found out HPE was downsizing. Before they laid off any employees, they offered early retirement severance packages for employees with over twenty years of service. I had already been considering leaving HPE, and now the company was offering to pay me to leave.

My entire life had been focused on family and work, but now neither family nor work required me to live in Northern California.

Additionally, I had made such strong friendships though USC and the MBV program that I was confident that if anything were to go wrong in Southern California, there would be plenty of people I could call on.

Accepting the position at USC meant taking a hefty pay cut. I would earn half of what I made at HPE while living in a city that was not known for its affordability. But with the HPE severance package, plus retirement investments enough for two people, my financial situation was such that I could afford to begin my semi-retirement career.

I accepted the job offer to become a member of the DSO faculty. I sold my home, packed my belongings in an orange U-Haul storage container, and before the end of August, I was in the classroom. What a whirlwind of a summer!

Five Years Later

Today, I am still teaching at USC's Marshall School of Business, where our undergraduate program was recently ranked number one in the nation, and our graduate school MBA consistently ranks among the national top twenty programs against competitors like Harvard, Stanford, Yale, and Wharton. For the undergrads, I teach courses in operations, project management, and supply chain management, and I am thrilled to teach operations for our MBV graduate students. I still can't believe I am living my lifelong dream of teaching at a university, particularly at one like USC. I couldn't be prouder of the institution I work for, and I couldn't be happier mentoring so many incredibly bright students. It's so rewarding to consider that maybe a small fraction of students has taken something they learned and gone out and changed the world. Back when I was an HP employee, I knew my work impacted the company and our customers, but that's nothing compared to the power of being a professor.

I was hired initially as an adjunct, or part-time professor, based on my decades of business experiences. Essentially, I was a contractor hired for each semester. Then in 2023, I was hired full-time as the assistant professor of the practice of data sciences and operations. I couldn't be prouder.

It's rewarding teaching the MBV graduate students after sitting in their seats not too long ago. I teach in the last half of the second semester, when the students have six weekends to go. I remind them to cherish their time together. Some students who hold full-time employment spend their lunch hours and breaks on their phones or laptops catching up on the latest crisis back at the office. Others get so caught up in schoolwork that they spend too much time on studying and writing papers instead of spending time with their peers. I remind them of the importance of networking with their classmates, particularly when they are together in person, and getting to know their classmates as well as possible. I tell them that the next few weeks will fly by fast and soon they will miss the every-other-weekend grind. They don't believe me.

Invariably, after graduation, I hear how the alumni are experiencing withdrawals from the program, missing their comrades. This ten-month program is an incredibly special time during which each cohort develops a close bond. We become a family, much like my comrades in the military. Even after thirty years in the business world, the close friendships I developed with my classmates from the MBV program were like no other.

I love teaching the undergrads, too. I frequently tell them that I don't do it for the money. I teach for the love of sharing my experiences and to mentor the incredibly bright students at USC Marshall. I am passionate about the courses I teach and truly care about the success of all my students.

Many of them have shared with me how much they love my classes and how they have considered changing careers because of my class. Undergrads often fear taking the required operations classes, which have a reputation for being some of the more difficult quantitative classes at Marshall. It's always rewarding when students take the time to let me know how much they enjoyed the class. I tell them that means more to me than my paycheck.

I find things to fill my time outside the classroom, too. I mentor and lead fellow faculty, and I am the beneficiary of mentoring, not only from experienced colleagues in my department, but also from the USC Marshall dean's office.

I love to volunteer. Much of my volunteering is focused on helping veterans and veteran causes. I am the faculty advisor for Marshall's Military and Veterans Association and an active supporter of USC's Student Veterans Association. This past spring, I volunteered with the USC Veteran Alumni Associate to clean-up and restore the USS *Iowa*.

One of my passions is helping women veterans, who often face more challenges than their counterparts. Since graduating from the MBV program, I have led panels of MBV alumnae and women business leaders who share the benefits of education when transitioning out of the military.

My most significant volunteering happens with the MBV program. I am so thrilled with what I got out of the program that I am happy to give back. I have been on the MBV applications committee since 2019. When I consider applications, it's important that the applicant is someone we can be proud of when they graduate. They need to demonstrate not only a strong academic background but also a professional trajectory that tells us that, after graduation, with new business skills and the Trojan network, they will go on to do great things.

If I could give anyone advice from what I have seen over the past thirty years, I would say be prepared. You never know what opportunities may present themselves. Be prepared to capitalize on every opportunity, make the best of every single situation, and do what you love. Because, as Mike has taught me, life is short.

My financial advisor tells me I can retire at any time. But I love what I do, and I get personal satisfaction out of finding opportunities to grow and contribute even more to this wonderful institution, to fellow veterans, and to my community.

In ten years, you may find me still doing what I'm doing. I hope to get promoted to associate professor and stick around long enough to become a full professor. In the end, what more can you ask? I tell my mentees to find their passion and do what they love. I found mine and continue to live the dream.

"Bigger and Brighter"

Kathy Takayama reflects on telling her story

I was honored when Professor Turrill asked if I would be willing to contribute to a book about MBV alumni. The MBV was and continues to be life-changing, so I answered Professor Turrill whole-heartedly, "Yes!" When you discover something amazing, you want to share it with others so they, too, can reap the rewards. I told him I would be happy to share my MBV story with anyone who might benefit from our program.

Then I thought about my fellow alumni. Many have overcome unspeakable difficulties or had interesting and incredible military careers. That's not my story. My upbringing was what I thought everybody had: I never went to bed hungry, and I thought gangs and drugs existed in the movies only. I had a loving and supportive family that encouraged me to go to college and start a career. My time in the military wasn't fraught with drama. I didn't travel to foreign countries, nor did I fight in a war. I don't know what it's like to fear for my life or for the lives of my team members. My naval career happened at a chalk board and from behind a metal desk.

I thought, who would want to read my story? Getting started was a little challenging. I felt like I didn't have a lot of interesting things to share. I struggled when faced with a blank piece of paper. Jotting down a bulleted list of main points got me started. After my first draft was complete, I had trouble finding time to get back to finishing. It wasn't until I had hard deadlines, bumping up this task's priority, that I finally sat down to finish.

Maybe my story gives a different perspective on veterans. I proudly served my country and would do it again in a heartbeat. I didn't suffer from PTSD or face many challenges transitioning out of the military. There are many veterans like me who followed their military careers with successful civilian careers. Maybe someone joining or considering joining the military, or maybe a veteran wondering if the MBV is worth the investment, can read my story and follow my path. If I did it, why not them? The story may end with you living a dream that turns out bigger and brighter than what you ever imagined.

Chapter 13

Fail Fast, Fail Forward
by
Lawrence Abee

Lawrence Abee works in corporate finance, executing financial planning and analysis and strategy services for a 110-year-old business based out of Southern California.

A veteran of the United States Marine Corps, he spent eight years in active and reserve duty obtaining the position of platoon sergeant in the intelligence sector. He specialized in providing remote sensing and geospatial capabilities for Marine Air-Ground Task Forces during a critical period in an unconventional theater. He was meritoriously commended twice for leadership and honorably discharged in 2015.

His interests, curiosities, and passions intersect in business, strategy, and design, which allows him to navigate dynamic situations, solve problems, and craft tangible experiences collaboratively. Lawrence resides in Southern California and enjoys playing the piano, building things by hand, tinkering in the garage with motorcycles and machinery, and most importantly, spending time with his wife and three boys.

The Minority

From an early age, I struggled with my family. My mother, a Filipina, married an American man. My birth father died from glioblastoma, terminal brain cancer, when I was two years old. My family spoke Tagalog, and the color of my skin signified to the culture at large that

I was a mixed American. Being born an American but not understanding what it meant to be American created a dissonance in my identity.

My birth father was a Marine with an illustrious family history of military service. My father's ancestral lineage traces back ten generations to Switzerland in 1645 and arrives in America around 1717. After his military service, my birth father became a machinist, engineer, teacher, and real estate private equity funder. I grew up listening to stories about him, vicariously living through storytellers, never knowing him, but maybe having two or three memories—or figments of my imagination.

My mother came from a prominent family in the Philippines whose seven generations include the president of the former Republic of Negros, a short-lived revolutionary government that was part of the First Philippine Republic; Philippines senators; mayors of Manila; and survivors and guerilla fighters of the Bataan Death March. To date, her family operates one of the largest private security contracting firms in the Philippines, employing several thousand people. When dictatorship and civil war broke out in the 1980s, and political families were attacked, my mother moved to America for safety. And that is how she met my birth father.

When my birth father passed, I was sent to the Philippines to live with my grandparents. I don't know who decided that was a great idea, or why. Now that I'm older, I believe my existence was too painful for my mother and she could not look at me after being widowed. Perhaps she thought I would gain a level of anti-fragility by experiencing a culture where people lived in shanty villages rampant with hunger, murder, and a dearth of utilities infrastructure.

But damn, just imagine being the only white kid speaking fluent Tagalog, surrounded by a high density of people who were all tan and brown. Let that sink in and really imagine being the object of social interaction and amusement, being loved and hated, being judged and taken for granted by simple circumstance. As a child I had no resilience.

To this day I remember the sights and smells of the Smokey Mountain garbage dump where people worked, slept, ate, and bathed amid burning heaps of filth. People who smiled showed their resilience. Others who were broken often died by violence, either self-inflicted or otherwise. Eating once a day and pumping water from who-knows-where was common.

I did not know our family was prominent and wealthy when I was a toddler, and I have wondered later in life why I had to experience living in a country that was not my own. I had no idea what an American was despite being one, and living in the Philippines as a toddler without my parents for five years divided me cognitively. I believe my family thought if I were exposed to their homeland, appreciation for my life in America would be much more impactful—it was meant to ingrain humility. My mom visited a few times and my lolo (grandpa) had the great idea that if I practiced singing karaoke then she would take me home. Filipinos love karaoke. It is one of our love languages.

When I returned to America, I had missed out on all the very formative years of parent-child interactions. All I knew was how my grandparents, born in 1926, raised their seven Filipino kids, which is how they raised me, too.

My mother eventually remarried. My stepfather was an amazing man, and he was my dad. In my teenager years I didn't believe it, but time heals. Although my stepfather was also Filipino, he was born in America, did not speak Tagalog, and was a Marine. I appreciate all men and women who can genuinely migrate into a family and raise non-biological children as their own.

I had just started high school when 9/11 happened, and Dad was involuntarily activated in the counterintelligence service. I saw my dad for only eight months cumulatively the entire time I was in high school.

After the Battle of Fallujah in 2004, I knew I wanted to be a Marine. I enlisted through the delay entry program in 2006 and subsequently earned my eagle, globe, and anchor at seventeen years old. I entered the intelligence field and deployed in support of Operations Enduring Freedom, Iraqi Freedom, and New Dawn. By 2009, my younger brother followed and enlisted in the Corps. The men in my household all wanted the same things: to bring back those who were over-deployed; to contribute to a calling which those who came before us had answered; and to fight for a country worth fighting for.

And at the peak of the surge strategy, all the men in my household were absent from home and serving their country. It came with a great cost. We lost a lot financially, but more importantly, we lost the time we could have spent building our family or growing professionally. When I got off active duty and entered the reserves, I took advantage of my educational opportunities and studied at UC Riverside to remain close to my family business endeavors. I worked on a few successful businesses and a few ideas that went nowhere.

The military gave me structure, opportunity, an appreciation for chaos, and filled many other voids in my life. Entry into the armed forces gave me a platform to separate from family history and create a path with a blank slate.

I don't care to talk much about what I was doing in the military because it was a completely different environment. I served during the period of counterinsurgency, supporting important unconventional specialized ground and aviation operations, and they are practically unrelatable with no real civilian use cases.

Historically, the military was oriented around conventional warfare (and today it appears they are returning to conventional warfare). This is important to note, because there is an entire generation of veterans who worked specifically in unconventional warfare.

Because I exited active duty and entered the reserves, I never left the military. In the 1990s, the transition assistance program (TAP) was created. My experience with TAP was an exit interview with a career planner who gave me my fitness report, some annual test scores, clearances, eligibility for reenlisting, and a one-on-one with my commanding officer. TAP was caught between transitioning service members out of the military and trying to keep eminently qualified personnel retained—a few checkboxes on paper that took about a week.

TAP gave conventional checkouts to every servicemember. That was the problem: the unconventional military was entering the conventional civilian work force. There were obviously big issues with that in corporate America. It was intimidating for employers to take young military service members with leadership experience, developed skillsets, and track records of consequential decision making. Conventional career paths don't accelerate at that pace.

But startups do.

Embracing Vulnerability

When I learned of USC Marshall's MBV program, I was considering admirable professional and personal goals. I wanted to explore some of my own issues as they related to military, personal identity, and work.

What appealed to me about the MBV program was the average age and leadership experience of the students, and the program's emphasis on value creation through entrepreneurship. Joining the Trojan family was a plus. I entered the program by grace of the selection committee's belief that my role as a Marine platoon sergeant, coupled with my other experiences, granted me high potential.

The MBV program is about large social impact, transformation, and innovation in an era of change. My education was personalized and specific to my interests. The pedagogy was finely crafted and designed. Most of the curriculum was taught by tenured department chairs and gave me access to a world class education. The time, resources, and scale we engaged at were intentionally executed with purpose from the first day to the last. My cohort significantly added to my education by the multiplicity of their experiences. The things I cherished most were: the meaningful and authentic relationships I developed; gaining a social board of directors; being part of a program focused on problem solving rather than just profits; embracing vulnerability; and fine tuning my career trajectory and growth mindset with my mentors and friends.

Fail Fast, Fail Forward

I sat outside Fertitta Hall for a casual late lunch a few weeks before graduation with Professors Rex Kovacevich and Michael Coombs. They asked me, "So what's next for you?" I told them I had no clue, because the only thing I wanted to do at the time was something different than what I was doing. What I was doing no longer served me, but I had no real plan and no success with unfocused job searching. What I did know, however, was that I had gained tools that needed to be put into practice.

My gut instinct told me to roll with what Dean Jim Ellis said: "We're entrepreneurs. We create value." So I did exactly that. I leveraged my means with no real managerial or strategic goal in mind and effectually reasoned through a few incubators, and I raised scattered conditional commitments to explore autonomous unmanned aerial vehicle (UAV) drones in the first responder space.

In Las Vegas, I met with the CEO of the leading venture in the first responder space and asked them for a strategic partnership. The proprietary device we were developing as a minimum viable product required a patent allowance on the company's assigned claim. In forming a strategic partnership, we could be the connectors for rapid adoption of their autonomous platform across many Southern California agencies. The response was enthusiastic.

A few minutes later over text message the venture capital funders turned us down. Several months of momentum vanished with a single text message.

We pitched to Motorola's venture arm, and our investment thesis was highlighted as significantly substantial. A big takeaway from that experience, and an insight shared among the VCs covering the industry, was that ninety percent of startups and early-stage ventures fail and/or consolidate ideas.

You could do a million things right, but if you could avoid doing a million things at all, that was better. If you could avoid wasting effort and investment by collaborating, there are significant advantages that could propel innovation forward.

I will also borrow some wisdom from General David Petraeus's playbook: Get the big idea right, communicate it effectively, oversee the implementation, and have a process for constant refining.

There's also a saying by Lewis Carroll: "If you don't know where you're going, surely any road will get you there."

As 2020 came around, I pulled the plug on the venture, canceled my commitments with partners with no loss but my own time and skin, and accepted the fact that failure is one of life's greatest teachers.

Five years later, I learned that the leading autonomous drone company had been acquired by Lockheed Martin Ventures, and the Motorola venture partner had become an investment manager at Lockheed Martin. Other emerging industry players ended up consolidating despite hundreds of millions in investment losses.

On the Verge

Three days prior to the start of the MBV program in 2018, my son was born. With the amount of time I spent in the program, and the pressure to excel after business school, came a feeling of remiss that I could have spent my time differently.

2020 seems like a blur now. It was a tough year. In February, my son was hospitalized for significant breathing issues. We had no health insurance and were ineligible for subsidized coverage options. The hospital refused to take him because they didn't think it was a big deal, but my wife made a compelling argument, and I signed a blank check. What we didn't know then, and what everyone knows now, was that he had contracted the novel coronavirus. He eventually recovered. We were lucky to have been exposed earlier rather than later.

At the end of the MBV program, I brought home a small stone as a totem. It was called the final stone. The final stone is an intimate values clarification exercise that Philip Folsom runs. The exercise asks a person to list several abstract concepts, infinitives, or characteristics that define what they value in life. You can imagine how vulnerable it is to write five to ten words on stones and tell everyone in a group why those values are important, only to trade and eliminate them one by one. The rule of the exercise is to end up holding a singular value. Why that matters is because when you operate across several values, it can result in an imbalance or incongruent pursuit of something that is probably unobtainable. These are life's trade-offs. To be present in the moment, free of distraction, living for the here-and-now, requires a congruent application of values. My final stone had Love written on it. I love my family, what I do, where I live, what I make, who I befriend, what I decide... and I love why I do it all.

The totem is important to me because, as it relates to a time and context in my life, I failed fast with the drone venture, my son got sick, society felt on the verge of collapse, and my career was in a rut. The totems are reminders that a healthy mix of risk and doing things for love keeps me going forward.

What I did next was I used my skills in derivatives trading, real estate development, and networking to work independently. I made the next few years extremely rewarding.

My greatest joy came in 2021 when my wife gave birth at home to our third son. We contracted a midwife for the delivery and learned how to deliver the baby if the midwife was absent. Lo and behold, the midwife did not make it to the delivery. There I was helping my wife—filling up the hot tub, communicating on delivering the child—and while she did all the work, I did everything the midwife taught me to do.

Bold Asks

I started working for another family-owned-and-operated venture in 2023, providing management services to a transportation manufacturing company based out of the City of Industry, CA. I operate in finance and strategy, and I work directly for the board of directors and principals to maximize return on investment.

It's a very rewarding opportunity because our products and services enable the backbone of North and South American supply chains to continuously improve standards of living. Now more than ever, our industry is prone to disruption by novel ideas and societal demands that seek to solve challenges. The timing of why I am where I am and what needs to be done has purpose in my eyes. Not one person or company alone can change the world for the better, and allies with common ground are needed.

Remembering the past informs us of what we experienced, how it got us here, and why we did it. Living in the present can provide great purpose for the future. A few years prior to the MBV program, and even a few months after graduating, I had blind spots while turning some corners. High levels of self-awareness helped navigate me, but it was through letting go of things that no longer served me that enabled me to grow. You can't control what is out of your means.

There is so much pressure on military service members and veterans to conform to or avoid altogether the labels that are generationally inherited from the heroes that came before us. I am a veteran, but that gives me no more or less of a right to be an American than any other citizen. My identity as a veteran includes a vested interest in respecting the Constitution and democracy and in restoring the balance in social institutions which the rest of society may be losing faith in. At my core, my family is the most important thing to me. As an extension of my identity as a veteran, I consider family those people with whom I served closely and those who relate to that shared experience.

In my day-to-day engagements with friends, strangers, colleagues, family, and so forth, I seek to reinforce social connection, not disconnection. I choose to be the bridge builder that connects people's beliefs in reciprocity so that we can be part of something greater than ourselves and we can do it together. This means acting intentional and saying only what I mean, promising only what I can do, taking pride in the quality of my work, and operating ethically, even if it hurts. Wherever I am, I try to be the force that untangles conflict and empowers generations old and young to build character.

Lead by example for others to emulate in internal and external corporate venturing. Flip resistance into momentum with bold asks. Be purposeful in work. Generate energy that attracts executive sponsorship to recognize experience and aptitude, not class identity.

Post-military transition is all about my daily interactions to reconnect and be a good neighbor. I build coalitions of tribes that find common ground on defining what sacrifice and values-based performance mean to them, and to convert them to long-term partnerships that put the human-focused experience at the center of our universe. Above all, I strive to be kind and humble.

"Advocacy For Others"

Lawrence Abee reflects on telling his story

The process of revisiting my story for the second volume of *Transitions in Leadership* was challenging and rewarding. I came late to the project, having not checked my school email in a long time. And time was an initial constraint, given work responsibilities and work-life balance. Allocating work on this project required taking it in chunks a few hours at a time on Saturdays.

Revisiting the first story and adding a second was an opportunity to reflect on my personal growth and everything I had learned from school and put into practice. I'm lucky to have failed in some areas of life and bounced back even better than before. I feel that my association with the program and the support offered by Bob and Eddie have been tremendous and a testament to the shared successful outcome.

One thing I learned engaging with Bob and Eddie regarded the representation of cohorts for this book, specifically the lack of involvement in Cohorts VII and beyond. Those students may not have access or opportunity to be involved in this series, which is why I argue that we need some sort of bridge to include those cohorts. Cohort VI was the last group before COVID, and Cohort VII experienced the greatest disruption with Dr. Turrill's retirement and classrooms going virtual. The program I went through is different now than it used to be. This needs to be studied for its evidence-based outcomes. Comparing those outcomes and experiences will be interesting.

I think the individual contributors who have dedicated their time and effort have done this not for their own benefit but to really push advocacy for others, which is motivating for me. Their sacrifices do not go unrecognized. I am grateful for the opportunity, the platform, and to the partners and mentors who have given time to this cause. My hope is that cohorts of past, present, and future remain connected through this book series, through their continued involvement and willingness to share their stories.

Chapter 14

Not Everyone Was So Lucky
by
Lester Ciudad Real

Lester Ciudad Real is a Marine veteran, entrepreneur, and dedicated family man. He joined the Marine Corps in 2007 and completed two combat tours in Helmand Province, Afghanistan. In 2014, Lester transitioned from military service to pursue a Master of Business for Veterans (MBV) degree at the University of Southern California. After completing the MBV program, he joined Goldman Sachs, where he supported real estate development projects.

Lester returned to USC to pursue a master's in real estate development. He has held positions in asset management, acquisitions, and development, and he started his own company, LL Real Estate Group, focused on real estate development.

In addition to traditional real estate ventures, Lester identified a niche in EV charging infrastructure development. He is now focused on developing EV charging hubs with retail amenities.

Everyone's Got a Story

A lot of what I have gone through is tied to my time in the service and to the transition back to civilian life. But at its heart, my story is about the ups and downs we all face. It's about getting back up when life knocks you down, finding ways to change yourself when the same old paths don't work anymore, and never stopping until you reach your goals. My journey is not unique because of the challenges I have faced but because of the way these experiences have shaped my understanding of the world and myself. I am a father, husband, military veteran, and entrepreneur navigating the complexities of life post-service. However, beneath these labels lies my struggle with self-doubt, a search for meaning, and the battle against unseen challenges.

Like many of my fellow veterans, I felt a calling to serve in the aftermath of 9/11. I witnessed in person the collapse of the World Trade Center towers and inhaled the burning debris for weeks. I witnessed firsthand the devastating impact terrorism had on my community. As a first-generation American, I believed it was my duty to serve, and I considered it a profound honor to stand and fight alongside the world's best.

The morning of September 11, 2001, began like any ordinary Tuesday. I followed my usual morning routine and, around 8:30 a.m., turned on the TV in the living room, filling the room with background noise as I prepared to head to school. As I walked past the television, I noticed that a plane had crashed into the North Tower of the World Trade Center. Initially, I didn't think much of it, assuming it might be a tragic accident, a small plane experiencing a navigation error, for instance. However, my thoughts changed abruptly when, while watching a reporter describe the crash, a second plane struck the South Tower. I knew our nation was under attack. It was a moment that forever altered the world and my future.

I remember the shock and disbelief that washed over me as I left my home and walked down the road to Boulevard East in North Bergen. There, I witnessed the towering infernos and, in a matter of time, watched in utter shock as those iconic structures vanished from the skyline. The air was filled with the sound of screams and crying. It is a day etched permanently in my memory.

As the hours passed, my initial disbelief resolved to a deep well of anger, and I recognized that I needed to take action.

Soldiering

I read once that children play by pretending what they want to be when they grow up. As far back as I can remember, I wanted to be in the military. I played soldier with my friends on the block. We shot bottle caps from guns made from wood, nails, clothes hangers, and rubber bands. We played for hours in stairwells and alleyways.

Later, when I learned the Marines had the toughest boot camp and the best-looking uniforms, I had to be a Marine.

I faced many challenges early in life and they prepared me for what was to come. I grew up in Bergen and Hudson County, New Jersey, across the river from Manhattan. I grew up in a blue-collar neighborhood that was predominantly Italian. I remember being one of the few Latino families in the neighborhood. I learned at a young age that someone like me was not wanted in the neighborhood. We grew up poor. My mother was a housekeeper and worked sometimes seven days a week cleaning houses. Eventually, she got a job as a hotel housekeeper. My father worked in a manufacturing plant for many years until he got a job as a doorman at a luxury high-rise building. My parents divorced when I was nine years old for reasons I am not ready to discuss. It broke our family and may have been the beginning of my out-of-control behavior as a teenager.

My mom was always working, and my father was not around. What does a young teenager do? He finds solace in the streets. I hung out with kids that had their own issues at home, and I guess I found comfort in that. But the reality is, if you spend too much time in the street, trouble will find you eventually, and it found me.

However, I had people who cared for me. I had some good high school teachers, coaches, and principals. I am truly grateful for the educators in my life. They didn't let me quit. If it was not for them, I would not have graduated high school.

I wish I could tell you that right after high school I went into the military, but that was not the case. After high school I got a job working in a warehouse stocking shelves. It paid minimum wage, and it was mind-numbing work. I could not believe that that was my life. I remember one morning I got on the bus, dreading to go to work, and at that moment I said to myself there had to be more to life than that job. I hopped off the bus, crossed the street, and jumped on another bus to go home. I quit the job that day and enrolled in community college.

In high school, I was not a good student. Education didn't come easy. When I was at community college I was placed in remedial math and English since I did not do well on the placement test. Early on in my college career, I struggled, and I had to force myself to study. Over time, I improved. Little by little I gained confidence that I could succeed.

Three years later, when I graduated community college with an associate's degree, I enrolled in a four-year school in New Jersey. I had to work a full-time job while going to school full-time, and I commuted to school. I was working twice as hard as my peers.

Life became increasingly difficult. Jobs were not working out and I ended up losing my apartment. I had to sleep in my car for a few weeks until I found a place to stay, but I did not stop going to school. And my age was becoming an issue. I was a twenty-seven-year-old senior in college with one semester left to graduate. I was cutting it close to the age cut-off for joining the Marines. I had to take a deep look at myself. I was struggling financially and emotionally. There were many things going wrong in my life. I needed to decide if I wanted to fulfill my goal of being a Marine.

I called an officer selection office (OSO) recruiter to speak about becoming an officer in the Marines. He told me that because I had a criminal record, becoming an officer was unlikely. The recruiter recommended that I join as an enlisted and work to become an officer on active duty. The next day I went to the recruiting office in West New York, NJ. Four months later, I shipped off to bootcamp in Paris Island, South Carolina. Enlisting in the Marines was one of the best decisions I ever made. I truly believe it saved my life.

Footprints

On June 3, 2007, I stepped on the yellow footprints at Parris Island. It was the start of a new life. I remember the bus ride. We got to the gate at midnight. Parris Island is a four-mile-long, three-mile-wide peninsula. There is only one way in and one way out. The drill instructor told us to keep our heads down, and as the bus stopped, he yelled, "Get off my bus." I ran out and stood on those yellow footprints. My journey in the Marine Corps had begun.

Twelve weeks later and thirty pounds lighter, I shipped off to combat training and then to the schoolhouse. My job was an individual material readiness list (IMRL) asset manager. I managed and oversaw aircraft equipment. It wasn't exactly the wartime job I wanted.

After the schoolhouse I was sent to my first duty station in Kaneohe Bay, Hawaii—not bad for a kid from New Jersey. I was sent to the Marine heavy helicopter squadron HMH-363, called "The Lucky Red Lions." I heard from other Marines that the HMH squadrons took volunteers for aircrew/helicopter machine gunner. I volunteered as soon as I got to the squadron. I passed the physical and other qualifications. They sent me to San Diego for swim qualification and survival training. I successfully passed the training, and I received my airwings and began flying as a helicopter machine gunner. A few months later our orders came in—we were going to Afghanistan.

Helmand Province

We flew into Camp Leatherneck in February, 2010. Our squadron was part of the Marine Aircraft Group 40 and Third Marine Aircraft Wing. During deployment, our squadron completed thousands of combat flight hours, thousands of sorties, and transported thousands of personnel and millions of pounds of cargo between battlefield circulations, inserts, and resupply missions.

We worked sixteen-hour days. We flew constantly day and night. The weather was hot with constant dust storms, sand everywhere. I had two responsibilities. The first was to manage the equipment and ensure the birds were in working order and ready for issue at any moment. My other responsibility was my role as aircrew member. I flew more than 220 combat flight hours that first tour.

I was part of forward operations in Marjah and Sanjin, where we flew Marines and soldiers in and out of hostile areas during enemy engagement. We were often shot at with small arms. RPGs were common. Luckily, I never suffered an RPG attack while on the aircraft. Not everyone was so lucky. Our replacement unit had their sponson (a wing-like structure attached to the fuselage) blown off by an RPG on their first mission out. The RPG blast missed the vital parts of the aircraft, and they were able to land back at base safely.

During missions in heavily-populated Taliban strongholds we were always on the lookout for the Taliban's favorite anti-aircraft weapon: a DShK .50 caliber machine gun mounted on the back of a Toyota pickup truck. I encountered it one night.

We were pulling out packs (troops) in a forward operating base (FOB) in Marjah. Marjah was not far. There was no electricity in the village, and it was a dark night, so we wore night vision goggles. The enemy could not see where we landed inside the FOB. The tail gunner and I pulled gear off the aircraft and reloaded to fly out. The packs, the troops and cargo we carried, were taking a long time gathering themselves and getting into the aircraft. When they finally got aboard and strapped in, we had been on the ground for several minutes. The CH-53 helicopter was extremely loud, and it was a large target. When flying into FOBs with high enemy engagement, you don't want to be on the ground longer than you have to be.

All I could think about was that the enemy could hear the helicopter. They knew where we were. The pilot thought the same because I heard him over the radio telling us to hurry the fuck up. "We are taking off!" I rushed back to my .50 cal. I locked in and made sure I had a round in the chamber. The aircraft lifted off. Once we were twenty feet in the air, we began taking fire from every angle.

Through my goggles I saw the enemy's tracer rounds flying through the sky. Off in the distance I spotted a truck with a DShK mounted on the back. The Taliban didn't have night vison googles and they started firing in the wrong direction. Our pilot picked up speed, took evasive maneuvers, and after a short while we were out of danger. We got lucky that night.

Before I knew it, I was hopping on a C-5 transport plane going home. I arrived back in Kaneohe Bay. I walked out of the aircraft onto a red carpet and welcome parade. I turned in my rifle. My wife came running out of the crowd and jumped on me and kissed me. The local paper caught the image, and it was on the front page of the local newspaper.

In August of 2011, HMH-363 redeployed to Afghanistan and suffered casualties. On January 19, 2012, one of the helicopters crashed and killed the crew on board. They were all good men: Captain Nathan R. McHone, pilot; Captain Daniel B. Bartle, pilot; Master Sergeant Travis W. Riddick, crew chief; Corporal Joseph D. Logan, aerial observer; Corporal Kevin J. Reinhard, crew chief; and Corporal Jesse W. Stites, crew chief.

Camp Leatherneck

I redeployed in 2012 with VMGR-352, a C-130 squadron. This deployment went smoother than the first. I did not fly with the squadron as they did not need volunteer flyers.

On September 13, 2012, Camp Leatherneck was attacked by fifteen Taliban on a suicide mission. They succeeded in damaging several aircraft. Two Marines were killed. Lieutenant Colonel Christopher Raible was killed when an anti-personnel RPG struck the side of the building that housed the squadron's workspaces. Sergeant Bradley Atwell was killed while taking cover behind equipment on the flight line. The attack has been described as the worst loss of US airpower in a single incident since the Vietnam War.

It was a different experience than getting attacked while flying. Rounds snapping and cracking overhead was frightening. As a sergeant, my Marines were my responsibility—to make sure they did what was expected of them, and to keep them alive. The Taliban had been killed within a few hours of the start of the raid, but since we did not know if there were any more, or if more were coming, we stayed on high alert. The ordeal lasted thirty-six hours.

Like many of my peers, leaving the Corps was one of the hardest decisions I made. Being a Marine was a dream. The decision was truly stressful. I made a list of the pros and cons of staying in. At that point I had put seven years into the Marine Corps, and if I reenlisted, I would be committing to twelve years of service in total, and at twelve years I might as well stay until I hit twenty and retire. However, I saw my future, my career, and I did not like what I saw. I decided to leave the Corps and go to graduate school and earn an MBA. That's when I came across the MBV program at USC.

My initial thought was that I wasn't good or smart enough for USC. But then I thought, what do I have to lose? A few weeks after applying, I got a call from James Bogle, the program director. James told me I was a good fit for the program. I was ecstatic. Getting into the program made my decision of leaving my beloved Marine Corps that much easier.

Skill Issue

On my first day of the MBV program I remember wondering if I had what it took to succeed in graduate school. I scanned the room and looked at my peers. I was impressed with the quality of my classmates. Among the class were enlisted, officers, business owners, first responders, and pilots. All were very well accomplished in their fields.

After the first day, we got out of class around 7:00 p.m. I walked across campus with feelings of joy and gratitude that were so overwhelming I had tears. You have to understand, people who grew up like me didn't go to top universities. We were lucky if we went to school at all. I took the opportunity and vowed to make something of it.

Anyone who has served can understand the pressure, fear, and doubts that come along with leaving the service we had committed to for so long. Serving in the military was our identity. Wearing the uniform, the demand of discipline, and the camaraderie—there is nothing like it in any organization or company in the world.

What made the MBV program exceptional was its emphasis on leadership. In the Marine Corps, we were ingrained with the belief that every Marine, regardless of rank, is a leader. This belief holds true, as even young Marines are entrusted with significant responsibilities, including the lives of their comrades. It's a unique experience not found in any other industry or job where an eighteen-year-old can have such a high level of responsibility. Most of us perform admirably.

When we transition to the civilian world, we find that our concept of leadership needs adjustment. In the corporate and business sectors, we are not hunting down the enemy, blowing things up, or supporting those that do. Learning how to adapt leadership skills became a crucial aspect of the MBV program.

In particular, I recall the lectures given by one of the founders of the MBV program, Professor Robert Turrill, who served in the Army during the Cold War in the late 1950s. I learned that leadership is a skill that can be improved, like any other skill. The case studies in the program were instrumental, and I still draw upon the discussions and lessons in my professional life today. During my time in the program, the course on leadership was undoubtedly the most valuable.

Another remarkable aspect of the program was the camaraderie it fostered. There's nothing like it compared to other graduate programs. When you bring together a room full of high-performing veterans, the results are outstanding. We were a group of men and women eager to learn and support one another, and, of course, we partied hard. I forged lasting friendships that I deeply cherish.

I am immensely grateful for the opportunity to be a part of this the, and I take pride in the accomplishment of Cohort II. The professional successes of many of my former classmates are a testament to the impact of the MBV program.

Preparation

I was not a confident person growing up. My background, environment, and influences did not allow me to have the confidence others had. My parents, God bless them, were immigrants with a sixth-grade education. They came to this country to work hard and nothing else. My parents did not know about finance or business or how to succeed in any of those endeavors. My drive and hard work ethic come from them.

During the MBV program I applied to a veteran internship program at Goldman Sachs. I knew two things: it was an investment banking firm, and they liked to hire pedigree. Other than that, I had no idea what Goldman Sachs did. But I took a chance. A few weeks later I received a response from the recruiter for an initial interview. I was happily surprised. After the initial interview I was asked to fly to Dallas where I had an in-person interview. I was extremely happy and nervous. I didn't think I would get the internship, but I decided to give it my best. I focused on what they could ask me in the interview. I studied and practiced and studied and practiced until my responses were polished.

I went into the interview prepared. There were other veterans there to interview for the internship. I spoke with some of them and realized that they were unprepared. That added to my confidence. The interview process itself was tough. I had four back-to-back interviews with different departments. In one of them, I was asked to tell them something about the company. I had studied that question, so I answered with confidence. Later, I learned that out of all the interviewees, I was the only one who provided information about what Goldman Sachs was and what they did. The interview was a success. I left knowing in my gut that I would get the offer simply because I was prepared. About a week later I was offered the internship, and that internship eventually led to a full-time offer.

I lived in Dallas for the internship and all the while commuted to Los Angeles every other weekend to finish the last semester of the MBV program.

At Goldman Sachs I worked in loan servicing operations and supported a team working on real estate development projects that utilized new market tax credits and term loans. I did not love what I was doing. Frankly, I did not love the corporate world. The job was repetitive, reviewing spread sheets and loan contracts. The office environment was not that great either. It was a typical cubical environment, not appealing, and it sometimes drove me crazy. Jumping from the military, where you were always on the move, to corporate cubicles was challenging, but I tried to adapt.

It wasn't all bad. I was introduced to real estate development, and I got to engage directly with developers and construction managers, since I was the one processing their loan draws. I quickly began to find an interest in real estate development and realized that I wanted to become a developer.

My wife had her own career ambitions in education. She got a teaching job at an elementary school in Irving. Then, one year after moving to Texas, she was offered an assistant principal position back in San Diego. She took the job. This gave me the opportunity to move back to Southern California and focus on my real estate development career.

As I looked for roles in real estate development, I faced many challenges. First, I set up as many informational interviews as possible. I reached out to anyone who would talk to me. I found that real estate development shops run lean. Development teams are usually small, three to six people. Most of those people never leave their jobs, which makes it very difficult to get positions at their companies. Adding to the challenge of job availability, real estate development companies want to hire experienced people. Since they run so small, they usually don't have time to train someone. I did not have the skills to get an analyst or associate role, so I decided to go back to school. I applied for and was accepted into the Master of Real Estate Development program at USC.

The MRED was amazing. I went in not knowing much about real estate and I exited a real estate professional. The program leaned heavily on the financial analysis side of the business. I learned how to model from scratch, how to underwrite, and how to understand the capital markets. And of course, I gained another great network.

After graduating, I set out on my real estate career. My goal was still to get an associate or analyst role for a multifamily developer. Unfortunately, development shops were further limited in San Diego. I had a difficult time finding a job. I networked and got several interviews, but no offers came. I looked for roles in other asset classes such as office and retail.

I found a job posting for an analyst in an office owner/operator. The manager was a former Navy SEAL who had also previously worked at Goldman Sachs. He had gone through the same veteran internship program a year before I had. I interviewed for the position and was hired as an asset management associate. I was part of a team that oversaw 2.5 million square feet of office space.

I learned a lot as an asset manager. I learned how the leasing process worked from the landlord's perspective, how lenders played a role, and how to add value to the assets I managed. All were extremely valuable skills that have helped me today.

A year or so into my asset management role, I received a call from a company I interned with during the MRED program. The company's managing director offered me a position on one of their teams developing industrial projects in Utah and Tennessee. It was an opportunity to work on development deals. I instantly said yes. The role was a great learning experience. I ran point almost every deal. I mastered the skills I had learned in MRED and as an asset manger.

Risks of Entrepreneurship

When the COVID-19 pandemic swept the United States, countless individuals found themselves unemployed or facing layoffs, and I was no exception. My initial response was to scour job websites, especially those that catered to the real estate industry. However, the available roles and salaries on offer were disheartening and, frankly, infuriating. I reflected on my accomplishments and the valuable skillsets I had acquired over the years. It dawned on me that I didn't have to work for someone else. I could establish my own real estate development venture. That's exactly what I did.

A fellow MBV alum and I identified a potential business opportunity in the EV charging space. Leveraging my real estate experience and working with a major EV manufacturer, we cofounded

StackCharge, envisioning it as the "fueling station of the future." Despite the corny tagline, our objective was clear—to support the growing influx of electric vehicles by creating charging hubs with retail amenities.

The venture was unconventional in the United States compared to the UK and other European countries where EV charging hubs were more common. Developing the hubs was challenging, particularly in acquiring power for the sites. My background in real estate development became instrumental in overcoming the challenges, because our charging hub concept was similar to a traditional gas station model.

While we encountered enthusiasm and support from EV-charging manufacturers and retail operators, raising capital posed a significant hurdle. Numerous private equity firms and high-net-worth individuals admired the concept but hesitated to invest until we had operational locations. This repeated rejection emphasized the importance of having well-established relationships. A hard lesson we learned early on.

We also encountered failures. Our initial acquisition attempt resulted in financial loss. But our second attempt led to the property we now own in Baker, CA. The property is positioned along the I-15 freeway between Barstow and Las Vegas. The Baker site is California's first operational StackCharge, boasting forty DC fast-charging stations and a national coffee brand as the retail amenity. Despite facing challenges such as power delays, fluctuating interest rates, and a year-long search for the right retail partner, we successfully raised capital, secured debt, designed the site, and partnered with a large EV-charging operator. The Baker site is set to be fully operational by mid-2024.

Embarking on this entrepreneurial journey has been one of the most challenging endeavors I have had to overcome. In entrepreneurship, the path forward is not always clear, and learning through trial and error or seeking guidance from mentors becomes essential. Mistakes, some costly, have also been my teachers. I have learned the importance of continuous learning, and that adaptability is paramount to success.

Patience has proven to be essential. Rushing decisions can lead to mistakes. While errors are inevitable, minimizing them is crucial. I also learned that choosing business partners wisely is extremely important, as it can make or break your business. Ensuring clear roles and responsibilities and having transparent conversations early on in the process are critical for long-term success. Raising capital remains a challenge and a skill I strive to improve. The ability to sell your vision or product is paramount. Above all, the most important lesson is learning how to deal with emotion. The highs and lows of being a business owner can put you to the test, and if you don't get a handle on your emotions, they really affect your business and personal life.

"Moments of Self-Discovery"

Lester Ciudad Real reflects on telling his story

From a New Jersey street kid to a Marine serving in Afghanistan, then onto the classrooms of USC and the challenges of real estate development, my journey proves that success is possible through perseverance, community, and the strength found in embracing vulnerability.

My experiences have taught me that growth is a continuous process, shaped by victories and failures both. Each step forward or backward has taught me a lesson in resilience, adaptability, and the pursuit of purpose.

Serving in the military equipped me and my fellow veterans with unique strengths and a spirit that many do not find on their own. Embrace the challenges as opportunities to learn and grow. The path may be unpredictable, but it is yours to shape with the same courage and commitment you showed in uniform.

Telling my story has helped me understand and recognize my strengths and weaknesses while reinvigorating my values. Writing my story motivated me to continue achieving my goals and has been a crucial part of my healing process.

Joining fellow MBV alumni to share our stories was driven by a desire to help other transitioning veterans. Many veterans lack a clear guide to assist them in their transition. My hope is that my story will help at least one veteran or active-duty member navigate their transition successfully.

The most impactful elements of this experience have been the moments of self-discovery. Through this process I gained a deeper understanding of my own transition and journey. I believe it is important to share our stories with those who did not serve. Our stories can help bridge the gap between veterans and the broader community, fostering empathy and understanding. Writing revealed a lot about myself. Often, we forget what we accomplished in our lives. My story reminded me of the grit I need to succeed. During this season of my life, those reminders have helped me push through the hard times. We all go through tough seasons and need reminders of our strength. While crafting my story, I encountered emotional hurdles and at times struggled to revisit certain memories. However, the support from fellow veterans and the therapeutic process of writing helped me overcome the challenges. The camaraderie of my fellow veterans has been invaluable.

Participating in this project was a transformative experience. It highlighted the importance of sharing our diverse and complex experiences, which can, in turn, help fellow veterans in the future. As I move forward, my goal is to take what I learned through this process and continue to help the veteran community. My storytelling journey has been not only about sharing my experiences, but about understanding myself better, healing from past traumas, and inspiring others.

Chapter 15

A Better Version of Myself
by
Lucas Lenhert

Lucas Lenhert left active duty after five years of service and two deployments in the army. While pursuing a BS in geological science, with an emphasis on environmental resources, from Cal Poly Pomona, Lucas joined the Army National Guard (ARNG). He went on to serve six years in the ARNG.

Lucas graduated with honors from the Master of Business for Veterans (MBV) program at the Marshall School of Business.

Since then, Lucas cofounded a nonprofit organization (The Sea Hut Foundation); volunteers for and holds an executive position with the Global Business Incubator (GBI); and founded his own business, VetHut Consulting LLC (VHC).

Currently, he operates EnneaPro Coaching, which guides small businesses into government procurement through targeted research, business development, team training, and management services.

The Logical Choice

Even at an early age I was planning out my future. In middle school I thought about what I wanted to do for a living, and by the time I was a freshman in high school I was eager to get away. But before I left, I needed a master plan. I needed to know where I was going, how I was going to get there, and what I wanted to do. I wanted to get out of the house and see the world, experience everything I could, make some money, and see what life had to offer.

I was a junior when I figured out the perfect career. I wanted to be an underwater welder. My favorite classes in high school were welding and metal working. I could quickly finish the projects, and I was allowed to create my own projects and develop my welding skills further.

I am not driven by money, but growing up in a single-parent home, money was a constant struggle. Choosing a path that enabled me to be financially secure, like being able to buy a house, to travel, and to spend quality time with my family, was constantly on my mind. By the end of my welding class, the teacher told me I would make a great welder. I didn't think I could earn a decent living doing it, and I was surprised when he told me that underwater welders make a lot of money. Initially I laughed because I thought he was kidding. But I began researching it. Soon enough I had the next six years of my life planned out. First, I would graduate and join the military. I would serve four years in the service and then ETS out. Then, starting with year six, I would use my educational benefits to put myself through a commercial diving trade school and become a commercial diver and underwater welder.

The military seemed like a logical choice considering the educational benefits I would receive. There weren't any family resources to pay for commercial diving school, and I didn't want to be burdened with debt. I gravitated towards the Navy. Common sense told me the Navy was where divers would be. But that's not where I enlisted.

I was at a farmer's market having a conversation next to a US Army recruiting booth. I was just about to walk away when the staff sergeant caught my attention and got me to talk to him. He said, "You look like you're in high school and about to graduate. What are you going to do for a career?" I didn't think the army was for me, since I wanted to be a diver. He told me the US Army also had divers, and he happened to have a brochure on that very occupation.

The conversation with the staff sergeant altered my future. I made three attempts to talk with a Navy recruiter, but every time I went to their office nobody was there. I took it as a sign from God that He had given me my path.

I Failed

My first enlistment was rocky. I had started with a blueprint for the next five years, and after five months I had lost all direction.

I integrated into the military easily. Following the rules made sense. I started basic training in July, 2001, and for the most part I had fun. Really, I was looking forward to it, like an adventure.

On the first day of basic training, when all the new recruits from all the companies stepped off the buses, I stood in formation. Multiple drill sergeants walked around and took turns yelling at us. One of them caught my eye. He bore a striking resemblance to Damon Wayans in the comedy *Major Payne* (1995). He even had the gold tooth to match. It was just too funny. Major Payne saw me smirking, and he stopped an inch from my face and started yelling. I had a hard time holding it together. I had to bite the inside of my cheek to keep from laughing or smiling, but it didn't work.

I soaked up everything basic training had to offer, and I was excited to move on to dive school. Dive school had an eighty-five percent dropout rate. Eleven days into the fourteen-day weed-out phase, I became part of that statistic. It was heartbreaking, and it had a dramatic effect on my confidence. I was convinced that all my plans had turned to dust.

When I failed dive school, I was given a few days to decide what to do next. They gave me some choices for jobs, some mechanical and some combat arms roles. Initially, mechanic seemed like the logical path because I would learn new skills, and when I got out of the army, I could use those skills to get a job. But my intuition told me to do combat arms. I didn't know where I was going, and I panicked. I had not talked with anyone about my feelings about my failure. Had I been able to lay my feelings out on the table and sift through them, I could have faced the pain of failure and not let it derail my plans.

Regardless, when the Army made me a mechanic, I was fine with it. I picked up my feelings of disappointment and defeat, put them in my emotional lockbox, and threw away the key. A part of me just wanted to move on instead of dealing with the defeat. I was embarrassed, and because I was embarrassed, I didn't talk to my family or friends about the situation. I kept my feelings to myself. Lack in confidence had taken over my mind.

Now I can see that losing confidence restricted me for years from using my full potential. I started moving backwards in life, and the hurt I felt caused me to turn to alcohol.

I learned that running from painful situations leads me down the path towards destruction. As a young man, my poor decision-making took the forms of hanging out with the wrong crowd, spending too much time and money on alcohol, and being lazy. I can't go back in time and repair those mistakes, but I can learn something from those experiences.

I spent a lot of time going at things alone. Unfortunately, in my childhood and young adulthood, I was let down so many times that I found it difficult to trust others, so I went through life not seeking help and making important decisions on my own. The problem with relying on myself was that I didn't get other points of view. It was hard to think outside the box.

Three years into my military service I realized I had made a mistake. I didn't want to go back, because I'm forward thinking. So, I applied for the Green to Gold scholarship, but I didn't get it. Then I applied to warrant officer school. My ambition was on full throttle and the only thing that stopped me was a second deployment.

After working as a mechanic, I knew I would not continue after the Army. I pushed to become a noncommissioned officer (NCO) so I could learn to manage people and operations.

Sometimes You Just Need a Good Shake

I need others to help me become a better version of myself. Sometimes those people have been right in front of me, and I haven't reached out. I often don't want to ask for help. Other times when I have been on the verge, and I needed to feel like someone really cared, I have tested them, but it didn't get me what I needed. Sometimes people come along who know what to do when they see someone in need—those people are amazing. Other times I have had organizations pull me in, rough me up, and spit me out a new man.

It was my second deployment (Kuwait, 2004) when I was called to the command tent and told that my dad had passed away. I never really knew my father. When I was a toddler, he was involved in a terrible accident and suffered severe brain damage. He was never a presence in my life. But still, when I heard of his passing it punched me in the gut. I immediately turned around and walked out of the tent. I felt the tears coming. I did not want to cry in front of my first sergeant.

Sergeant Henderson happened to be walking by and saw the tears in my eyes. He didn't say anything, he just grabbed me and hugged me as if he knew what was going on. If he had asked if I wanted a hug, I would have pushed him away or told him to fuck off. But he did the right thing. You see someone who could use some help, and you shake them. Later that evening I remember feeling comforted by the hug. Suddenly I had someone I could trust with my feelings, and we talked several times after that. Sergeant Henderson was an example of great leadership. He had no responsibility to me. In fact, we didn't really know each other outside of giving or receiving paperwork. As he walked by me, he could have very easily looked the other way. But he took action. I'm grateful that he came along and shook me.

Expiration

I posted at the same permanent duty station from 2002-2006. I ended up becoming an NCO there.

I had to make the hard right choice to leave over the easy wrong choice to stay, because something unexpected happened that was enough to tip me towards letting my expiration term of service (ETS) date come and go without renewing.

I had a strong desire to serve my country and there were aspects about service that really appealed to me. I valued the fact that I could change my career, that professional development was included, the tools were paid for, and there was the potential to see the world. The career potential was particularly intriguing for me. The idea of being able to stay an NCO and go up the ladder, go warrant, or go to college and go officer seemed like a fun journey.

On my last deployment, I held many positions as an NCO, including squad leader, scheduled maintenance section sergeant, and unscheduled maintenance section sergeant. One of the soldiers in my platoon was—to put it lightly—not a good soldier or team member. He had stolen from soldiers in our platoon, consistently violated rules and regulations, and was constantly disrespecting NCOs and other soldiers. He had been moved from one squad to another several times and placed under different NCOs in our platoon. Nobody could control him. Finally, he was placed in my squad. He came with a UCMJ judicial punishment and a field grade Article 15 non-judicial punishment. This soldier had literally been sent to a military prison in Kuwait while we were on deployment, and when his time was up, they sent him back to our platoon.

The time and effort this soldier required, and the lack of effort he reciprocated, plus the stress he put on the whole platoon, far outweighed any work he produced. I had to confront him several times, and the conversations ended with me fearing for my life. Eventually paperwork was filed from a different command for the soldier's separation from duty. I had to testify in a court of law while on deployment.

By that point I had lost respect for many in my command, and I was willing to go around them for my own health and safety. Even though what I did was legal, I was indirectly punished for my actions. The leadership pulled several opportunities from me, and it made me angry enough that I gave the middle finger to reenlistment. I had several people tell me to not let my first duty station determine whether I would stay in the military. But as I weighed becoming a commercial diver against continually having to deal with similar situations as an NCO again, I wanted to let my ETS date happen.

Commercial diving was on my mind again because a buddy told me of his plans to go into commercial diving school when he ETS'd. I had written off underwater welding after dropping out of Army dive school, but for the last six months of my deployment, it was back on my mind. It had suddenly occurred to me that just because I didn't become an Army diver, that didn't mean I couldn't become a civilian diver like I had originally planned. A seemingly random thought reignited the old flame and it further tipped me into ETSing. When I went on leave for my father's funeral, I visited the commercial diving school to check it out. I signed up and had a start date for the next year. I got my confidence back.

I worked as a commercial diver for three years, and I loved it. If I hadn't also wanted to start a family, commercial diving is what I would be doing today. I only got out because I knew I wanted to be a father. I wanted to be around for my child, and working in that industry would hardly give me the balance I needed for my family.

My family is a value of mine. What that means is that, in nearly every decision I make, I consider if it helps or hurts the family. Not if it helps or hurts me, but the family. If I want to make better decisions for myself, that is a good place to start.

Vet Groups

Initially, my transition out of the Army was easy. I had a goal going in and plans coming out. The plan didn't go exactly as desired, but I still achieved the end goal.

After graduating dive school, I was at my celebration dinner with friends and family, and I couldn't help but think that just five years earlier, my feelings of failure had consumed me. It used to feel as if all my plans were crashing down. But just look at me, I graduated dive school. I had succeeded, not failed.

The 2007 and 2008 recessions impacted the diving company I worked for. Work slowed enough to the point I thought about using my VA benefits to attend the local community college. I left the commercial diving industry and began pursuing a bachelor's degree. I never thought I would see myself in college.

I also jumped at the chance to go back into service, and I joined the Army National Guard. I loved stepping back into the uniform. I changed my military occupational specialty to infantryman (MOS 11B), and off to school I went.

With the money from the GI Bill, I didn't need to take on extra jobs. I had plenty of time for studying, exploring opportunities, and experiencing what the university had to offer. The National Guard culture was quite different from active duty, but I adapted and had fun. I fell back into military life seamlessly. I reverted to the same attitudes and behaviors I had in the military—most notably I became a champion cusser.

I was a year and a half into my bachelor's degree when I transferred from Washington State University (WSU) to Cal Poly Pomona, and there seemed to be some trouble setting up my educational benefits at Cal Poly. Simply speaking, I was lied to several times by the staff member whose responsibility it was to certify VA benefits. As a result, more than two months into the school year had passed before any payments came in. I had to take out loans. At a certain point in the process, I went into the office and got the staff member's attention. In a not-so-polite way I asked why she was lying to me and not getting my educational benefits paperwork processed. I turned some heads in the office. After that, the college took the responsibility of certifying VA benefits away from that individual. It was not the first time a veteran had had issues with the particular staff member. I was glad to see that the university recognized the flaw in their system and took steps to fix the problem.

Before the end of the year, I found a good community of veterans at the college. Working at the Veteran's Resource Center (VRC) made a huge impact. Working there really helped me navigate between civilian and military life. The VRC coordinator was heavily invested in helping veterans. She put high expectations on her staff, and we got excellent training which helped us in mentoring other student veterans. The VRC built an environment where people felt comfortable asking questions, sharing, and participating in events. Part of our training included meeting the department chairs, so I got to know everything about each department so I could efficiently help anyone who came to the VRC.

One day, I was talking to a former Green Beret, and during the conversation he told me he had toured many schools in the area and what led him to choosing Cal Poly was how welcoming and knowledgeable we at the VRC were.

The VRC was not just a place to study or get information about educational benefits. The space allowed veterans to openly talk about their struggles, and the staff were well equipped to mentor them. For the first time I was able to discuss my own issues and struggles, including the struggles I had with the college environment and culture.

Most college students just wanted to have fun. Working hard and learning didn't seem to be a high priority. More often than not, I was frustrated with my group project members for not doing their part. I always made efforts to try and help them by doing some of their research and providing them with notes. In the military, if I got frustrated, I yelled, and the work got done. (That only happened a few times.) I had a hard time not letting a few people get me so frustrated. That frustration caused stress, and that stress pulled time and energy away which I otherwise would have spent on my studies. I could see that these situations had something in common—anger—and that I had handed my decision-making abilities over to my feelings. Luckily, I had a place to talk to other veterans.

Talking to other vets helped me better communicate with college students, especially working in group projects. I learned to be frank or honest but not rude.

Skill Sets

Near the end of my bachelor's degree I decided to go to graduate school for public administration or business administration. Through the VRC, I got in touch with a member of MBV Cohort III, and we discussed the program. Before that discussion, I knew nothing about USC. I didn't know what I was getting myself into. I took away from the discussion the idea that the MBV program was going to be a great journey. I wanted a ticket on the MBV train.

Immersing myself in the veteran community had been a crutch for me in transitioning to civilian life or just dealing with life in general. But when I first got out of active duty, I did not seek out a community of veterans, though one did end up finding me. After researching the MBV program, I remember thinking to myself, veterans really know how to build cool shit, and that's where I need to be!

On orientation day, a guy walking in the same direction as I was introduced himself and asked if I was attending the MBV program. How did this dude know I was a veteran? I thought, "It's fate that I'm here, attending this program, with these awesome people!" It wasn't the degree itself I was so excited about. It was the experience and journey I was about to have with these awesome people.

I looked forward to every weekend and everything it brought because the cohort seemed ambitious, motivated, and everyone was excited to be there. A huge difference between getting my bachelor's and my master's degrees was that I never felt any anxiety in my group projects. I was always comfortable within my group. I had confidence in the people, and we worked like a team. We did what we said we would do, and I trusted my team. In the MBV, my fellow students had careers, families, had been deployed, and much, much more.

Working in small teams is where I thrived. I felt at ease giving my thoughts and opinions, asking questions, taking initiative, and challenging the ideas of others. When I say, "challenging the ideas of others," what I mean is that I am less likely to people-please. I have learned to listen more attentively, because I can appear condescending, patronizing, not present, or even aggressive. Even though I don't mean to be, I occasionally come off in those ways and it has limited my capabilities as a leader and team member.

As I looked back and inspected which of my attitudes and behaviors led to less than desirable outcomes, I recognized that my anger and frustration were driven by impatience. I have an inverse relationship with patience. Where it comes from, why having patience is so difficult sometimes, and how I can improve, these are things I have identified since the MBV program, and I'm working on improving my interpersonal communication skill set.

Becoming a Better Version

During the MBV program, I wasn't just learning about business concepts and strategies, I was learning about myself: how I operated, how I saw myself, how I communicated, and identifying my strengths and weaknesses. Some of the things I learned about myself and about my relationships were eye-opening. I not only learned about emotional intelligence, but also how to apply the concepts to my relationships.

There was a moment in the MBV program that made me do a lot of thinking about self-development and self-awareness. We read *Working with Emotional Intelligence* by Daniel Goleman. The book includes an assessment about leadership styles. After taking the assessment, I felt a little disappointed. I wanted to be a "commanding leader" but tested as a "democratic leader." After thinking about how many past situations matched this leadership style, I could not deny that the "democratic leadership" style was a perfect match.

That assessment, and the others, really got me thinking about how I could be different. I wanted to get my points across in more desirable ways, I wanted to be more persuasive, I wanted more patience, and I wanted better listening skills. One specific weakness I identified was a lack of patience. And not just with others; I can be really impatient with myself. My lack of patience has been a limiting factor throughout my life. It has been a constant battle. It contributed to the end of some relationships. I have pissed off countless people. It has cost me opportunities. I have received tickets because of it. In short, I have made others feel bad.

I finally realized how badly my impatience had been limiting me when I started having constant arguments with my wife. Yelling was frequent. It came from my struggle with listening. I assume I know what the other person is going to say, or a point they are going to make, and because I'm angry or frustrated with them, I stop listening. I cut them off, talk over them, or take what they are saying personally. I don't listen well because I assume I already know the answer. It comes from a place of anticipating the future. When someone is talking to me, I might be thinking about the end of their argument while they are mid-way through their point, and I interrupt them.

After gaining this insight, I devised a plan. When I'm communicating with my wife and getting frustrated, I remove myself from the situation. I work on centering myself and separating my feelings from objectivity. Then I come back to the conversation with a fresh perspective. Instead of reacting to my feelings, I trying to respond to them. So, I might tell myself, "I'm feeling impatient and that's not going to lead to a win. I want a win, so here is what I can do to not let my impatience get the best of me."

Instead of being impatient I can instead be persistent, thorough, and focused. Depending on the situation and what is needed to get a win, I tell myself to do the other things. My system isn't perfect but it's a start. Since employing this strategy, it has completely changed communication within my marriage for the better. Prior to having this insight about myself, my wife and I had gone through a year and a half of marital counseling.

The tools I was introduced to through the MBV program showed me that I had more potential than I was currently accessing. Through self-reflection I was able to see how my impatience limited my communications skills. If I'm not listening well, cutting people off, and getting angry, then others are not going to feel comfortable around me. They are not going to enjoy working with me, and they are less likely to want my input. That closes off important lines of communication in both my personal and professional lives.

For example, I loaned some training tools to a colleague with a verbal agreement that I would get the materials back. Later, I made several attempts to get the tools returned to me, with no success. I started getting impatient. I recognized that my impatience was turning into frustration, and that frustration was turning into anger. I knew that if I didn't try a new approach, I might not get my tools back at all, which would complicate the relationship I had with the colleague. I asked myself what I needed and wanted from the situation. I needed my things back and I wanted to maintain a good relationship with my colleague. Would my frustration and anger get me where I wanted and needed to be? No. Then what actions could I take? I put aside my feelings and focused on being friendly but persistent. I made sure to communicate in a way that reflected positive energy. Positive persistence got me what I needed, and my colleague and I could still be friends. Because of my awareness of the situation, it turned out to be a win-win, when it could have easily been lose-lose.

The MBV program provided a comfortable environment for me to explore my future potential. It challenged what I thought about myself, my future, and my capabilities. The idea of having a goals partner (aka battle buddy) was one way in which the MBV shook me. Partnering with a person whose responsibility was to ask me what I wanted to do, what my plans were, what my aspirations were, what possibilities were open to me; to have someone who would listen, follow up, and hold me accountable was refreshing, challenging, and inspiring.

MBV in Action

My son was born a couple months before I started the MBV program. My wife was already well into her career as an attorney, and she did not want to stay at home. We both felt it was best for one of us to be a stay-at-home parent, so I quit my job as a field engineer. Not having a full-time job allowed me to immerse myself in the MBV material, and I was able to participate in many of the extracurricular events. However, not having a job outside the home limited my ability to test out the leadership concepts I was learning in class. While I do not regret the decision I made for my family, in some ways I missed out on a professional growth opportunity. A silver lining was that instead of applying my leadership capabilities in the workplace, I used them to work on my marriage.

After graduating from the MBV program, I had a lot to think about. I wanted to do something with all my ambition. But I couldn't go and get just any job because I was still a stay-at-home parent. If I started working, it would have to be part-time from home. One night I sat down with all of my ideas. I looked at all the factors that were affecting my life and work and began going through some future career possibilities.

Part of the reason why I chose the MBV program was its focus on entrepreneurship. I hoped to get clarity on and explore my entrepreneurial ideas. Learning about small business certifications led me to research government procurement. I attended government procurement workshops, and I volunteered for nonprofits in the government contracting industry. All of this led to a business plan with a marketing focus on government contracting. The MBV program was pivotal in molding those entrepreneurial desires.

During the program I built a feasibility analysis around my idea for a nonprofit that would bring better educational and career knowledge to transitioning veterans. After graduating, I began building a business blueprint out of that analysis. One day, I met up with an old buddy who was also a veteran. It turned out that he was working on a business plan for a nonprofit to provide a scholarship for student veterans. We talked about our ideas and decided there could be some synergy between them, so we began building one business plan together.

A year later we started the business and began the process of turning it into a nonprofit. My partner wanted to hire a lawyer to help us file the 501(c)(3) paperwork, but a lawyer would have cost thousands of dollars. My business degree from USC had given me some big balls, and I'm a tenacious person by nature, so I convinced my partner we should give the paperwork a go ourselves. The application was twenty-eight pages long. After we had finished putting it together, it ran about seventy-five pages. We assumed the application would get rejected a few times, but we were determined to keep submitting. Our 501(c)(3) status was approved the first time. By doing it ourselves, I was able to learn the ins and outs of nonprofit regulations, standards, operation, and management.

After cofounding the Sea Hut Foundation (SHF), I began building another business plan and doing some side gigs building plans for others. Cofounding the SHF was a way to get my feet wet before starting my own business. I was inspired by a presentation I saw during the MBV program. One of the presenters talked about taking advantage of his veteran status, specifically the veteran-owned small business (VSOB) certification he had. The presentation focused on government contracting and leveraging those certifications. I had some experience with purchasing on the government side but not on the business side. I started educating myself about government procurement, and at a certain point I decided to find a client, one who needed help with navigating government contracts (as opposed to going after government contracts through my own business), at least to start. In other words, I wanted to work as a subcontractor. That way I could learn the business side of things before allowing my own business to take on risks of signing a government contract.

In 2019, I signed my first contract/client agreement with a construction company to do business development and estimating. I researched and estimated government contracts that fit my client's niche. For the next two years I put in twenty-five to forty hours a week for the client. I estimated tens of millions of dollars in contracts, including capturing two MATOC for that client. (MATOC, or multiple award task order contract, is a special type of contract with multiple recipients.) I did most of the project management, which included managing projects in California, Oregon, Idaho, and Hawaii. At the height of activity during this period, I signed contracts with two other independent contractors. It was an exciting time.

Then it wasn't. A couple projects were pulled and suddenly I was my only employee again. It's crazy how fast things can go from good to bad.

I founded VetHut Consulting with the intention to service the training industry. My first client was construction business development, estimating, and project management. I adopted an opportunistic business strategy. Although it wasn't my original intention to go back to the construction field, a good opportunity came along, and I didn't want to say no. My bachelor's degree in geological science, along with a decade of construction experience and many certifications, made me more marketable in the industry.

My ambition, experience, and education put me on a path towards entrepreneurship, but my health issues proved to be a major roadblock on that journey. I suffered from chronic pain in my back, neck, and shoulders, the result of too many accidents, hard jobs, and living a rough life. The discomfort and chronic pain limited the time and effort I could put into my career and family. These health issues forced my career to slow down, and as a result my business became a hobby. I had to lower my stressors and cut out responsibilities in my recovery. Repairing my body became top priority.

I want more than I have time for. I often can't accomplish all the different ideas that float around in my head. But because I still try to do it all, often I don't meet my own deadlines, and many of my projects go unfinished. It's not that I can't accomplish each project on its own, but when I take on too much, my ability to put forth one hundred percent on any of the projects goes away. My bandwidth diminishes. My responsibilities and projects suffer. That's what happened at work, and it had a negative impact on my home life. That's why it's extremely important that I take care of my own health so that I can properly take care of my family. I have a moral obligation to really be there for my family and be fully present in their lives. I'm OK with that outcome. Transforming my business into a hobby was a hard decision to accept at first, and it was hard to implement. But the only other option was to close the business. Calling it a hobby has made it easier for me to take a step back and prioritize my health and my family over my career.

I wasn't sure how it was going to work out, turning my business into a hobby. I returned to business development, as that was what I did initially, and then I came across a phrase I had never heard before: business coaching. It didn't take me very long to realize that business coaching would be a perfect hobby. I could do all my work and marketing from my computer, and setting my work hours to be appointment-only would be easy, and I could merge coaching with what I was doing before.

Over the last three years, I have helped business partners build stronger communication skills so they can have more efficient, more effective, and more profitable businesses. I help business owners and their spouses develop stronger communication and improved routines, so they can be more present in their business. Working as a business coach has been very rewarding. Not only has it allowed me to continue working, but I get to spend quality time with my family and I'm doing something fun for my career.

"A Lot of Mixed Emotions"

Lucas Lenhert reflects on telling his story

I graduated from the MBV with Cohort IV in 2017. I still miss being in the MBV program. I enjoyed the program so much that I would do it again. Any chance I get to reconnect with other MBV alumni, to share an experience and make memories with them, and to give back to the program, count me in. So writing this was a good experience.

One element of the MBV has been most impactful for me—the love of personal and professional development. After being introduced to Daniel Goleman's leadership assessment, it sparked a flame inside me, and seven years later I'm still seeking wisdom through self-awareness and constantly seeking out new emotional intelligence tools. Learning more about myself and how I operate humbles me. I'm more willing to receive criticism on how I'm falling short, when I make a mistake, or how I could have done something better, without getting uncomfortable or letting the criticism derail me.

In my story I look back on some past experiences that I have a lot of mixed emotions about. I could have told a story that focused on improving my strengths over my weaknesses, which for me would have been more fun. But I'm on a self-improvement journey, and I wanted to be intentional about that improvement, to be real with myself, so I can really learn something, and so I can move on.

When I first started writing about my journey, I thought I would be finished after a few months. Along the way I kept getting stuck trying to perfect a sentence or trying to figure out specifically how I felt about different events within my story, whether I remembered the events clearly and correctly, or whether after sitting on some of my memories for so long I now had a different impression of events.

The clarity I got about the events leading up to when I chose not to reenlist has been interesting. There were several influences on me during that time, causing me confusion and anger. I see now that I was not in the best place to make such an important decision in my life. I got out because I was stressed and angry. Had I been in a better place mentally during the reenlistment period, I probably would have reenlisted.

Looking back on events from my life, trying to figure out what I learned from them, I realized how I was turning my decision-making abilities over to my feelings. I don't identify as someone who uses their feelings a lot, so to see how I just let myself be controlled by my feelings, especially during such an important decision in my life, was profound.

Chapter 16

Selfless Service
by
Maria Maciel

Maria Maciel served for six years as an intelligence analyst in the US Army, earning the rank of staff sergeant and the title of senior instructor at the Joint Intelligence Combat Training Center in Fort Huachuca, Arizona. Her service included a fifteen-month deployment to Baghdad during Operation Iraqi Freedom.

Maria's academic achievements include a BS in business administration with a concentration in management from University of California, Riverside. She continued her studies at the University of Southern California, earning a Master of Business for Veterans (MBV) degree. Maria also obtained a TESOL certificate from the University of California, Riverside Extension.

Maria enjoys giving back to her community by volunteering at a local animal shelter with her daughter.

Serving Others

My earliest memory is an accident I had when I was five years old. I came home from school with an apple I had saved for my brother and sister, who were three and four years old, respectively. I wanted to give them each a half, and I wanted to do it on my own, so I didn't ask the babysitter for help. I walked into the kitchen, grabbed a knife, and sliced into the apple while holding it. The blade didn't stop in the apple, and it sliced my palm. I ran to the restroom, grabbed tissue, and packed the wound. I fell asleep waiting for my mom to get home from work. She took me to the emergency room where they stitched up my hand. When I felt the pain of the tissue being cleaned out, I realized I had not had the best idea.

I was born in Madera, CA, to Mexican immigrant parents. As the oldest of six children and twenty-seven grandchildren, I was no stranger to serving others. I was raised by a single mother who spent a lot of time out of the home working, so I often had to look after my siblings. Independence, caring for others, and the ability to tough it out became second nature to me.

I don't have a lot of good memories of my childhood. Although my mother did her best, we went through some very tough times. I read somewhere that when you experience a lot of trauma, your brain rewires itself to protect you from the bad memories. I try not to think too much about it for fear that memories I don't want to relive will be unlocked. And I wish I could forget some things my brain still holds onto: watching my mother endure physical, verbal, and mental abuse from terrible men; the times we didn't have anywhere to live; and all the yelling and fighting we grew up with.

Shortly after I turned nine, my father was arrested and spent the next twenty years in prison for a nonviolent crime. I was embarrassed of being judged for this, so I lied and told my friends that he lived in Mexico. Now, I am proud to share it as a part of my story.

I spent a lot of important milestones, like my wedding day, college graduation, and the early stages of raising a child, for example, wondering what it would have been like to be raised with a father. Times were good when he was around. I have great memories of him taking us out to the lake on his boat and for rides on the quads. It's hard for a single mother to do it all on her own, and it would have been a lot easier if she had had support. But I wouldn't be where I am today without the impact of his absence.

I didn't have a clear role model. I had examples of what not to do or what not to become. That's not to say I don't have good people in my family; I just didn't have anyone I aspired to be. I got most of my dreams and aspirations from television. I wanted to be a doctor, then a detective, or an FBI agent. Fortunately, I was surrounded by strong women whom I learned a lot from. From my mother to my grandmother and my aunts, I had great examples of women who made something of themselves despite the many challenges they faced. For that, I will always be thankful, and I credit them for giving me the mental strength I needed to succeed in the military.

Joining Up to Run Away

One of the most common questions we hear as veterans is why we decided to join the military. Typical responses often include, "I didn't know what I wanted to do after high school," or "I wanted to serve my country." For me, I thought the military looked like fun. I was twelve years old when I made the decision to join. I watched the commercials of soldiers running through fields carrying rifles and jumping out of planes. I'm scared of heights so jumping out of planes didn't appeal to me, but I was still drawn to the intensity. The catchy slogan "Be all that you can be" really spoke to me. I couldn't see it at the time, but now looking back, I think my need to escape my less-than-ideal circumstances pushed me to join. That I may have joined up in order to run away was confirmed about ten years later when, during a heated argument, my brother accused me of taking off and leaving my siblings behind. It was never my intention, but I understood how my brother perceived it that way, especially since I haven't returned home. I didn't waste any time—as soon as I was eligible, I started the enlistment process.

The recruiter presented me with job options that had huge sign-on bonuses, but I didn't care about how much money I got. I grew up with little. Money was never a big motivator. I chose a job that didn't come with a bonus, but which sounded cool—intelligence analyst. From the start, I told the recruiter that I wanted to serve four years and go to college. He added the Army College Fund to my contract, which is college money on top of the GI Bill. At the time, I didn't know how valuable this would be. I signed up when I was seventeen, with my mom's signature accompanying mine, and left for basic training eleven days after graduating high school.

The Way I Was Designed

The recruiters didn't tell me that basic training in the summer was not a good idea. I flew to Fort Leonard Wood, Missouri, and spent nine weeks waiting for the torment to end. Basic training was the hardest thing I had done. I was naturally athletic, but nothing could have prepared me for the constant physical and mental exhaustion we were put through.

I remember the drill sergeants shouting as soon as our bus pulled up. We exited with only a duffle bag filled with our belongings, and after filing into the auditorium, the shouting continued. The screaming didn't faze me because I had grown up with screaming. That was my childhood, lots of screaming and fighting. From that moment on, I made sure to fly under the radar. I didn't do anything to make myself stand out. In many ways, I'm still like that. I do what I am supposed to do, and I don't like the spotlight on me. I have been like this since I was a child. I don't know if it's the way I was designed. Maybe it was because I was always the new kid in school. We moved around a lot when I was little, so I constantly started at a new school. I hated walking into class and having the teacher announce my name. Everyone stared while I fought the urge to run.

One of the worst things about basic training was the mass punishment we received when other people made bad choices. There was always someone who couldn't follow basic instructions. It didn't matter how hard I worked to do the right thing. Sometimes, the drill sergeants marched us to an open shed filled with shredded tire mulch and made us exercise to exhaustion. We kicked up so much dust it was nearly impossible to breathe. I remember their favorite drill: mountain climbers to kick up the dust, then push ups to breathe it all in.

The Army's values of loyalty, duty, respect, selfless service, honor, integrity, and personal courage quickly became ingrained into every fiber of my being. These values resonate so deeply even today because they put into words the type of person I have always strived to be. They describe an internal truth I have always known, and I do my best to live by them.

Advanced Individual Training

My next stop was advanced individual training (AIT) at Fort Huachuca, AZ. There I learned the basics of being an intelligence analyst. We learned to gather intelligence from multiple sources and paint a picture of what the enemy was doing. Then we presented that information to the person in charge so they could develop an action plan to stop or capture said enemy. Fort Huachuca was also where we got our first taste of freedom and, for many, it was a true test of character. The freedom was easy for me, and I followed the rules to the letter. The value of integrity, doing what was right even when no one was watching, was always at the top of my mind.

After a few weeks, we were allowed to wear civilian clothing in the evenings and on weekends. I saw groups of friends who were so different from each other they most likely would have belonged to separate cliques in high school. There was the preppy jock hanging out with the punk in baggy pants with chains. It didn't matter what your background was, the uniform brought us all together. This would hold true even after service. Being a veteran is a brotherhood and sisterhood like no other.

First Duty Station

Towards the end of AIT, I was notified that I would be stationed in Fort Riley, KS. Never in my wildest dreams did I think I would end up living in Kansas. No offense to anyone from there, but my first thought was, "What the heck is in Kansas?" I spent a year training with my unit (Fourth Brigade, First Infantry Division) before we deployed to FOB Falcon in Baghdad. It was 2008 during Operation Iraqi Freedom and the peak of the surge, so our deployment extended to fifteen months.

I don't remember much from my time in Iraq aside from the extreme heat, powerful dust storms, and long twelve-hour shifts. I do vividly remember the hero flights that left the base with the body or bodies of fallen soldiers. Saluting the Black Hawks as they took off was a very somber experience, especially when you knew the soldier. It was a constant reminder that our days weren't guaranteed, and we could be gone in an instant. I was nineteen when I left for Iraq and twenty-one when I returned. Living such an intense experience during the formative years of early adulthood greatly affected who I grew up to be. I do my best not to take life for granted and I don't hold resentments because you never know if a negative interaction with someone may be your last.

I served six years of active duty in the Army, three of them as an instructor at the Joint Combat Intelligence Training Center at Fort Huachuca. I earned my senior instructor badge and my promotion to staff sergeant before leaving in 2011. Throughout adulthood I have pushed myself to meet the highest standards possible. I continue to set goals and do whatever it takes to accomplish them. Looking back on my military career, I think I did an excellent job, and my service is something I am extremely proud of. If I could go back and speak to my twelve-year-old self, I would tell her she chose a great path for me.

I learned the importance of meeting deadlines and having a strong work ethic. Showing up late was not acceptable. Forget about telling the commander you didn't complete a project. The values instilled in me during my time in the military have been instrumental to my success.

After marrying a fellow soldier, my daughter was born two months before I left the army, and I divorced approximately two years later during one of the most crucial times in my adult life.

A New Life

My transition was delayed as I focused on raising my daughter on my own while her father was stationed in South Korea, and after my divorce I had to figure out what I wanted to do with my life. That period marked a lot of firsts. As I began experiencing what it meant to exist outside of uniform, I also explored the intricacies of single parenthood, all while completing college courses in pursuit of a business degree. Going to college after the army had always been part of the plan. Life took a detour when I had a child first, but I never lost sight of college as a goal. I started by taking community college classes before my divorce. By the time my divorce came along, I was ready to apply for admission to a university and I was accepted to University of California, Riverside. At the time, my daughter's father was stationed at Fort Irwin, CA, so I chose a university close enough for him to see our daughter.

Going to college full-time while raising a child proved to be the most challenging thing I had ever done, even harder than breathing tire mulch in the shed. I used to go to bed late and wake up extremely early so that I could do homework while my daughter slept. Then her father was sent to Fort Polk, Louisiana, shortly after I started school, so I didn't have his help. I was too far from my family for them to be of much assistance. However, one of my sisters drove over on multiple occasions to help me when I really needed it. I also convinced one of my sisters to stay with me for a summer so I could complete a summer session. My sense of duty didn't allow me to slack off or request extensions to deadlines, so I had no choice but to figure it out.

I remember pulling my daughter in a wagon through my school one day because she had a fever and couldn't attend daycare. I held her in my arms as I took notes in accounting class. I could have easily skipped that day, but accounting was a struggle for me, and I couldn't afford to miss the material. She also sat with me during a finance final one night. My professor had a portable DVD player and headphones for her. My daughter sat next to me watching *Frozen* as I took my final. I remember she started singing at one point before I reminded her that she had to be extra quiet. Everyone in class got a nice laugh out of it.

Although I was only twenty-seven when I started at UCR, I felt decades older than the other students. Probably because I had lived through a lot by that point and had the responsibility of raising a child. Because of this, I naturally gravitated to the students I related to, the other veterans. I became a member of the Student Veterans Organization (SVO). It felt like I had found my community. I was a transfer student, so I only had two years there before graduating, but I made the most of it. During my senior year, I became president of the SVO and joined the university's Diversity Council to help the voices of veterans be heard on campus, and to push for more school resources.

The executive assistant to the vice chancellor of business services was looking for a student assistant. They wanted a veteran since the executive assistant was a US Air Force veteran and the vice chancellor was a Marine veteran. I wasn't looking for a job, but I saw an opportunity to make connections that could help UCR's student veterans while adding to my bare resume. I was selected for the job and found my way into as many conversations about student veterans as possible. Eventually, we convinced enough people that we needed more space and resources, and a veterans' resource center was approved. Although I wouldn't benefit from the center since it was going to be built after I graduated, I was happy that future student veterans would have access to the services they needed.

Towards the end of my time at UCR, someone mentioned a new program at USC called the Master of Business for Veterans (MBV) program. I was a business major, so it piqued my interest. I attended an informational session and decided to take a shot at applying. I didn't use my GI Bill during community college, and I had enough benefits to get me through the program, so I figured I had nothing to lose. It would also buy me some time while I decided what I really wanted to do. After submitting my application, I didn't have high hopes and kept my expectations low so that I wouldn't be crushed if I wasn't accepted. I focused on the goal at hand, which was to graduate from UCR, and I became the first college graduate in my family.

I was notified a few days after my graduation that I had been accepted to the MBV program. I can't put into words what I felt at that moment—a mix of excitement, nervousness, and pride.

A Family Photo

I was extremely nervous walking into the MBV program on the first day. The feeling was completely different from my first day of undergrad. Unlike when I was one of the oldest in my class at UCR, here I was one of the youngest. I was surrounded by people who had already established careers. It was a bit intimidating. My nervousness and uneasiness quickly subsided when, like good veterans, everyone began to make service rivalry jokes. I was back with my brothers and sisters, but this time it was better. This time, they sat next to me in class, they worked with me on group projects, and they had a direct impact on my education. There is nothing like working on yourself and advancing your education surrounded by people who hold you accountable and push you to succeed.

The classes were great, especially the entrepreneurship class, and I still reflect on the things I learned there. Although I feel like most people go into entrepreneurship with the desire to make millions, I was able to walk away with some ideas on how I could apply some the principles to nonprofit work. For our final project, we had to develop a business plan for a company, and I created a fictional nonprofit which I still dream of starting.

On the day we had to do a negotiation exercise and give our final presentation, I didn't have anyone to watch my daughter, so she joined me. A friend took a photo of us from behind the one-way glass used during the negotiations. There were four of us sitting at a table discussing whatever business deal we had to make. It had something to do with coffee. Behind us in the photo is my daughter, standing at the whiteboard filled with pictures she drew. The photo fills me with so much joy because it beautifully captures my educational journey as a mother. I had to figure out how to balance motherhood and being a student. The photo also highlights the importance of community and having people around who support you. In this case, the professor allowed me to bring my daughter along, and my classmates welcomed her into our negotiation room.

The last day of the program still plays vividly in my mind. We started the morning with a team building exercise which included writing a few values on small stones. Later in the day, the entire cohort gathered in a room, and we took turns being vulnerable, sharing what we would leave behind and what we would take with us going forward. We filled a jar with the stones that had our most important values written on them. My stones read "loyalty" and "selfless service." I decided to keep "selfless service" as the value that I would always carry with me.

The MBV program was a great opportunity to surround myself with successful veterans I looked up to. I used to downplay my service, thinking I had "only" served and hadn't had a real job. Then I saw all the veterans with amazing careers they had built after the military. They helped me understand the value of my experience in the service, and they taught me that I had actually accomplished a lot.

There is a special bond you develop with other veterans. Not many people really know what serving entails, and having to provide all the prerequisite details to give context to stories gets tiresome. I found friends for life in the MBV program, and although we might not speak regularly, I have a circle that I can count on to have my back because they understand me in ways that those who have not served never will.

As the first person in my family to graduate college, and the first to obtain a graduate degree, I held my head up high as I walked across the graduation stage. In that moment, I felt that I brought some honor to my family, especially to my parents.

After graduating from the MBV program, I enrolled in and completed the Teaching English to Speakers of Other Languages program at UCR Extension. I proceeded to earn my teaching credential and worked as a substitute teacher part-time before I learned that working with children was not the right fit for me. Don't get me wrong, I love kids, but trying to get them all to sit quietly and follow instructions was a lot harder than my time serving as an instructor in the army.

Present and Future

I currently work in commercial real estate property management for a high-rise in Los Angeles. Like many of my coworkers, it's not an industry I was looking for, and I sort of fell into it. My mother spent most of my childhood working in real estate. Since she wasn't home very much, real estate quickly became synonymous with an absent parent. For a long time, I swore that I would stay away from anything even remotely related to it. Now, I understand that it's not the career that determines how present we are. It's how you prioritize your life and what you choose to give your time to. I have made the conscious decision to always put my daughter first.

Ironically, I do a lot of tasks related to accounting, which happened to be my least favorite subject in college. Now that I put it into practice, a lot of the material I learned has clicked (but I am still not a fan). I don't love what I do, but I don't wake up dreading my workday. I consider that a small victory. I am still in self-discovery mode until I find the job and the career that really moves me.

During my free time, when I'm not driving my daughter around to her extracurricular activities, I volunteer in the community. While I was stationed in Arizona, I started volunteering with Big Brothers Big Sisters and with a local church youth group. In Riverside, I assisted with immigration workshops, translated for Spanish-speaking patients during health fairs, and spent time as an adult English tutor. Now, in Los Angeles, I volunteer as a kitten foster with a local rescue, and I volunteer at an animal shelter where my daughter also joins me, so she can learn the importance of selfless service.

As I write this, I am thirty-six years old and have spent exactly half of my years away from my family, trying to build a life I can be proud of and providing the life for my daughter which I would have wanted—a life of peace without constant turmoil, filled with love. My daughter is twelve years old, the same age I was when I outlined my future. She is far wiser than I was. Unlike me, her brain hasn't been wired by trauma, and she has a wonderful ability to retain information after just hearing it once. She loves history and learning about the world. We have in-depth conversations about various world issues that many kids her age would probably be bored with. I have done my best to keep an open line of communication with her, something that I didn't have with my mother when I was growing up. I am confident that, unlike me, she will never feel the need to flee.

I have been out of the army for more than twelve years, but I feel like I just got out. I am still searching for my true calling. I am destined to serve others, but I need to figure out in what capacity. I don't measure success by the amount of money I make but rather by the life I provide for my daughter and the memories I create for her. It's strange having grown up with trauma that makes your primary goal as a parent to never pass your childhood on to your own child. I also measure success by how much I give back to others. It has never been enough for me to just make money and live comfortably. If I am not doing what I can for others, I am failing.

After reflecting, I think these feelings mostly come from my human nature to serve others and the examples I witnessed throughout my childhood. Remember the strong, resilient women I mentioned before? Those women also help others any time they can. They don't help because they want something in return, they help because they want to see others do well. My siblings and I have carried on this quality. Sometimes, all someone needs is one person to reach out and help. It makes a world of difference. I want to be able to extend my hand to anyone that needs it.

My ultimate goal is to do nonprofit work serving the Latino immigrant community, the veteran community, or both. I'm still not sure how I will get there but I am determined to find a way.

"I Have the Right to Tell My Story"

Maria Maciel reflects on telling her story

Writing my story acted as a sort of therapy. Each time I sat down to write was another session where I took a deep dive into my past and really focused on the choices I made throughout life. Analyzing why I chose certain paths was the most impactful part of this project. I feel more pride for where I am today. I know that I still have work to do but this has been a great start in my healing process.

When I was first approached about the project, the thought of being vulnerable and sharing my story with complete strangers made me shy away from it. Then, a friend suggested that this project could be something I write for my daughter. I was drawn to the idea that my daughter could learn about me on a deeper level and learn things that might not have come up in our day-to-day conversations. In reading my story and learning about the different challenges I overcame, I hope she is inspired. I want to show her that through hard work and perseverance, she can do great things. The fact that I may inspire anyone else reading this is an added bonus.

I had never given much thought to my decision to join the military. After some deep reflection, I came to the realization that I was probably trying to escape my home life. Of course, I still don't know if this is completely true, but it would explain a lot of my choices. I don't want my family to think that I ran away and turned my back on them. Since doing this project, I decided that I am going to put in the extra effort to visit them and stay connected.

One of the biggest obstacles I faced during this process was finding the balance between sharing my story while not revealing too much that might hurt the people I love. I know that I have the right to tell my story, but I wanted to be respectful of everyone else's right to share or not share their own stories. Another obstacle that presented itself was how I doubted the impact of my story. I didn't think it was impressive or important enough to share. I had to stop comparing my challenges or accomplishments to others. I focused on the reason for writing it and reminded myself that everyone is unique, and we all measure success differently. We all have something to offer by sharing our journey.

Chapter 17

Serving My Purpose
by
Michael Chavez

Specialist Michael Frank Chavez enlisted in the army in 1993. At Fort McClellan, Alabama, he trained as a nuclear, biological, and chemical specialist. He was stationed at Fort Wainwright, Alaska, in a light infantry unit, and at Fort Carson, Colorado, in a mechanized infantry unit.

Chavez started his first business repairing pagers and cell phones. He worked in the marketing department of a local Colorado hospital selling health insurance, and for twenty years he worked for engineering firms in their marketing and business development divisions. Currently, Chavez is a mortgage loan originator and national educator for Veterans Lending Group in Colorado.

Chavez runs a YouTube channel, "Finances with Mike," where he teaches personal finance, from credit optimization and budgeting to home ownership. He runs another channel, "Uncle Mike Reads," where he promotes literacy and critical thinking skills through reading books to children.

Lessons

At the age of six, I started working for my dad for two dollars a day. He was a painting contractor with some light carpentry skills who specialized in residential interiors and exteriors. In the summer months, when I spent most of my working days, I was thankful to wear shorts and a T-shirt while Dad was covered from head to toe indoors to spray the walls and ceilings. I recall how, after he was done spraying, as he took off his gear, I saw he was drenched in sweat. I learned that hard work is honorable. Do it with pride and without complaint.

Dad loved the oldies and conservative talk radio. Working with him, I also grew fond of both. As we sat down one day with the lunches Mom prepared for us, I casually said to him, "I really like this radio host, I agree with everything he says!" Dad finished his bite, looked at me and replied, "I like him too, but it's important that you hear multiple opinions about something so you can make up your own mind." I learned not to blindly follow someone you agree with. Seek out other sources to be better informed.

Dad didn't advertise his services. He worked purely on referrals. I would occasionally accompany him when he gave estimates. Frequently he was the most expensive, and sometimes he got pushback from the homeowner. He responded with two things: his quality of work and his reputation. He liked to tell them that he had no problem being outbid. Go ahead and hire someone else, he told them, and when the work didn't live up to the quality they expected, he was happy to fix it for them. He was only there because someone they trusted referred him. Would they rather have someone highly recommended who used quality materials, or the cheapest bidder and hope for the best? I learned that if you take pride in what you do, and do exceptional work, you should stand up for what you are worth.

Dad believed to his core that everyone deserved respect. He was always kind to strangers, tipped generously, and had a charisma that drew everyone to him. Gossip angered him greatly, and he was quick to put us in check whenever we started talking bad about someone. I learned that kindness and decency were two of the greatest virtues.

Sowing Seeds

To say Dad was influential is an understatement. He not only exemplified how to be a good man but also taught me how not to be like him. We always struggled financially, and for the most part it didn't matter that much to me. I had a dad who was the envy of all the neighborhood kids. I knew from an early age that I didn't want to struggle all my life with money, and I worked my rear end off to earn the little money I had. I refused to let all that effort be wasted on trivial things, so I taught myself to save up for what I really wanted, which at the time happened to be a Jabba the Hut action figure that cost $29.99. I gave myself a budget on what I was willing to spend on candy at the liquor store and video games at the pizza place just down the street.

The day came when I finally had enough to buy the action figure. As I eagerly waited for Dad to come home to take me to the toy store, I distracted myself by riding my bike with the neighborhood kids, my life savings in my pocket. I headed home when it got dark and the streetlights came on, and that's when I realized my pockets were empty. I had lost everything! I searched to no avail. I was crushed. When Dad eventually drove up, I ran to him in tears, explaining through sobs what happened. He hugged me and consoled me. He didn't offer to buy the toy for me anyway, and he didn't try to push a lesson on me about being responsible, either. He simply sympathized and let me express my sorrow. I learned a valuable lesson that day, and I never had an incident like that again.

There is nothing worse for a kid than to stand out in a negative way among your friends. While the neighborhood kids joyfully showed off their Christmas hauls, I would buy myself toys and pretend they were gifts. This was only possible by being consistently disciplined with saving and spending, which solidified my perception of money and its usefulness when budgeted correctly.

A Mother's Love

My mother has been the one constant in my life—my cornerstone of security, love, and belonging. She was there for every scraped knee and all the high jinks I got into. She cheered me on in sports. My parents had very traditional roles, and Mom did not compromise about religion. Not only did she insist on enrolling us in private Catholic schools totaling more than 100 years of private school tuition between us all, but she was insistent that we go to Mass every Sunday. I had a short-lived gig as an altar boy. It was cool to experience the services from the other side.

Tension grew between us when I started to question our religion in the second grade. The catalyst came when I discovered the truth about Santa. Once I understood that I had been misled into believing something that wasn't true, regardless of the purest intentions, I began to question everything I was told. Combined with Dad's advice to seek alternative viewpoints, I became a handful.

Perhaps the most difficult pill to swallow coincided with a particularly rough day of being bullied. My emotions were overwhelming, and I gathered all my courage and went to Mom. I told her I was thinking of ending my life, and her response was, "Oh no you're not," and she resumed doing whatever she was preoccupied with. It was crushing in the moment, and it forever dampened my willingness to confide in her, but in a way it actually helped. By immediately dismissing it as not even a possibility, I began to assume that same outlook.

The lyrics of "One" by Metallica tell the story of a man who goes off to war, gets blown up by a landmine and loses his limbs, sight, hearing, and ability to speak. He is kept alive with medical technology and suffers in horrendous, constant pain and isolation. I thought about the song for a long time, and I asked Mom what she would want in that situation. She told me that she would use that time to pray for others. It was her way of continuing to make a difference in people's lives. It really put in perspective for me not only how strong her faith was, but how giving and selfless she could be.

Rebellion and Independence

Growing up, I was small for my age. It is no surprise that being bullied and harassed comprised much of my childhood. I focused on doing well in school and having a core group of friends that accepted me. Reading comics was my gateway drug into reading novels for entertainment. With reading and math my natural strengths, school wasn't much of a challenge. It was the social side of things that was my Everest to climb. Being small didn't help much when it came to playing sports, but the one physical thing I had going for me was my speed. I was fast. In soccer tryouts, I left everyone in the dust. It was the first time I was ever recognized for excelling at something physical, and it solidified my status among my newfound team.

"Why?" remained a common utterance for me, and through reading I was exposed to so many different viewpoints and perspectives. I began to challenge my faith, politics (recall the years of conservative radio), and who I was and wanted to be. I dove headfirst into hard rock and heavy metal when I discovered it was a cathartic outlet for the repressed anger and frustration I carried from being bullied. I grew out my hair, wore heavy metal T-shirts and my signature black leather jacket in the sweltering Southern California heat.

When I turned eighteen, I was more than ready to apply for my first credit card. When I was initially denied, I discovered I already had a few outstanding accounts with balances that had all defaulted. Someone had applied for credit in my name and maxed out the accounts. After doing some digging, I found out it was my eldest sister. Fortunately, I was able to remove every account from my credit report and start working on building credit, but the experience had already taught me that even family can betray your trust.

To Leave a Mark

As a senior in high school, like many seniors, I was chomping at the bit to move out of my parents' home and be on my own. With my aptitude for learning, it was a no-brainer to apply to the local colleges. I had no clue what I wanted to do or who I wanted to be, I just did what I thought was expected of me. As it happened, I met an Army recruiter who was skilled at talking up the positives with a clear spirit of service and purpose, and it resonated with my core unlike anything else.

I had heard many stories about Dad in the Navy getting into bar fights with Marines; my older brother's adventures in the coast guard before joining the seminary; and my uncle's time as an Army Ranger. Joining the service was never at the top of my mind, particularly because of my diminutive stature, but I had always admired and looked up to those in my family who had served.

It is amazing how palpable it is, for some young men, to want to leave their mark on the world and make a real difference. My girlfriend at the time was adamantly against me joining the service, threatening to break up with me if I joined. With these conflicting motivations, I made what I feel was the best choice at the time and signed up. After a few rocky months, we did indeed break up, then we got back together shortly before I left for basic training.

To say that basic training was a culture shock would be to put it too mildly. The mental and physical stresses they put us through were substantial. I was still small, about 125 pounds soaking wet, but I was in relatively decent shape from my years playing soccer and flag football. Once I realized that the mental stuff the drill sergeants put you through was just a game, the rest came easy. It was a year before *Forrest Gump* came out, and the part of the movie that resonated with me is when Forrest, in basic training, realizes that all he has to do to stay out of trouble is to just do whatever they tell you and "Always answer every question with 'YES, DRILL SERGEANT!'"

After basic and advanced individual training (AIT), I was stationed in Fairbanks, AK, in the middle of January. Picture this: a native Southern California boy, no body fat for insulation, for whom 70 degrees Fahrenheit is chilly, was dropped into a place that was 30 degrees below zero, the cold hitting him like a truck, and he wished like heck he could get back on a plane. Ironically, it ended up being my favorite duty station. It was there I met some of the most exceptional people it has been my pleasure to come across.

My first sergeant is the person that most quickly comes to mind. Not only would he be out there in freezing temperatures with us every morning for physical training, but as we ran in formation, he would run around the formation, taunting us with things like, "I'm twice your age doing twice as much. The very least you can do is keep up." And to a man and woman, we did. He was fond of telling us he didn't want us there if we didn't want to be there, and he worked to send us somewhere warmer anytime we were ready to leave. He also told us that if we did stay, he expected us to excel in everything we did, so we could not only reach our personal goals in the service but make our unit the envy of the military. He supported anyone willing to put in the work it took to be promoted, and he was true to his word.

The other individual that stands out during my time in Alaska was a lieutenant colonel I had the pleasure of driving around and for whom I essentially acted as an administrative assistant. She was quite simply a force of nature. No nonsense, with clear motives and an iron will to see them through. She didn't shy away from her femininity and was one of the most honest, assertive, loyal and compassionate people I have ever met. She shattered the misconceptions I had of what a woman could achieve in the military.

To be clear, not all my experiences in the military were so positive. I blew out both knees completing a six-mile run on icy roads. That was almost thirty years ago, and since then I have lived with chronic pain.

Favoring one knee or another began to impact my back, causing more chronic pain.

To put it kindly, not everyone in the military was as overwhelmingly exceptional as the two examples above. It was a lesson in ensuring trust is earned and not freely given.

I saw that many of my brothers and sisters had been taken advantage of by predatory lenders. A close buddy fell into the trap of payday loans. I ended up creating a budget for him and helping him break free of that vicious cycle.

During my time in the service, I married my girlfriend. We bought our first house and had a son together. The homebuying process was a nightmare, as both my realtor and the lender were not very experienced with VA loans. The VA appraisal was ordered very late in the game, which delayed our closing date. We had already given notice to our landlord, so we had to scramble to find somewhere to live until we closed on the purchase.

In general, my wife did not adapt to military life very well. Being away from her family and friends took a toll that eventually became intolerable for her.

Splits

Before my Army expiration term of service (ETS) date, I partnered with a coworker who was a significantly higher rank than me and started a pager and cellular repair business. This further alienated my wife as I was constantly working after hours and on weekends to get my new venture off the ground.

In addition, I experienced difficulty finding employment after the service, eventually leading me to sign up with a temp service. I landed a gig in the marketing department of a local hospital that sold its own health insurance. Again, I had the good fortune of some wonderful coworkers and leadership, and I thrived in that environment.

Then the rug was pulled out from under me. My business partner, being older and far more knowledgeable with the laws, essentially screwed me over. He seized our assets and froze me out. Since his name was on everything, I had little recourse. That was the final straw for my wife. She said she was leaving to go back home to California, and she was going with or without me. Not wanting to break up my family, I agreed. We sold our home and moved back to California, and our relationship limped along for a couple of months before finally dying.

Hitting Bottom

The period afterward was simultaneously the most difficult and the most rewarding. My wife maxed all my credit cards on purchases for her teenage boyfriend and on pet items, all of which I was on the hook to pay for. I was heartbroken, deeply in debt, sleeping on my friend's couch, with all my remaining worldly belongings stuffed in a trash bag in the closet, and stuck with a $12.50 per hour job. There was only one thing to do—soldier on.

Finally utilizing the hard-earned lessons of my youth, I cut my spending down to the bone and threw every extra penny at getting out of debt. After six months, I had completely recovered financially. My friend and I rented a small two-bedroom apartment, and I focused on saving everything I could for my next goal: purchasing my own home. After just another year, I bought a small two-bedroom condo and retained my good friend as a roommate. I continued to save money while making extra payments on the mortgage.

My career in marketing and business development began to take off, and I finally found myself in a stable financial position. When my friend eventually married and moved out, I was still living far below my means, making double mortgage payments and still significantly contributing to my savings each month. I had finally reached my goal of being financially stable and not worrying about money anymore.

It was also during this period when I began to experience pain flare ups so intense I couldn't sit up in bed. It not only impacted my professional and personal life, but it wasn't soon after that I was diagnosed with two herniated discs, acute anxiety, and chronic depression. Since then, I have been on medication and learned to cope with it, but I became quite practiced at hiding it all. I hate the idea of being a burden on others or bringing them down with my problems. I would much rather be the entertainer and focus on helping others get through their own challenges.

Primary Custody and Tutoring

During this period, my ex-wife had custody of our son during the week, and I had him on the weekends. I got permission from my boss to modify my work week to nine-hour days Monday through Thursday and four hours on Friday so I could leave by noon. I used that extra time to drive to my son's school and volunteer in his class.

I got to know his teachers and classmates, and I got a better feel for how he was doing in school. My son seemed to always struggle with academics, which was surprising to me and something I couldn't relate to. Things came to a decision point at the end of seventh grade when he had four Fs and two Ds. I had a number of conversations with his teacher, and he indicated that my son almost never turned in homework and was frequently absent. His self-esteem was very low, and he believed he just wasn't smart enough for school.

I proposed an idea to his mom: we swap schedules and let our son live with me during the week so that I could help him with school. After some initial resistance, she agreed. We ended up butting heads when I tried to tutor him, and he protested significantly. I couldn't understand it. We had always had a close relationship, and I was a fun and effective tutor.

Determined to make it work, I tried different approaches. I kept him accountable for every homework assignment and conducted study sessions. The poor guy didn't even know you were supposed to study before tests. It was a completely new experience for him. There were days he was at it from the time I picked him up until he went to bed crying. I felt like a failure as a parent.

The turning point came one Friday after school. My son came to me and said something to the effect of, "Dad, I need to tell you something. Today at school, my teacher was handing back our writing homework telling us she was very disappointed. She said they were so bad that I was the only one who got an A. I didn't think I heard her right, and she repeated my name and congratulated me." He barely got that last sentence out because we were both in tears, and I was overwhelmed with pride in him. He said that he finally got what I was trying to do with him and apologized for fighting it. He said that he only got an A because of me. I told him he deserved all the credit. I showed him the way and believed he could do it, but he was the one who put in all the work.

After that, while he didn't completely stop resisting (he was a teenage boy, after all), things went far more smoothly. Sure enough, his next report card was proof of his efforts, with one A, three Bs, and two Cs. While he never quite reached straight As, he nevertheless retained a belief that he could pass his classes. And he did.

Back to School, Part One

At the age of thirty, I felt that I had reached the limit in my career without a degree, so I went back to school to obtain my bachelor's degree. Since it had been quite a while since I had taken any classes, I was a bit rusty. Nevertheless, I welcomed the challenge and did well. Since I used the GI Bill and was awarded $10,000 in scholarships, I obtained my undergraduate business degree with no student loans and with a pay increase at work. I entertained the idea of going right for a master's degree, but quite honestly, I felt burnt out from school and decided to revisit that option later.

It was during this period that I found love again and married an amazing, supportive woman.

YouTube

One hot summer weekend, I was clearing out the attic. I came across a box of old children's books which I used to read to my son. I sat and went through them all, recalling so many great memories of reading them together. It also happened to be my niece's birthday coming up, and I had a flash of inspiration. Why don't I give her my favorite book of the bunch and read it to her to give her a fun memory, too?

My sister ended up recording me on video reading to her daughter, and we had an absolute blast. On my drive home, I felt so energized, and I wanted to share that feeling with my other twenty-four nieces and nephews. Since my family is spread across the US, I decided to make a YouTube account and post the video my sister took. While I was putting it together, I thought, "Why limit it to just my nieces and nephews? Why not put it out there for anyone to enjoy?"

With that, "Uncle Mike Reads" was born. Since I love reading, and because I still feel guilty to this day for failing to instill a love of reading in my son, I felt it was a great way to promote literacy to beginning readers. However, I didn't want to just read the books, I wanted to get my viewers to think about what they read. I wanted the viewers to not only enjoy reading along, but to develop their critical thinking skills.

Since then, I have made almost eighty videos. I have a great time creating them. I have slowed down considerably the past few years as other projects demanded attention, but I love every time I come back to do another one. In one of my recent videos, I cut my hair to donate to the charity Children with Hair Loss. It felt good to have another way to give back. The experience of growing out my hair also taught me another valuable lesson: friends, family, and coworkers can be quite intolerant when you don't conform to their ideas of how you should present yourself. Since then, I focus my time and attention on those who accept me as I am, not for what they expect me to be.

Back to School, Part Two

With each year that went by, the desire to get my master's faded. One of my good friends announced he was graduating from a master's program at USC. At his graduation party, I asked a million questions about the program and how he did it. He told me the degree was a Master of Business for Veterans, or MBV. He said it was a one-of-a-kind master's program only available to veterans and active service members, which leveraged their military experience to shorten the length of time needed to obtain a degree. My interest in going back to school rekindled. I attended the next information session and was hooked. With the full support of my wife, I decided to go for it.

I submitted my application on the first day of the next enrollment period and waited. And waited. For two months I was stuck in limbo, anxious to hear if I made it. Finally, I received an invitation to do a virtual interview. I was ecstatic and nervous. The interview was far more relaxed than I anticipated, and it was an overall pleasant experience. Shortly after, I received an acceptance letter.

Orientation was an amazing experience. The USC campus was awe-inspiring, and the energy and positivity of the faculty were palpable. I never dreamed that a poor kid from a big family would ever have the opportunity to go to USC. Yet there I was, surrounded by my fellow veterans.

I did have one major concern. While I was able to use my GI Bill and obtain my undergrad degree without debt, I knew this time getting a graduate degree would be different. I no longer had the GI Bill as an option and had to take out student loans to cover the cost. However, the program had a robust scholarship program based on the amount of out-of-pocket costs incurred. After they calculated that, they allocated the funds accordingly. It covered more than two thirds of my tuition. I was absolutely floored that I received such sizeable assistance, and I was so deeply grateful.

To say the program was intense is an understatement, as it challenged me in numerous ways. It was fast-paced and left no room for procrastination. Doing the reading ahead of time and preparing for each class were critical to keep up with the workload. This wasn't some stretched out undergrad program where you could easily miss a week's instruction without much impact. They crammed so much into the eight-hour Friday and Saturday sessions that missing even one day was challenging to recover from.

In addition, it was extremely difficult and painful on my body to drive to and from campus as well as to remain sitting during the long instruction days. Some days were worse than others, and there were several times when I was tempted to stay home because of my pain level and/or depression. But I was determined to see it through.

Rising to the challenge, our cohort came together to support one another. Those who had a strength in a particular subject held review sessions for those who were struggling. With accounting and finance being my strengths, I started my own sessions to help, and I created Excel spreadsheets to explain and break down the formulas. We shared tips, notes, and study guides. It really was amazing how many people stepped up to help others. This was, I believe, one of the most special and impactful aspects of the program.

The other aspect of the program that was far different from my undergrad experience was the cohort model. For the first semester, we were split in two groups in order to fit in two classrooms. Every class, every day, the same group was together. We also had the opportunity to socialize with the other group during breaks and lunch. This resulted in bonds forming quickly between us. Not every personality was compatible, and some of us had psychological injuries from service making compatibility even more difficult, but by and large, we coalesced into a supportive and cohesive group with a common mission.

One of the hardest aspects of the program was trusting others in group projects. I was trained to trust myself only to accomplish anything, so relying on others pushed me far out of my comfort zone. I had been disappointed so often that I had reduced my expectations to zero. I had decided that I was better off doing things on my own.

Fortunately for my personal growth, I had no choice but to rely on others for the numerous group projects to succeed. This presented quite a challenge for me. But I was repeatedly amazed at how consistently the others came through when it mattered.

One incident made me question whether to continue with the program. For a group project, we were separated into small groups and given a negotiation scenario. Someone posted one of the negotiating side's details to the entire class, which I felt undermined the entire point of the exercise. I, along with a few of my classmates, sent a group email to the instructor of the class the day before the exercise was to commence, outlining our concerns.

The professor changed the assignment that morning, much to the irritation of some. I took the brunt of the heat for it and refused to give up the names of the others in that group email. This was right before the winter break, and those last two days were filled with tension.

During the winter break, I emailed both Professor Turrill and James Bogle, explaining that I felt overwhelmed by the negative feedback from my cohort about what we had done, even though we had only followed our conscience. I told them I was seriously considering dropping out of the program. I received a very prompt response asking to meet in person, which I agreed to. Both Professor Turrill and Mr. Bogle drove all the way to meet me for lunch near my work. They were so considerate and understanding of my situation and asked me to consider the situation from other positions. That took me out of my own head and forced me to take a more nuanced view.

Swallowing my pride and feelings of righteousness, I reached out to the people who were most directly offended and asked to meet with them. I took a Saturday to drive down to San Diego to meet one on one to resolve things. The two individuals most impacted by the situation agreed to meet separately with me, and we were able to find common ground and work things out. Each commented that they respected that I not only reached out but followed through by driving out to meet them face to face.

My biggest takeaway from the situation was that Professor Turrill and Mr. Bogle empowered and encouraged me to act and find a resolution to the conflict without directly instructing me to do so. Their understanding, support, and gentle guidance empowered me to resolve it on my own.

In addition to forging lifelong friendships during the program, it also forced me to reassess my career path and motivations. While I achieved some success in my career, I never felt like I was making a positive difference in people's lives. I didn't want the consolation prize of only making a difference after hours. I wanted to make a difference with my actual career, and I discovered that I now had the will to face the risks and make it happen. I didn't know what form it would take, or what career would help me achieve it.

Volunteering

The MBV program inspired me to invest in my community. I ran for a spot on the board of my homeowner's association and discovered we had a fiscal emergency. With a budget of $93,000 per month, we were running a $30,000 deficit. It was simply not sustainable. During the first year of being on the board, I asked many questions, but the majority of the board was resistant to making any meaningful change. The following election resulted in a majority of the board willing to make necessary changes, and I was elected president to shepherd those changes. Together, we found numerous ways to cut spending, resulting in an average monthly surplus of $35,000.

My next opportunity came when USC hosted the Festival of Books on campus. During lunch between classes, I walked around and met several children's book authors and spoke to them about my YouTube channel where I read books to kids. I instantly connected with an author who loved what I was doing. She informed me she was the president of a nonprofit called Reading Is Fundamental of Southern California, and she asked if I would like to join the board. She explained that they provide brand new books for free to children in underserved school districts, and I was hooked. I am now the current president, and I made expanding our reach a top priority.

After graduation, I volunteered at the local library bookstore, selling donated books to raise money to support city youth literacy programs. It was a joy interacting with the kids and parents, and it's one of the things I miss the most since moving to Colorado last year.

Veterans Resource Team is a nonprofit that provides free education and resources with a focus on transitioning veterans. After being recognized for my passion for educating and serving active military and veterans, I was invited to join the board. This was instrumental in expanding my ability to reach and assist veterans and their families.

Journey to a New Career

My first step towards a new career came when I posted on social media about how I bought an investment property for cash. Friends, family, and social media connections reached out to ask me how I did it. Having learned to recognize opportunities in the MBV program and drawing from my experience with my "Uncle Mike Reads" YouTube channel, I decided to make a video explaining how I did it. I created a new channel called "Finances with Mike." From there, I have posted episodes covering budgeting, personal finance, and credit repair. I finally felt like I was making an impact, falling short only because it was still outside of my actual job.

The next step came one year into COVID. I finally got to the point where the need to make a difference in my actual job exceeded the need for the high salary I commanded. I had a lengthy discussion with my wife, sharing everything weighing on my mind, and her immediate response was something to the effect of: "Just quit. We'll be OK with the other sources of income we have. Take the time to figure out what you want to do and then go for it." Unconditional. I was floored by the sincerity and support. She didn't have to tell me twice. With the courage I had to embrace calculated risks, I resigned from my twenty-year-long career, which included fifteen years with the same company, without hesitation.

Now that I finally had the room to breathe and the time to think, I did some soul searching to find that elusive career, one that played to my strengths and made a difference in peoples' lives. I continued to produce episodes for my YouTube channel, and individuals were messaging me asking how much I charged for one-on-one coaching. Again recognizing an opportunity and seeing a need, I set up a sliding scale fee structure and began financial coaching. I hoped that it would not only provide me income but make a real difference in others' lives.

I found financial coaching to be deeply rewarding. I cannot express my reaction to seeing a sixty-five-year-old Marine veteran in tears, telling me that I finally gave him hope and a clear path forward after decades of seeing no way out of his financial struggle. While I did reach my goal of making a real difference in peoples' lives through my career, it wasn't taking off like I hoped it would.

In 2021, my wife and I began looking into residential investment property with her brother in Colorado Springs. The entry price point was far lower than Southern California, and we figured it would be beneficial to have someone we trusted be on-site to quickly remedy any issues. As time went on, we reconsidered, recognizing that financial entanglements with family almost always ended up straining relationships, which was a risk neither of us wanted to take.

My wife was the one to float the idea of buying a new home, which had always been her dream, and moving into it instead of just buying rental property. That was the polar opposite of everything I knew about real estate, where you buy low and sell high. The market was significantly overheated, but my wife explained that she was already working remotely full-time due to COVID and could easily move, and I had no work ties to California anymore. As luck would have it, we happened to find a new development in the location we were looking for, selected the perfect model together, and through a lottery system secured the last plot. We took that as a sign, signed on the dotted line, and it became ours.

After that monumental decision, I knew the clock was counting down on my new career. I brainstormed what was most important to me. Among the top requirements were: my work had to have a clear benefit for others; it had to play towards my strengths; and it had to bring in a decent income. Bonus points for directly helping veterans or kids. Since Colorado Springs has five military installations in or near the city, veterans seemed to be a logical choice. The next step was determining how best to serve them.

As I analyzed different careers, I couldn't find any that checked all the "must have" boxes. I recalled the extraordinary difficulty I had in purchasing a home when I was stationed in the military, mostly due to the inexperience of my agent and lender. Being an agent seemed a bit too salesy for me, but being a lender? Crunching numbers and solving problems for people looking to get approved for a loan? That sounded much more like me. With that decided, I began to focus on lending firms that catered to the military. I found two main local contenders: Veterans United (VU) and Veterans Lending Group (VLG). Each had their strengths, with VLG edging out VU with their free education program. I walked into both locations with resume in hand and told them I was willing to do whatever it took to join their teams. Both companies directed me to their online application system, but only Veterans Lending Group was willing to give me a shot.

I went through three rounds of interviews and finally landed the gig. I had to take classes and pass the certification exam before they officially signed me on, but all of that was no match for the rigorous coursework I had completed in the MBV program. I was prepared for the challenge and knocked it out.

Since then, I have worked as a loan officer specializing in VA home loans and catering to active service members, veterans, and their families. I became one of the few national educators in the company, traveling around the country teaching veterans for free about their VA Home Loan benefit and how to leverage it toward their goals of homeownership and real estate investment. I also teach real estate agents how to serve and grow their businesses with the military community. I'm currently expanding our curriculum with new classes about first-time homebuying, credit optimization, and budgeting.

The most rewarding part of the job is assisting those who have been unable to qualify for a loan and were turned away by other lending companies. I use my experience as a financial coach to create a customized budget for them and prioritize their efforts to increase their credit score. If they are willing to do the work, I stick with them for as long as it takes to get them to a stable financial position. The feedback I have received has been overwhelmingly positive.

Bringing it All Together

From my dad, I learned the value of hard work, taking pride in what you do, and doing it to the best of your ability. Decency, kindness, and compassion are the greatest virtues. Earning money through hard work, it taught me to value and maximize its power through budgeting and discipline. I also learned to think critically and not accept as fact what I heard and read.

From my mother, I learned the value of being a constant presence in the lives of your children. The stability of a loving home was a core reason why I thrived. It was through her example that I strive to be of service to others.

My teenage years gifted me with some confidence in my physical abilities, which began the process of discovering who I was and what I believed, and the hard lesson that even family could betray your trust.

My time in service introduced me to a handful of exceptional people and taught me valuable lessons in leadership (especially leading by example), supporting the goals of those you lead, and being honest and assertive. Through watching my fellow service members get taken advantage of, I found ways to be of service by helping them get out of financially challenging situations.

My transition out of the military, and my business partner's subsequent betrayal, reinforced the fact that even people close to you can fall short of your trust.

Hitting rock bottom pushed me to step up to the challenge and build myself back up. It gave me a new appreciation for those struggling financially and for those suffering with intense chronic pain. I discovered that focusing on others and finding ways to make their lives better gave me a sense of purpose and meaning that helped me power through my low moments.

My time in the MBV program molded me and pushed me to re-prioritize. As a cohort, we faced challenges by supporting each other and working as a team. I graduated with the courage, motivation, and revitalized purpose of finding a career for myself that would help people directly. I also became more confident in my ability to resolve conflicts without the intercession of those in charge.

MBV instilled in me a drive to become more involved in the community. Not only was I able to pull my HOA out of debt and become financially stable, but I am part of nonprofits that provide desperately needed services.

Without the MBV program, I would most likely still have found a rewarding career, but it vastly accelerated the timeline. Completing such a rigorous program changed my perspective on what I was capable of. It gave me the courage to reach beyond what I thought were my limits. The lifelong friendships I forged in the program are connections I treasure to this day.

"Valued, Supported, and Honored"

Michael Chavez reflects on telling his story

I did not expect to dive so heavily into this project and share very personal information that few know about.

Dr. Turrill held monthly Zoom meetings for us to come together and discuss our progress and what we were struggling with. These meetings helped enormously in keeping us engaged, feeling supported, and building the camaraderie throughout the shared experience. Some who were ahead of schedule opened their drafts to the rest, and I was impressed and a bit intimidated by their accomplishments. Imposter syndrome was a very real force, but I soldiered on regardless.

During the most recent USC tailgate weekend, Dr. Turrill held a gathering and dinner on campus. I made it a priority to fly into town to attend with my wife, and I am so thankful I did. Meeting up with those I hadn't seen in years, and meeting so many new contributors to this project, was so motivating to me. I felt valued, supported, and honored to be in the company of such exceptional men and women.

The ten-month experience of going through the MBV program was indeed transformational to me. It helped me redirect my life onto a path that is far more personally fulfilling. This process of telling my story allowed me to lower my walls and share some deeply personal parts of myself, and I feel completely unburdened as a result. I'm so thankful to have been asked to be a part of this, and I fervently hope those reading our stories understand how life-changing the MBV program can be.

Chapter 18

Service Never Stops
by
Michael Guadan

Michael Guadan is a Marine veteran who worked for a decade in local municipal law enforcement in Southern California before transitioning to the private sector. As a Marine, Michael served with Fourth Force Reconnaissance Company as a reconnaissance team leader and deployed twice for Operation Iraqi Freedom.

As a police officer, Michael worked in uniformed patrol and was awarded his department's Lifesaving Medal in 2009. Michael holds a BA from Arizona State University, a master's in international studies from Concordia University, Irvine, and a Master of Business for Veterans (MBV) degree from the University of Southern California.

After the MBV program, Michael joined a public safety software startup. Michael was then recruited to another public safety company focused on Next Generation 911 solutions. For the past five years, Michael has worked in multiple revenue-generating roles at the company and holds the position of senior director of sales enablement.

The Calling

I always felt a calling to serve my country and community. Even though my family has a history of military service, serving was never forced upon me. On my mother's side, military service dates back to WWI. More recently, my father served in Vietnam as a Marine. My earliest memory of the military was attending an air show at Marine Corps Air Station El Toro that featured a B-2 stealth bomber. I saw the B-2 from afar, then after we returned home, I happened to be outside when it departed, and the B-2 flew directly overhead. A remarkable memory I will never forget. The youngest of five boys, I was the only sibling who ended up serving. Military service is not for everyone, and I understand that. My dad used to say that giving back in any way is essential before taking advantage of this country's opportunities. I never forgot that lesson.

My desire to serve in the military grew stronger each year as I got older. I grew up in a large, blended family in Orange County, CA. My parents had been married before and had children from their previous marriages. For me, the only difference was the holiday schedule. Luckily, my brothers and sister never made it an issue. Everyone in my family deserves credit for maintaining as much normalcy as possible.

My parents gave all of us what they could and left a lasting mark on the family. My immediate and extended family never wavered, always providing me with genuine support and encouragement. My parents always supported me, even when they did not agree with my decisions or when I did not have a plan or know what was next. Love and support gave me a safety net for life. Home stability gave me the confidence to seek out new adventures. I grew up in a middle-class environment, and I was given everything I needed and, at times, was fortunate to receive more than I felt I deserved. However, my family's consistent love and unwavering support shaped my upbringing. I realize their intangible gifts were priceless. Without my family's foundation of love and support, none of my life's experiences and successes would have been attainable.

Growing up in Southern California was a blessing. Only later did I fully appreciate the nearly perfect weather and proximity to the beach, which ended up playing an essential role in my military service. I remember having access to numerous world-class attractions that could keep any '90s kid occupied for days. As a young student, I was more drawn to the social aspects of school, seeing the educational side as more of a hassle. It was not until later in life that I discovered my love for learning and reading after realizing it had to be aligned with my interests.

Back then, I focused solely on doing what I wanted. I spent time at the beach instead of studying. The beach holds a lot of my adolescent memories. I always found an excuse to go. I could relax on the sand, swim in the water, or play volleyball. I just loved being there. Even today, the beach is one of the few places I feel relaxed. As high school graduation approached, I lacked any noteworthy academic or athletic achievements to help me with college applications. Academic rigor did not suit me during those years, and while I was athletic, I was not a standout athlete. If I was going to college, it was simply a decision of which community college to attend, but I cannot say that I was even interested. College was just one of many steps for my generation then. It weighed heavily on me, as I saw college as a hindrance to my true calling.

While many of my friends were focused on their college applications, I was determined to pursue a career in law enforcement. In California, you had to be twenty-one years old to become a police officer. I decided to join the military to fill the gap between high school and being eligible for the police academy. It was not a difficult decision for me, I just followed my instincts. But as I always did with major life decisions, I sought advice from my parents. Even though I was technically an adult, their input was important to me, and I was grateful for the opportunity to have their guidance.

It was unsurprising to have my parent's support. What surprised me was that my dad, a USMC Vietnam veteran, encouraged me to think twice about joining and to not lose focus on college and graduation. One of my dad's special qualities, which meant a lot to me, was that he was loving and supportive even when he questioned my ideas. Now, after multiple deployments and having my own kids, I can certainly appreciate and understand his position much more deeply than I did back then. Because I valued his opinion, I took it upon myself to carefully consider and research my options. I dedicated myself to evaluating and exploring every available option. Eventually, I decided to join the US Marine Corps Reserve. I saw the Reserves as a win-win scenario, where I could still join the military and become a Marine while working towards a college degree.

When I met with the recruiter, I told him I wanted to become a Marine. I placed more importance on the title than any specific job. I was an overly motivated and naïve recruit who just wanted to enlist: I made a perfect recruiting target. It did not take much for the recruiter to convince me that being a radio operator was a great role. In fairness, he had a convincing argument.

My experience in the military would far exceed my most optimistic expectations.

Yellow Post-it Note

In January, 2001, after signing on the dotted line, I was sent off to Marine Corps Recruit Depot, San Diego. Then Uncle Sam blessed me with military occupational specialty (MOS) school in Twenty-Nine Palms, CA. Although I was not a recruit anymore, most of the time it still felt like I was. Even though the radio operator course was excellent and incredibly interesting, I struggled to find fulfillment in the role. Fate intervened on one of the last weeks of the course when our regular classroom instruction was put on hold for an unscheduled presentation. No one knew what the presentation would be about. I sat anxiously. The presenter was a Marine with a commanding presence—an experienced staff noncommissioned officer who spoke with authority. The presentation was my first time learning about Marine Reconnaissance. It captured my attention like nothing thus far in the military. Like most pitches, the slideshow showed only the "coolest" parts of the job. The presenter skipped over the harsh realities and challenges. Like many others that afternoon, I imagined jumping out of planes and fast-roping from helicopters. It was precisely the challenge I sought when I joined the Marine Corps. How can anyone realize their true potential without ever being pushed?

I volunteered and passed the initial screening. My experience of spending days at the beach in high school paid off, as the pool training part of the screening process was challenging for most men. Then the next day, my life took an unexpected turn. At a break in class, the seasoned staff sergeant who had given the presentation called my name and told me to report outside. I felt confused, rigid, locked up. As a private at parade rest, I stood before the staff sergeant who wore shiny pins on his uniform. With my eyes locked forward, the staff sergeant said, "Hey, Marine, relax, you are a reservist. You passed the screening for orders to an active-duty recon company. If this is what you want to do, here is my buddy's number at a reserve recon company in Reno. Give him a call to set up a screening there." The number was written on a yellow post-it which he handed to me. This seemingly small gesture completely changed the trajectory of my life. I never saw that staff sergeant again. I do not remember his name; he certainly would not know me from a stranger.

There we were, Marine to Marine, and he had gone out of his way to give me a chance at improving my life. He could have easily ignored a random, foolish private. Sending the elevator back down is something we should always do. Seemingly simple acts have significant impacts we can't imagine.

"One Weekend a Month"

I called the number written on the post-it note. My inquiry was well-received, and I was invited to attend a screening session. I had a few weeks to prepare, and I was given a list of the required gear and informed where and when to report. The second day of screening focused on my skills and what I could bring to the unit. I was lucky to meet the requirements of each phase, and I received my orders. That was just the beginning. I was assigned to the Recon Indoctrination Training Platoon (RIP). I had only been a member of the RIP platoon for a few months when the United States was attacked on September 11, 2001. My commitment to the military remained unwavering. The attacks were a stark reminder of why I chose my path. After the attacks, I found myself in Reno, NV, at the Fourth Force Reconnaissance Company, entirely dedicated to becoming a qualified Reconnaissance Marine.

Recon training tested me beyond measure. I failed quickly and often. I was challenged so much that I lost my fear of failure, and I developed a willingness to embrace challenges. I persisted because I never quit. The best lessons focused on the basics—to not fear failure, to stay humble (because the best idea always wins), and to make contingency plans. These lessons stand out the most, and they remained essential to my transition. Meanwhile, I continued working towards my original goal of starting a career in law enforcement and taking college courses between USMC school slots.

I was lucky to be part of a unit that provided support and resources to pursue my career goals. The unit allowed me to balance my military responsibilities and even postponed some of them when I was hired as a municipal police officer in Southern California. The military had prepared me for life at the police academy, despite the differences in environment and organizational missions. After six months of academy training, I graduated and began the field training program. Over the next few years, I worked full-time as a police officer, attending community college, reserve drill training, and taking more recon-specific school slots as they became available. It was nonstop, and I loved it.

In 2005, we activated to support Second Force Reconnaissance Company for Operation Iraqi Freedom. My first deployment was a transformative experience. I embraced my junior role and witnessed many lessons in leadership. I acclimated quickly to recon culture, incorporating those values and standards into every facet of my life. Reconnaissance culture demands discipline, teamwork, and an unwavering focus on mission accomplishment. The overarching goal is the mission and the team. The impact of reconnaissance culture inspired me to seek challenges, think unconventionally, and embrace a unique understanding of human camaraderie. After a seven-month deployment, I returned home and, without delay, jumped right back into police work and school.

Back at work, I hoped to experience the same deployment highs. On deployment, the operational tempo was high, and we focused on the job, collaborating and doing what had to be done to get home safely. In contrast, civilian police work was disappointing and slow. Despite a few bad days in deployment, I wanted to go back.

The bad days were terrible—friends killed and wounded in action. Losing such committed and brave warriors is unfair. And it's always a risk. I try not to dwell on it. Instead, I focus on what I would have wanted if I had not come home. I would want my brothers and sisters to be the best versions of themselves, live their best lives, and take advantage of every day. The experience taught me to value each moment and strive for excellence in all aspects of life. Understanding how fragile life is, I became more determined to make meaningful contributions professionally and personally. The sacrifices of others inspired me to honor their memory by living with purpose and intention. This change in perspective empowered me to face challenges with renewed energy, embracing opportunities for growth and cherishing the relationships and experiences that truly matter.

I went from a fast-paced mission set to routine police work, homework, and everyday personal tasks. The contrast was stark between an intense deployment and the daily rhythms of civilian life. Although I kept myself busy, I experienced an unexpected sense of boredom. The contrast between the two worlds was deep and growing daily. Military service provided me with more personal and professional satisfaction in a way that police work could no longer. In 2007, I volunteered for a second deployment to western Iraq to support combat operations with Third Recon Battalion.

After returning home, I once again started working full-time. I completed college and earned a bachelor's degree. Even with that goal accomplished, I still felt like something was missing. Everything seemed to be changing around me, and I could not shake the feeling that more change was inevitable. I decided to leave the military despite not knowing exactly what I was missing or why I felt the way I did.

Cooling Down

By 2010, I had finished a decade of nonstop activity—years of fast-paced work, college courses, intensive training, workups, and two deployments. There was never a break. Once I slowed down to working full-time only, boredom and resentment flooded into my life. I knew I had no reason to feel anything other than gratitude and appreciation. Activities that used to excite me now seemed dull, and I felt increasingly restless as I continued with routine police work. I missed the strong connections I had formed with my unit and our collective sense of purpose. I found myself feeling constantly on edge and out of place. The desire for change and a fresh start became impossible to ignore. Transitioning from the high-stakes world of overseas operations to the more familiar yet still challenging environment of municipal law enforcement was my new reality.

Unlike before, I had more time to process and be more present in this significant shift. I often drew parallels between my current colleagues, the workplace culture, the job itself, and my military experiences. This habit led me to set unreasonably high expectations for the people around me and the leadership. As a result, they consistently fell short of these unrealistic standards, which fueled a growing sense of disappointment and resentment within me. That is when I began contemplating a life beyond my role in the police department.

The truth was, I had changed. And a lot about the job was also changing. The changes were perhaps inevitable and too significant to overcome. Though incredibly rewarding in its own right, policing couldn't replicate the military and life overseas. Moreover, separating two very different training methodologies became a distraction and occupied a lot of headspace. Going from speed, surprise, and violence of action to de-escalation and using the minimal amount of force necessary got me thinking about a different future for myself. I didn't believe constantly switching contexts was optimal for my professional competence, especially considering the stakes. Even though I was proud of my law enforcement experience, my passion for the job diminished. For the first time in my life I was open to trying something new. There was also a noticeable change in community support. Because of these reasons, I felt it was best to find new challenges and a new environment. In 2011, I voluntarily left law enforcement and started a new adventure.

I decided to go back to school. My past experiences had shown me there was still much to learn. Before beginning a master's program in international studies, I took a few months to travel to various countries and historical sites. At the time, I believed in living without regret and fulfilling the promises I made to myself during my deployments. Looking back, spending four months traveling alone was a much-needed break. Traveling allowed me to release tension built up from years of pressure. I had put high expectations on myself in my military and law enforcement careers. Without those burdens, I gradually felt the tension lessen. My focus shifted inward toward long-term well-being rather than outward for the safety of my teammates and the community I served. I immersed myself in unfamiliar cultures during my travels, often feeling like an outsider. And I was fortunate to feel accepted and understood by many, which contributed to a deeper self-awareness and empathy for others. Traveling before returning to school really helped me. It gave me a new perspective and a profound understanding of different ways of life, which formed the basis for cultivating compassion. After I traveled, I thought more broadly and challenged myself intellectually, especially once I was a student again. The vibrancy and unpredictability of my travels gave way to the structured environment of academia. While I relished the opportunity to expand my knowledge, I often grappled with uncertainty about my future path.

Completing a new degree was a journey of discovery and self-reflection. I navigated complex subjects and collaborated with peers from various backgrounds. Yet, beneath the surface of academic rigor, I was still searching for clarity and direction in my post-military life. During this educational period, I acquired new skills, struggled to understand my place in the civilian world, and worked to apply my experiences and passions to the next chapter in my life. Later on, the MBV program built upon this foundation.

A Promise

Near the end of my master's program, my dad experienced some poor health, which put my journey into sharper perspective. Fortunately, he received the care he needed from specialists at the Keck Medical Center of USC. My family and I spent months around the hospital and staff. It was not easy, and the stress of the unknown was hard to handle. In a dark and unpredictable time, I felt the caring and professionalism of the doctors at Keck. Even during a difficult time, USC made an impression on me and my family. I remember making a promise to myself then that someday I would contribute to or give back to USC. I will forever be grateful for the care and attention USC gave my father, who eventually made it home after a major surgery.

About a year later, after graduating with a master's degree and entering a stagnant economy that had not recovered since the 2008 collapse, I found myself at a crossroads. I still did not have a clear direction or the prospect of a viable plan. I spent over a year figuring out my next move, sending out resumes and looking at different programs, still trying to figure out what I wanted. Asking for help was not an option for me at the time. I had had such a command of my life before. I wasn't sure why I couldn't figure out the next steps. I kept associating personal fulfillment with a job, maybe because, for the first time in my life, I did not have one. More so, I did not have anyone in a mentorship role to help me understand what specifically about my work and military experiences I loved.

One late, restless night, I browsed USC's website and learned about the MBV program. It immediately appealed to me, as it seemed tailored to leverage my capabilities and offered the opportunity to surround myself with fellow vets, many of whom had faced or were facing similar situations. As a new husband and father, I was motivated to pursue a new opportunity. MBV seemed like the perfect choice for me to channel my potential into a new mission. I paid special attention to the application process and put my best foot forward, as this was both a personal and professional endeavor. I was fortunate enough to be accepted into the program, which allowed me to embark on a rigorous business education and a formal transition journey.

Even Higher Education

The MBV program's small cohort became my new team. My classmates and I bonded through shared military traditions and humor. Soon, we became partners in tackling military transition together. I found inspiration in learning about my classmate's wide range of goals and dreams. Everything about the program aligned perfectly with my renewed sense of purpose. The MBV program offered more than academics; it provided a community of like-minded veterans and a sanctuary where my experiences were understood and valued.

The MBV program was two programs in one for me: an academic program and a personal program. On a personal level, I had the opportunity to rediscover myself and shape the next chapter in my life. The concept of strategic goal planning, informational interviews, coaching, and career preparation was foreign to me. I came from public service, where promotions are often less about meritocracy and more about tenure and experience. MBV helped me peel back the layers of my interests, goals, and priorities. It was gratifying to find correlations between what I liked most about the military and my previous work experiences. Even though I was new to these techniques, coaching resources were available. The best part of the MBV experience was approaching the program with an open mind and staying open to any opportunity that came my way.

The most impactful aspect of the MBV program was what I left behind: the military operational identity that my service had formed inside me. While it was helpful in the military, it limited me in civilian life. It led to periods of merely transactional relationships between family and friends. But I learned how to temper the intensity of the more instinctual behaviors that no longer served me. The past is important, but it didn't have to be my primary focus as I moved forward.

The catalyst to moving forward came during one of the MBV program's many distinguished faculty and guest speaker presentations. One speaker in particular changed the trajectory of my life: General David Petraeus. During one of General Petraeus's talks, he asked what types of companies one of my classmates invested in. My classmate explained his investment process and methodology. General Petraeus mentioned he had recently invested in a police software company whose mission was to innovate police technology. I was captivated and intrigued. For the first time, I realized the possibility of merging my law enforcement background with my interests in technology and innovation.

After hearing General Petraeus, I started researching. I found the company he referred to. My research opened my eyes to the world of startups, specifically startups in government technology (GovTech). These niche companies were mission-driven. Through an MBV introduction, I was connected with the company's CEO, who encouraged me to apply. That kicked off a seven month-long process with multiple interviews and negotiations. I leveraged the MBV career coach whose insights and experiences proved invaluable. I had never participated in such a rigorous process. A couple of months after graduating from the MBV program, I accepted a management position at the company. I credit MBV for preparing me to demonstrate my capabilities beyond my service record. The faculty's and the administration's investment in my success opened the door to a life-changing opportunity and provided the resources to help me navigate the other side. Soon after I graduated the MBV program, my first startup job awaited me.

Not an Overnight Process

Working for a startup was exciting and intriguing. The opportunity to work in a small team with a mission-focused mindset was familiar. Solving complex challenges, predictable and unpredictable, was the type of work environment I had been looking for. Turns out that startups share some similarities with the military. As I was still new to the private sector, I needed to integrate the vocabulary of business into my daily routine. Adhering to and understanding meeting etiquette was entirely new, too.

Being made late because of meetings exceeding their scheduled time often created stress for me. I was used to being on time. I initially thought being late was a big deal, even sparingly, and people would think less of me. A sense of urgency was another problematic concept for me in the private sector. Yes, startups are fast-paced. But that does not mean people do things quickly, like answering emails. I struggled when someone took a few days to respond to a simple email. In the beginning, I took it personally. Thoughts kept lingering in my head that I could have explained the situation better or articulated the question differently. I could not fathom procrastinating on things so simple and quick to resolve. Of course, this got better over time, especially as I became more and more the culprit of email procrastination. There is a lot to the private sector that I do not understand or, perhaps, do not want to understand. The point is that transition is not an overnight process. It takes time and it may never feel complete, just better.

In the years since the MBV, I expanded my skills scaling public safety technology startups. I worked for what is called a first-mover startup, in particular, a company that was the first to deliver cloud-native digital transformation to mission-critical software systems within public safety. In the US, cloud software had not penetrated areas such as computer-aided dispatch (CAD) and records management systems (RMS). From that company, I moved to another startup focused on digital transformation for Next Generation 911 systems. I played a small part in bringing data, accessibility, and innovation to first responders.

Initially, I took business development roles that utilized my relationship-building and strategic planning abilities. Then I pivoted into strategic and technical positions as I gained more insights into the technologies and users' needs. Now I serve as a senior director in enablement and revenue acceleration. The urgency of our projects and the teamwork they require relate closely to the most meaningful aspects of military service. Leadership principles I learned in the Marines still guide me, but much about my approach has evolved. Motivating talented and willful technologists requires more nuance than barking orders at them. My process is not perfect, but I try, and I care. That is more than a lot of people can say. I focus on enabling everyone to work at their full potential, together. This collaborative spirit consistently drives results and the innovation we need to be successful in our industry.

The Anchor of Family

My family's unwavering support and grounding influence were fundamental to my life's journey. It is strange to think that at one time, naively, I thought I would be happiest traversing the world alone and presumably without burdens, responsible only to myself and my own adventure. But the same purpose that motivated me in the military became renewed with starting my own family.

Since my children were born, their presence made a sanctuary for me amid the many uncertainties I faced, before and during my transition and while attending the MBV program. My wife and children have been my anchors. They are my constants, and they teach me invaluable lessons in empathy, patience, and unconditional love. Parenthood, in particular, has been a transformative experience. The days have been long, but the moments are powerful. The daily joys and challenges of raising children have instilled a more profound sense of responsibility, and a renewed sense of purpose has followed. Every smile, every question, and every milestone has been a reminder of the importance of being present and nurturing not just my children's growth but my own as well.

In the early days of my transition, when I felt unsure about my direction, my family's support was my refuge. They helped me to look beyond my past and find fulfillment in my roles as a husband and father. My wife's partnership and understanding, and my children's pure, unfiltered zest for life have been the foundation on which I have recentered my life. Their love reminded me that my value and purpose were not solely tied to a job or to the military.

My wife has been a pillar of support. We met after my first deployment, and throughout the next decade, she supported me during many of life's challenges. In 2015, we got married and immediately started a family. Since then, she has provided us with unwavering support. She has become the emotional rock of our family and friends, especially when I lack the emotional capability. When I look at problems and see only missing solutions, I often lack the empathy to maintain the correct stability. But my wife's ability to fill that gap and give acceptance and love exceeds any accomplishment I will ever have. Since becoming a parent and completing the MBV, she has provided the stability I desperately needed. Overall, her patience and empathy have taught me the true meaning of partnership, and our marriage has become the foundation for the environment we seek for our family.

The arrival of my children added a new dimension to life. Each one has a unique personality, boundless energy, and a pure innocence. They have become the lens through which I see the future I want for them. Every play session, bedtime story, and shared moment of discovery has taken me another step toward focusing on the future rather than dwelling on the past. Their curiosity and enthusiasm for life remind me to stay in the present moment and pull me away from thoughts of "the good old days." My happiness comes from witnessing their joy, and I find comfort in their laughter and a purpose in their dreams. I realize that my role as a father is the most important mission I will ever have.

The parenting journey has been as challenging as it has been rewarding. It constantly pushes me to grow and discover more about myself. It has tested my patience, which was already in short supply. However, with the help of my kids, I have grown more patient over time. (Perhaps they wouldn't agree with that statement.) They drive me to be better and do better as I navigate the highs and lows of raising them. When it comes to the lows, I recognize that I am not good at giving them the empathy they deserve when trying to understand their fears, joys, and perspectives. In a world filled with devices and distractions, the best gift I can give them is my presence. As I seek to build a better tomorrow, balancing spending time at home while also achieving my professional aspirations is a constant battle. Family life has been the influence I needed in order to find that balance.

Now I consider how my loved ones perceive my words and actions. I have started to incorporate the idea of other people's perceptions into my way of thinking. I realized these values were not limited to just one aspect of my life; they could equally apply to all of life.

As I continue on this journey of personal growth and professional development, I carry the lessons learned from my family. They have taught me that true strength lies in vulnerability, that real success is measured by the love and connections we cultivate, and that our most profound purpose is serving those we love. My family's influence has been a cornerstone of my transition and growth. They have been my constant source of strength, guiding me through the complexities of reintegration and helping me to build a life rich with purpose and joy. As I look to the future, I am inspired by their love and support, ready to embrace new challenges and opportunities with the knowledge that I am never alone on this continuous growth journey.

Continuous Growth

From when I was inspired by my family's rich military history to navigating the transformative experiences in the Marine Corps, graduate studies, and family life, my journey has been marked by significant milestones. Each phase of my life has shaped who I am today, teaching me the enduring importance of embracing change, nurturing connections, and striving for excellence in all aspects of life.

Traveling abroad and immersing myself in new cultures and academic environments was a profound turning point in the journey. It pushed me out of my comfort zone and broadened my perspectives, allowing me to reflect on my past and consider my future more clearly. The solitude of travel provided a much-needed break, helping me decompress from the intensity of my military service and redefine my identity outside the uniform. These experiences were crucial in reshaping my outlook and laying the foundation for continuous personal growth.

A more significant milestone in my personal development was enrolling in the MBV program at USC. The challenging curriculum and the camaraderie of my fellow veterans provided a structured setting to apply my military skills in new and impactful ways. Surrounded by a community of like-minded individuals, I honed my leadership abilities and deepened my understanding of business fundamentals, bridging the gap between my military background and civilian aspirations. The program pushed me to think critically, communicate effectively, and lead with empathy—skills invaluable in my professional and personal life.

I have faced ongoing challenges with staying connected to friends and the veteran community, particularly to the MBV program. This disconnection is disappointing, as I owe much of my recent accomplishments and education to the program and its incredible staff. Recognizing this gap, I am committed to addressing it as part of my continuous growth journey.

I intend to pursue more advanced learning opportunities through professional courses and certifications that align with my career goals. Lifelong learning is a crucial aspect of continuous growth, and I am committed to staying updated with the latest industry trends and knowledge. This dedication to education will enhance my professional skills and keep my mind responsive and open to new ideas.

With time, I realized that the essence of transition lay in the ability to let go. Despite feeling unprepared and lost at times, I began to understand that my journey was defined by continuous change, significant moments, and unwavering support, all of which connected my past experiences to the present.

"The Necessity of Embracing Change"

Michael Guadan reflects on telling his story

The most influential factors in my life have been the support of my family, my children, and the transformative experiences I had in the military and law enforcement. My family's unwavering encouragement gave me a stable foundation and the confidence to pursue my goals and overcome obstacles. The military, particularly the Marine Corps, instilled in me discipline, resilience, and a strong sense of duty and commitment, significantly shaping my personal and professional growth.

I was motivated to join fellow MBV alums in sharing my story in order to connect with others who have faced similar transitions. Sharing my story was important because it was my way of expressing my feelings and sharing them with others. It allowed me to reflect and document my journey.

I wanted to convey the importance of resilience, the value of family support, and the necessity of embracing change; and, furthermore, I wanted to inspire fellow veterans to seek growth and transformation while highlighting that personal and professional transitions are continuous and evolving.

This process taught me more about myself, especially what I can do when motivated. Being vulnerable with the writing, my story presented several obstacles, including condensing complex experiences into a coherent narrative. I also needed help to get it done on time. To navigate these challenges, I employed strategies such as focusing on key themes and allowing myself the space to process emotions as they arose.

Chapter 19

Post-Traumatic Stress to Post-Traumatic Growth
by
Nikkea B. Devida

Nikkea B. Devida is a US Air Force veteran whose career has combined her military, education, and business backgrounds with cutting-edge science, music, and energy psychology.

Nikkea served for four years at Hanscom Air Force Base, Massachusetts, where she was responsible for negotiating and managing $200 million defense contracts. She was awarded the Secretary of Defense Superior Management award while working on the Technical Onsite Inspection (TOSI) program in support of the Intermediate-Range Nuclear Forces (INF) Treaty with Russia.

Nikkea's academic credentials include a BS from the US Air Force Academy in 1986 (the seventh graduating class of women), and a Master of Business for Veterans degree from the University of Southern California's Marshall School of Business.

Nikkea is the founder of Sisters Who Serve, a nonprofit organization dedicated to supporting the unique needs of women veterans. An accomplished singer, songwriter, and recording artist, Nikkea's has earned a reputation as a sought-after speaker, trainer, and consultant.

You Are Not Alone

It would not be an exaggeration to say that it has taken the last twenty-five years to heal and overcome the litany of trauma from the first twenty-five years.

I half-jokingly say that there's not a healing modality out there I have not written a check to. I have put many therapist's kids through college. I did a sobering exercise to calculate the money I invested over the years in healing myself—I stopped counting after $500,000. Now at the age of sixty, I have put my money where my mouth is, and I have done the work to get to the other side of the trauma to a place of peace, self-love, forgiveness and acceptance.

It doesn't mean that my life is now all angels, butterflies, fairies, and unicorns. Life is still in session—there are still challenges, struggles, and obstacles. It's just that now I have tools and resources to self-regulate and respond to, as opposed to reacting to those challenges.

I have had some extraordinary therapists, coaches, mentors, friends and guides along the way who helped me heal. Now I have the tools, experience, and wisdom to pay it forward and support others through their healing journey. I believe that healing begins the moment we share our stories and realize we are not alone in our struggles.

My greatest joy and fulfillment comes from sharing my gifts, talents, and wisdom to help others overcome trauma, become the best versions of themselves, and achieve their goals faster and easier. I am driven and called to serve to make a positive difference in the world. I have no interest in sacrifice or being a martyr. I believe in doing well and doing good. I can help more people if I'm sharing my message in a way that feeds and fuels my joy and gives me energy. I also learned it's a lot easier and more enjoyable to play to my strengths instead of working hard to develop my weaknesses.

Above Average Dysfunction

Since I was a little girl, I had a passion for singing and writing lyrics and poems. I loved music and I dreamed of being a singer performing for audiences around the world. I have always seen the systems, processes, and patterns of things. Both strengths set up an inner conflict, or tug of war, inside me. Which strength was going to win? Which would I pursue? My first love was to be creative with music, songwriting, and self-expression. But life led me to the analytical as a matter of survival.

At the same time, there was a lot of conflict with my family. Growing up, and through most of my adulthood, I never felt safe to speak my truth because it was always used as a weapon against me. Being silent was safe.

My parents were born in 1930 in the Bronx, NY, during the Great Depression. My father's parents were immigrants from Italy, and my mother's parents were immigrants from Ireland. They were very poor.

My father is the youngest from a family of six children. My mother is from a family of fourteen children. Two of the children died when they were toddlers. My mother was born after they died, and she was named after one of the deceased siblings.

My parents married when they were nineteen. My father had a Dr. Jekyll and Mr. Hyde personality. He put on a charming face to everyone outside of our household. It was very important that everything looked perfect to the outside world and everyone else thought he was one of the nicest people.

After twenty-one years of a very unhealthy and dysfunctional marriage, my parents divorced when I was nine years old. After the divorce, we were all left to fend for ourselves. Back in the 1970s, there was very little awareness of trauma and abuse. There was a stigma about seeking help from a mental health professional. Only "crazy people" saw a therapist.

I became my brother's personal punching bag. I was severely abused physically and emotionally (including hospital visits and stitches) nearly every day until I turned eighteen. Because my mother was mostly deaf and a single working mother, she didn't hear or see what went on. When I came crying to tell her, I was the one who got in trouble. My brother never got in trouble for beating me. Understandably, that sent messages to me that I wasn't good enough or worthy of love, protection, or caring.

My family was not exactly your average dysfunctional family. It was above average. As I got older, I was highly motivated to get away from that environment.

Whether by luck or divine protection, I chose academics and athletics as my coping mechanism and a way out of my family. I stayed after school for as long as possible to avoid going home. While both of my brothers were addicted to drugs and alcohol at an early age, I was involved with extracurriculars such as the Key Club, singing in the choir, school plays, and playing the flute. I was also a competitive athlete on the varsity gymnastics and diving teams. I built an excellent resume for college applications, which included applying to the military service academies.

In hindsight, I have discovered that people go into the military for one of two reasons: a call to serve their country, or to escape their home environment. My reason was the latter. I joined the military so I wouldn't owe anyone anything and could move forward on my own terms. I wanted to get as far away as possible from my family in New York. Although no one in my family had ever gone to college, I still knew that getting an education was the way out. I was an honors student with a 4.35 GPA and graduated sixth out of 450 students, but I couldn't afford college without a full scholarship.

Academy and Station

I applied and was accepted to the United States Air Force Academy (USAFA) in Colorado Springs, CO.

The core curriculum was largely science and engineering, but I pursued the closest equivalent to a business degree. I graduated from USAFA in 1986 with a BS in management, part of the seventh graduating class of women since women had been accepted to the academy for the first time in 1976. The guys weren't used to us back then, but we were all smart, tough cookies.

Over the years, I have been asked what it was like being around so many men. Frankly, it was comfortable. I have been surrounded by men all my life. I'm the only girl in my family of three brothers, four nephews, and two great nephews. I was the only girl in my neighborhood growing up. I was used to being the only girl.

Through it all, I never lost sight of music. I sang in the Air Force Cadet Chorale. We traveled all over the country putting on shows, including at the National Cathedral in Washington, D.C.; for retirement ceremonies of generals and other high-ranking officers; and at community events and celebrations. The Cadet Chorale got me through the academy.

In the '80s, there were several incidences of sexual harassment and hazing at the academies. As a cadet, I was raped by a military officer. I was the only female deployed on a summer program in the

Philippines. The reason I didn't report it was that because, back then, those who reported sexual assaults were given a diagnosis of borderline personality disorder and then a dishonorable discharge for having a pre-existing condition. The systemic silencing of reporting sexual harassment was reminiscent of Don't Ask, Don't Tell.

As an officer, I was stationed at Hanscom Air Force Base outside of Boston, MA, from 1986-1990. I was a contracting officer and negotiated and managed approximately $200 million in defense contracts, all at the age of twenty-two. I received high-level training in contracts and contract negotiations, project management, cost and price analysis, systems, operations, logistics, and leadership in the military. I started in research and development projects with local universities such as MIT, then physical defense security systems, and then the famous Strategic Defense Initiative (SDI) contracts, better known as the Star Wars program. In Washington, D.C., I was awarded the Secretary of Defense Superior Management Award by then Secretary of Defense Casper Weinburger for my work on the Technical Onsite Inspection Program (TOSI) that constructed the verification site in Russia according to the Intermediate-Range Nuclear Forces (INF) Treaty.

While I was active duty, I entered Air Force talent shows to earn a spot in the world-renowned Tops in Blue traveling musical group. It was my dream to perform in that show. But to get there, you had to win at the local base level, then at the command level, and then at the nationals. The Tops in Blue performers were selected at nationals. It was a tough competition with a lot of talented singers, musicians, dancers and performers.

I won the base competition four times, and I competed at the command level. I entered as a female vocalist, one of the most competitive categories. Only first-place winners were selected to compete at nationals. On my fourth attempt, I finally took first place in the female vocalist category. I was so excited that I was finally going to nationals. One by one, the names were announced: "For the first time in the history of this competition, we decided not to bring anyone from the female vocalist category." My name wasn't called. I wasn't going to nationals, even though I took first place, and I was never selected for Tops in Blue. I was devastated. When I got back to Hanscom AFB, things started to unravel.

On the outside, I had all the trappings of success. I had a good education and a good job. I had a new car, a condo that I owned, and I was in a loving relationship. But my soul was dying inside. All of the abuse and trauma was catching up with me.

Given the home I came from, then USAFA, and the military environment I went to, I was a hypervigilant bundle of anxiety. I felt like a pressure cooker in need of a release valve.

If there is one thing an Irish and Italian family from New York likes to do, it's eat and eat and eat. That was my socially acceptable release valve. In fact, my grandmother's favorite phrase, and the only Italian I know, is stai zitta e mangia, which means shut up and eat.

The Big Secret

While I was at USAFA, I developed a severe eating disorder. I struggled with bulimia, binging and purging 2-6 times per day, every day, for four years at the academy and four years on active duty. It was destroying my life, my health, my relationships, and my finances. I tried many things on my own in secret to get help and overcome it, but nothing worked.

In Boston, after spending tens of thousands of dollars of a junior military officer's salary on therapy, support groups, and treatment after treatment that didn't work, I finally went to the Air Force for help.

At first, they didn't want to acknowledge my condition. They said it was impossible for me to be active duty with an eating disorder and that I should never have graduated from USAFA with that condition.

Desperate for help, I contacted local Anorexia Nervosa and Associated Disorders (ANAD) groups and the local news media to inquire about how to share my story. I also researched what resources, facilities, and programs the military used to treat eating disorders. I learned of an in-person treatment facility outside of Washington, D.C., at Malcolm Grow Medical Center, located on Andrews Air Force Base. I was getting my negotiating chips lined up.

As a first lieutenant, I marched into the base commander's office and demanded to speak directly with the general. I informed them that if they didn't get me into the treatment center at Andrews, I was going to expose the story to the press. Further, I said I was at the end of my rope and was not going to go back to work until I got an answer. I have no idea what made me think I could get away with it, other than desperation. Within three days, I was airlifted and admitted to Malcolm Grow Medical Center at Andrews Air Force Base, MD. It was early December, 1989. I spent my twenty-sixth birthday and the holidays alone in the hospital.

A lot of good came from being in the hospital. I owe a lifetime of gratitude to my therapist, who taught me about mental illness and personality disorders. For the first time in my life, I learned that the trauma I had experienced wasn't my fault and I hadn't deserved it. I learned that nearly everyone in my immediate family very likely had mood and personality disorders: my father, my mother, and my abusive brother. Learning about mental illness, personality disorders, and trauma finally put things in perspective.

Making Do

I was hospitalized until I was medically discharged from the Air Force in May, 1990. My military career was over. I had to leave the place where I felt a sense of camaraderie, belonging, and purpose. In the military, I knew my job and I did it well. I knew what was expected of me. I knew the rules. I knew how to win and be successful.

There was no Transition Assistance Program (TAP). The only tap I got was the door hitting me on the butt. Once I walked out that door, I was a civilian. I had no idea what I was going to do. There was no one to help me. I was alone.

Because of the Mass. Exodus, in which it is estimated that almost 300,000 Massachusetts residents moved out of state between 1990 and 2002, I couldn't get a job after leaving the Air Force. I had many interviews, but no offers. There were hundreds of applicants per job opening. I couldn't even get a minimum wage job. To make ends meet, I took whatever part time work was available.

After eighteen months of miscellaneous odd jobs, I moved to Indianapolis, IN, where I got a corporate job as a senior buyer for a wholesale pharmaceutical company. Because of my experience in the military, I was used to a lot of responsibilities. Corporate America came as a culture shock. I wasn't used to the secrecy or the low level of trust. I learned the jargon fast and quickly became responsible for managing an inventory of more than 22,000 SKUs single-handedly, over three times the volume of any other buyer. I left that job after the company president grabbed my crotch during one of our conferences in San Antonio, TX.

I then became a purchasing and project manager for companies like Walt Disney Records and Macmillan Publishing in Indianapolis. Then, at Nova Development, a software development company in Calabasas, CA, I gained more experience and training in leadership, project management, systems, operations, logistics, contracts and contract negotiations, purchasing and inventory management, and supply chain management. I was building a solid resume.

Music

Over the years, I wrote song lyrics to express my feelings, but I never did anything with them. While living in Indianapolis, I completed a six-week continuing education course in songwriting at Indiana University–Purdue University Indianapolis (IUPUI). I met a great musician, producer, and composer there. Near the end of the course, I got the courage to ask him to look at my lyrics and songs. He loved them, and we started collaborating. I bought a home recording studio, and we started recording and producing songs together. The story about my half-brother became the inspiration for a song I wrote called "Do You Remember Me."

The underground single made it in regular rotation in some dance clubs in 1995. Unfortunately, my co-writer fell in love and moved to Missouri, ending our collaboration. I moved to California. I sold the studio and haven't found a good collaborator since.

My Healing and Spiritual Journey

I got married to my first husband when I was thirty-one. We were married for eighteen months. He was a talented chiropractor. I learned a lot about natural health from him. He also introduced me to feng shui. It immediately made sense to me, and I became a practitioner and trainer right away. To this day, I practice the Black Hat Sect of feng shui as taught by His Holiness Lin Yun. I was a full-time practitioner until 1998 when I moved to California. It became part time because of all the driving. I transitioned from doing feng shui consultations to doing subconscious belief change work, and I started gaining traction doing teaching workshops.

From 1990-1998, I tried many traditional healing modalities such as individual talk therapy, group therapy, support groups, and pharmaceutical interventions, etc., to overcome my eating disorder. Nothing worked. I tried numerous alternative, mind-body, complimentary, holistic techniques and spiritual practices in the form of books, tapes, CDs, DVDs, workshops, conferences, masterminds, rituals, ceremonies, and private mentoring. I learned and studied personal and spiritual development, dozens of different healing techniques and modalities, neuroscience, cellular biology, and psychoneuroimmunology. None of them healed me of the eating disorder.

I became a certified trainer of hypnosis, neuro-linguistic programming (NLP), Psych-K, feng shui, Reiki, Hawaiian shamanism, the Sedona Method, the Silva Method, chakra balancing, sound and vibrational healing, and many others. I thought that if I got certified, they would work better. I didn't get lasting results with the eating disorder, but I learned and received a lot of value from every single one of them. Each of them was instrumental in my healing journey.

While studying neuroscience, I learned that just being aware of and understanding your trauma does not heal you nor does it transform the trauma. I also learned that repeating affirmations over and over for days, weeks, months, or years does not actually reprogram your beliefs nor install new ones. Awareness and repetition are not necessary at all. I was eager (and a little desperate) to learn everything I could. I was cautiously optimistic that maybe this was the missing piece I was looking for. To my surprise, and for the first time, I got some relief from the eating disorder I had been suffering with for thirteen years. That got my attention.

True Calls to Service

My second husband taught me a lot about being a business consultant. Right after we were married in 2006, we got the contract for Peak Potentials. We lived in beautiful Vancouver, British Columbia. My husband was the COO, and I was the director of production of the largest department of seventy-five people running the 156 events per year that drew an average audience of a thousand. Harv Eker, founder of Peak Potentials, wanted to sell the company in five years, so we had a lot of work to do. We had to get the right people in the right roles, and have all the procedures identified, streamlined, and documented in order to be ready for sale. We worked sixteen-hour days, seven days a week. It was the toughest job I ever loved. I credit that role with giving me an opportunity and the responsibility to use all my leadership and operational skills.

In 2004, I attended a women entrepreneur conference where the founder of a nonprofit women veteran's organization gave a presentation. That's when I learned about the shocking statistics facing women veterans. Women veterans? I thought. I was a veteran, but it had never occurred to me that I was a woman veteran.

I learned that women veterans are the fastest growing homeless population in the US; young female veterans have an unemployment rate as high as 36.1 percent; approximately 33 percent experience military sexual trauma (but the percent could be as high as 60); and women veterans are eight times more likely to experience PTSD as their male counterparts.

I sobbed uncontrollably when she shared those statistics. In that moment, I felt a call to serve my fellow sisters-in-arms. For the first time, I realized that I had experienced most of the troubling statistics.

That inspired me to write and record a song called "Sisters Who Serve" as my personal tribute to women veterans. It was originally supposed to be used as a fundraiser for the nonprofit, but they ended up not wanting to use it. No problem, I thought, I would find another one to donate the song to. But after researching several nonprofits, I discovered that either their mission, their culture, or their leadership was not compatible with the song. As a result, I decided to start my own nonprofit for military women called Sisters who Serve in July, 2014. Sisters Who Serve is dedicated to securing and providing resources, training, and mentoring to military women and women in military families, with the goal of supporting their physical, emotional, spiritual, and financial well-beings. The nonprofit has six initiatives: Heal a Vet, Hear a Vet, Hire a Vet, House a Vet, Honor a Vet, and Help a Vet.

Years ago, I coined the term "sacred self-indulgence" to refer to something you love to do, and when you do it, you feel a sense of joy, oneness, and timelessness. It could be playing a sport or recreational activity, music or performing arts, a craft or hobby, learning something new, being in nature, playing with children or pets. My sacred self-indulgence has always been music.

I discovered that whenever I mention music, singing, songwriting, or recording and women veterans in the same sentence, magic happens in my life beyond anything I could have planned. For example, I was invited to be a keynote speaker for the Women in Military Service for America (WIMSA) memorial at Arlington National Cemetery in Arlington, Virginia. I was also invited to sing "Sisters Who Serve" at the Women Veterans Alliance and at the California Women's Conference during the opening ceremonies' tribute to women in military.

Transformation

While attending a military veterans conference, I learned about the Master of Business for Veterans (MBV) program at USC. The more I heard about the program, the more I wanted to attend.

A major complaint I have about my transition experience is that I didn't know that I was eligible for the GI Bill or vocational rehabilitation (VocRehab). Not only did I not know I was eligible, I also didn't know the education benefit expired after ten years and twelve years from date of separation, respectively.

By the time I realized I was eligible, my benefits had long expired. It never occurred to me that I was a GI. I thought it was just for combat veterans or enlisted people. For years, I tried to get my benefits reinstated or find a loophole in the regulations, and for years I was unsuccessful. Then I found a tiny little phrase in the VocRehab regulations called "extreme financial hardship."

I applied. To my surprise and delight, I got accepted into the MBV program. I temporarily closed my business in 2018 to go back to school. I was very nervous about it, too. I wondered if I still had the ability to learn and do well in school. Even though I had been continuously learning since 1986 when I graduated from USAFA, I hadn't taken additional formal college-level courses. I was determined to do my best and learn everything I possibly could from the opportunity.

To say the program was life-changing and transformational would be an understatement. The MBV program ranks among the top highlights of my life, higher than either of my weddings. I graduated with Cohort VI in 2019.

I would not be who or where I am today without the MBV program. One of the major nuggets I learned was that personal development and leadership development are synonymous. The program made me a better human, not just a more educated veteran, leader, entrepreneur, business owner, and consultant. The program was impeccably designed and executed from start to finish, as one would expect from the fine institution of USC. It gave me the opportunity to up my game on every level.

I applied myself to learn everything I could about leadership and team building. I worked to master the process of creating a business plan, a marketing plan, and a financial plan. I even entered two of USC's pitch competitions. I received so much guidance and support, including from my advisor who was a neuroscientist and a professor of entrepreneurship both.

COVID-19

In October, 2019, I moved from California to Las Vegas, NV. I wanted to continue my education while looking for employment and going on interviews. To keep the momentum going, I also completed my project management professional (PMP) course certification from Syracuse University. I couldn't complete the final certification exam due to the COVID-19 shutdown.

When COVID hit, I had just started a new remote job as a fractional, outsourced CEO and COO of law firms, so I was already working remotely. In fact, I had been working remotely since 2001, so I had very little adjustment to make. While the world shutdown, I was working seventy-five hours per week managing the leadership, growth, and profitability of as many as twenty-six law firms. I learned a lot, but it was intense, grueling, and exhausting. I really respected and enjoyed the caliber of my colleagues, but the leadership and culture left a lot to be desired.

My tenure there ended after eighteen months of employment. Almost instantly, a weight was lifted in my heart. At the same time, the pandemic restrictions were starting to lift in Las Vegas, one of the last cities in the country to reopen. I was finally able to start exploring the live entertainment scene, which was one of the main reasons I had moved to Vegas.

I found another job as an operations manager/COO of a company that did outsourced accounting and bookkeeping for law firms. It was a well-run, fast-growing company with a really positive, supportive culture where everyone worked as a team and supported each other. Part of the positive culture was the good work-life balance and reasonable work schedule.

I think about what I learned during MBV in order to grow and improve as a leader. We have doubled our team and tripled our revenue over the past two years. We have upgraded our internal infrastructure to be scalable, and we have documented our key processes and procedures.

Back to School Again

Because I became unemployed during COVID-19, I qualified for another scholarship through the Veterans Rapid Retraining Assistance Program (VRRAP). There was a list of approved schools and certification programs eligible under the program. I wanted to do something very practical and make my heart sing. I looked through the list for any online music production programs, and there were! I did my research, chose a school, and applied for the Independent Artist Program (IAP) at the Musicians Institute in Hollywood, CA. I thought the MBV program was tough, but in the IAP program I was learning an entirely new language of music. The program was extremely challenging and rewarding. I wrote and recorded several new songs. As a result of the program, I finally gained the knowledge, skills, and training I needed to become more self-sufficient as a musician and producer.

The IAP program provided the missing link I needed to complete the business and marketing plan I created in the MBV program. I synthesized all the work I did and updated my plan to include music. The IAP program equipped me with the tools and knowledge I needed to write, record, and produce my own music. In the past, I had to rely on (and pay for) other musicians and producers.

I believe I was put on this earth, and had to go through the experiences I did, for the purpose of helping others (especially military veterans) heal from the suffering of trauma, neglect, and abuse.

Post-Traumatic Growth

My true calling is to step onto a national platform and raise awareness and funding for military veterans and first responders. I want all the challenges I have experienced and overcome to mean something, and I want to share what I have learned in a sustainable, scalable way that can help people long after I'm gone. It's not about me; it's about serving and supporting others.

Each phase of my journey has been meaningful because I decided to make it meaningful. Healing is a choice that comes from within, and I'm grateful for all I have learned along the way to become who I am today.

My journey of post-traumatic growth has been a long but rewarding one. Instead of retiring, I'm more excited than ever to start a new endeavor. It's terrifying and exhilarating at the same time. My goal is to look back on my life and know I left it all on the field. I don't want to die with my music left inside me.

"Real, Raw, and Vulnerable"

Nikkea B. Devida reflects on telling her story

When Professor Turrill announced that he was compiling a second book and was looking for authors, I immediately volunteered. Any opportunity to connect with Professor Turrill and fellow MBV grads is always a valuable and transformational experience that never disappoints.

The most impactful part of the process for me personally was to reconnect with the MBV community again. They are always a source of inspiration, growth, wisdom, depth, caring, and sharing. Attending the regular check-in calls to reconnect, share insights, challenges, and stories was always a welcome break and oasis in a very busy schedule.

The hardest part of any journey is taking the first step. The first challenge was making the time to write and figure out what to write. My strategy was to reflect on the most impactful stories of my life, both positive and negative, which have shaped who I am and who I have become. As a seasoned project manager, I put myself on a schedule to write five pages per week to complete the assignment.

Because my goal was to get my entire story out on paper first, I did a brain dump and wrote about twenty-five different stories in a linear timeline sharing the good, the bad, and the ugly. We were encouraged to be real, raw, and vulnerable. I dove in headfirst, and my first draft was three times the length of the final draft. I was curious and nervous to receive feedback from Drs. Arámbula and Turrill about what I wrote. I genuinely appreciated receiving their valuable, direct, honest, and always constructive feedback. I edited the chapter at least a dozen times. It's a lot easier for me to have too much and edit it down than to not have enough material. Because I will be re-purposing my story to create a one-woman show, I took this as an opportunity to write a much more comprehensive life story, and then I developed the ones that felt most relevant to the stated purpose of the book. I'm grateful for the opportunity to share these stories, and I hope that they provide inspiration and courage.

I have been meaning to write my story for several years. I started a draft and never finished it due to time constraints. I have wanted to share my story for the purpose of inspiring others. We all have important stories to share. All have value and deserve to be heard.

It was valuable to hear others share about their challenges with writing, and to hear their stories as they were being written. At the age of sixty, I still forget sometimes that I have thirty years more experience, perspective, and healing under my belt than some of my classmates. I have so many goals and aspirations that I forget to give myself credit for how far I have come and what I have accomplished.

Chapter 20

Other Than a Soldier, I Was a Trojan
by
Rob Fortner

Rob Fortner served for ten years in the US Army as a medic in the 173rd Airborne Brigade combat team. His 2007 deployment to Camp Keating in the Landay Sin Valley, Afghanistan, was reported as part of Jake Tapper's 2013 book, The Outpost: An Untold Story of American Valor. *He was awarded the Silver Star for gallantry in action and the Purple Heart for wounds received in action, as well as numerous Army Commendation and Army Achievement medals.*

Rob has a bachelor's degree in finance from the University of Phoenix and a Master of Business for Veterans (MBV) degree from the University of Southern California. He is currently pursuing a certificate in business analysis from Purdue University.

He has served as vice commander and commander of the Military Order of the Purple Heart Chapter 604 in Kern County, and he remains active in helping veterans to this day, including in his work as an operations manager for Amazon in Dupont, Washington. He is married with two children.

Family and School

I live what many consider the American dream: I have a wife and two kids, a nice suburban house, a dog and a cat. I work as an operations manager at an Amazon facility. The journey that brought me to this location was the most terrifying and exciting adventure of two lifetimes. It was filled with the highest of highs and the lowest of lows. I lived through multiple life-threatening situations while also visiting some of the most beautiful and historical places imaginable.

I grew up with six siblings raised by a single mother. For being alone, and considering there were seven of us born across twelve years from oldest to youngest, my mom did an extraordinary job raising us. She instilled in us a sense of right and wrong. We were active in church, and that is where, personally, my own sense of right and wrong was forged. My parents divorced when I was in junior high school. Up to that point my mom had been a stay-at-home mom. After the divorce proceedings began, my father became completely absent and was what some call a "Disneyland Dad," but without the Disneyland. As my mom spent more time at work, I drifted a bit.

I wasn't a terrible student, but I definitely did not apply myself as much as I could have. I spent way too much time staying up past reasonable bedtimes and participating in activities that neither challenged myself nor lent well to how others perceived me. I really underperformed my freshman and sophomore year, finally waking up to reality before my junior year. It took me the bulk of my junior and senior year to make up the missing credits from my shenanigans. For example, in sophomore year, I walked around the neighborhood (as one does) lighting illegal fireworks. I threw one into the Kern County fire chief's backyard, and the chief himself caught me. I find it humorous now, because he was in charge of a task force that was looking for people using illegal fireworks. He went easy on me and didn't throw me to the wolves, which he definitely could have done. But he gave me the stern talking-to that was four years overdue. It was the talk that good fathers give their sons, covering topics such as expectations, fairness in life, results from hard work, and so on. What I took from it was that I wasn't unique. It reinforced a feeling that no one was going to feel sorry for me and give me what I wanted.

Despite my mistakes, I made up the credits and improved my grades substantially. The next two years, I finished with more than a 3.0 GPA each semester. Growing up in a single- earner household with very limited finances, combined with lackluster freshman and sophomore years, made me a subpar candidate for college scholarships. I did not have the means or the opportunity to go to college, even junior college, straight out of high school. I realized that if I was going to make something of myself it was going to be from my own sweat, from my own tears, and most importantly, from my own pocket. I decided I would join the military in order to take advantage of the GI Bill, which would pay for future college endeavors.

A Recruiter's Dream

I started my service in 1993 in the US Navy. I was way too immature at that time to really take advantage of the opportunities afforded to me. For the majority of the time, I lived paycheck to paycheck the way most teenagers did. Due to my military service I had more access to bars and nightclubs than one normally had at my age, and that is where far too much of my income and free time was spent. I did not have any semblance of a plan for my near- or long-term future when I was active duty. I didn't love my job. To be honest, I hated it. I hated the Navy, but I loved being in the Navy. The many countries I visited made it worthwhile.

I was a recruiter's dream. I wanted to join the Navy, but I had no idea what I wanted to do. All I wanted was to be on an aircraft carrier around airplanes. I don't know why I wanted that. I had probably spent a little too much time watching *Top Gun* and fantasizing that I was going to be the next Maverick.

I signed up for a two-year enlistment with absolutely zero training. I was called an undesignated airman. The way it was sold to me was as an opportunity to bounce from job to job and choose the one I found myself most attracted to. Really, it was a way to get somebody to do the jobs they couldn't get anybody else to volunteer for. I worked the supply department of an aircraft carrier. It's probably the hardest-working and longest-working department on any ship. We had to work even when we stopped at all the cool ports around the world. When the ship pulled into Hawaii, I worked two out of the three days we stayed there. Japan? I had to work three out of the four days. Singapore, Australia, Tasmania, you name it—if it was a fun place, I probably had to work. But, after all, I did get to spend time in many different countries I never would have visited had it not been for the Navy.

I didn't receive any kind of job-specific training that translated into a civilian skill set. But what I did develop during those two years was the foundation of what would make me successful for the rest of my life. It also served me well when I reentered the military years later. It was discipline! And it was respect for others regardless of their standing in the military or civilian world. I learned that I was representing not just myself or the Navy: I was an ambassador of something much larger and more complex. To this day, I don't fully understand the impact that young soldiers and sailors have on their surroundings, especially overseas.

A War Coming

After the military, I began junior college—a return to Bakersfield, CA where I grew up. Unfortunately, my GI Bill did not pay enough to live independently. I had grown accustomed to my independence. I picked up odd jobs here and there to make the ends met and ensure I had a roof over my head and a plate of food at the end of the day. Even so, living expenses started to outpace my income.

As the third semester began, I went to school less and worked more. I worked and went to school intermittently for two years before I realized that I really had enjoyed the military. I did not want to go back to active duty as an enlisted person, so I joined the National Guard. It was an attempt to continue my service while getting a financial reprieve from my education. I spent less and less time on my education and instead focused on my career in the National Guard. I built a skillset in telecommunications, which, in the late 1990s, was just beginning to bud into a lucrative industry.

Two years of active duty and three years in the National Guard gave me something my coworkers did not possess—a sense of duty and mission accomplishment. I took that mindset and applied it to my civilian job. I progressed quickly and in 1999 I was earning $55,000 as a twenty-four-year-old. However, I found the job dull. Despite traveling all over the state for jobs that lasted anywhere from a week to a month, I did not find satisfaction in what I was doing. That would all come to a head in October, 2000, when the USS *Cole* was bombed in a port in Yemen. When I saw the news, I was enraged. I felt called to serve immediately. I felt the threat, even though I didn't understand it. There was a war coming and I wanted to be part of that war.

I still had not completed my bachelor's degree. I was taking one class per semester. I wanted to join up so bad that I lost sight of my original goal to return to the military as an officer. I sacrificed my education chasing something I believed in. By January, 2001, I was on a flight to Saint Louis for Army boot camp at Fort Leonard Wood.

Return to Active Duty

The Army required I choose a different job when I returned to active duty. I wanted to be as close to the battle as I could be, but I didn't want to be infantry. I didn't know why at the time. I thought the best way to be with the infantry was as a combat medic. The Army had other plans for me.

I spent two years in Bamberg, Germany. I was assigned to the base's health clinic. This was disappointing because the entire reason I rejoined was to be part of the fighting force. At least the beauty of Bamberg and northern Franconia took the sting out. My tour in Germany gave me the opportunity to visit nearly every country with open borders. I was centrally located with easy rail and autobahn access to the greatest spots you could imagine. What for some might be considered the opportunity of a lifetime to see centuries-old architecture and culture, had for me become a near daily experience. Less than a month after I arrived, the September 11 attacks happened.

I felt everything quicken in the months after September 11. I still wanted to go and fight. My chance would come with the 173rd Airborne Brigade.

After leaving Germany, I went to the Army Basic Airborne Course. I was disappointed once again, as the Army put me in a combat support hospital that was attached to an airborne medical brigade. There would be no jumps for me.

In 2004, I re-enlisted for four more years. I was given options for my reenlistment. I choose to go to an airborne unit and to go back to Germany. Back in Germany in late 2005, I took over the senior medic role of B Troop in the 1-91st Cavalry of the 173rd Airborne Brigade.

Cancer

When I arrived in Schweinfurt I felt at home. I was finally in a unit I really wanted to be in. I had access to great weekend getaways, and I was in a combat unit. It was perfect. The unit was scheduled to deploy to Iraq in the spring of 2007.

While leading my squad of medics during morning training runs, I began to develop an odd pain and sensation which I assumed was a cramp. I didn't get the pain on medium distance runs typically between two and four miles, but I got them on short fast runs and on long slow runs. I brought it up a few times to our squadron surgeon and he assumed the same thing I did, cramps. It continued through the summer of 2006. In November, I felt another sharp pain while training, but I also felt it in my groin. These two symptoms led to a suspicion of something greater. After several exams and a trip to Landstuhl Regional Medical Center, I was diagnosed with testicular cancer.

I don't believe anything can really prepare you for something like that. There are feelings of despair, fear, anxiety, and anger. Things happened very quickly after my diagnosis. The plan that evening was to get an orchiectomy, meaning they were going to remove the testicle. My unit was scheduled to deploy in just over six months. The surgery finished without complications, and the recovery didn't hurt nearly as bad as I expected. Reality began to set in. I had cancer and I wouldn't be deploying with my team. On the one hand, I would get to stay with friends in Germany; on the other, I felt like I was abandoning my brothers when they needed me. But I challenged myself to be ready to deploy. I felt it could be done, and I was someone who could do it.

My treatment lasted for eight weeks. I decided to have radiation therapy opposed to chemotherapy based solely on the estimated length of treatment. Chemotherapy treatments could have been done in Germany, but they would have taken five months to complete. In comparison, radiation therapy was scheduled for only eight weeks. But it wasn't available in Landstuhl. I had to go to Walter Reed Medical Center in Washington, D.C. With radiation, I could finish treatment by the end of January and rejoin my team shortly thereafter.

I didn't know what to expect from cancer treatment, especially in regard to how it would affect me physically. I knew I needed to stay as fit as possible to return to my unit. When my treatments started at 7:00 a.m., I would wake up at 5:00 to run for forty minutes. I thought if I could maintain my stamina in the first weeks, then I could maintain strength training as the treatment progressed. That plan lasted about seventy-two hours once the treatments started. They floored me. After the first treatment, I was so nauseous I was spinning. I couldn't get out of bed for very long before I felt completely exhausted. I thought the best thing to do would be to get a good night's sleep and continue my plan the next day. The first morning after treatment I made it twenty minutes into my run before I couldn't move. I needed to walk back in order to change into dry and clean clothes. The short walk to treatment on the second day zapped nearly all the energy out of me. I took a five-hour nap. After waking up I still had no energy.

Every day I made fewer and fewer minutes into my run before running out of energy. By the end of the first week I had stopped running altogether, and my physical activity was reduced to the walk to and from treatment. After the first week, I lost more than ten pounds from a complete lack of appetite from the nausea. I started to notice lessening symptoms in the second week, and I started taking afternoon walks around Walter Reed Medical Center. I still had little energy waking up, and I needed time to recover after both the walk to treatment and the afternoon strolls. By the end of the second week, I was down nearly twenty pounds. You need to eat when you have cancer. But the food probably wouldn't have stayed down anyway.

After each treatment, a nurse who worked in radiation oncology delivered a sandwich to me in my hotel room. I quickly regained energy and started the fourth week able to jog to treatment. A childhood friend, who was stationed at then Fort Bragg, NC, came up to spend Christmas and the New Year with me. Having a goals partner gave me the extra motivation I needed. I got up earlier in the day and was far more active after my treatments. I also went on jogs in the late afternoon. I really felt that I had turned a corner. After my friend left, my energy levels increased steadily through week eight as I became more accustomed to the treatment and the radiation dosages decreased, focusing directly on the tumors. At the end of January, 2007, my CT scan revealed zero tumors, and I was cleared to go back to my unit. I had been gone for two months.

Afghanistan

Coming back to my unit in Germany was a very special homecoming. I felt like I was back with family. The adulation I received from that team really made me feel special. It was the first week of February, and the unit was headed to Grafenwöhr for a forty-five-day field exercise. There we learned that we would not be leaving in March for Iraq but rather in May for Afghanistan.

My unit deployed to the Regional Command East (RC East) zone of Afghanistan, in the Nuristan Province, part of the Hindu Kush Mountain range. The 1-91 Cavalry's operating area extended from the village of Naray north into the Landay Sin River Valley. B Troop took control of Combat Outpost (COP) Keating, which was situated across the river from a village called Urmul and at the bottom of a narrow, rugged valley surrounded by steep mountain slopes. (Two years later, in October, 2009, COP Keating was lost in the Battle of Kamdesh.) B Troop also manned two other observation posts: OP Kamu outside the village of Kamu, and OP Wahrheit outside the town of Kamdesh.

As the senior medic for the troop, I worked with every platoon in the unit. In July, 2007, I left Keating on a seven-day overwatch mission to OP Wahrheit. On the morning of July 27, we assessed a nearby village called Saret Koleh. As we left the village my commander, Major Thomas Bostick, looked me in the eye and said, "Get ready."

My Day of Days

The fighting on July 27 was different from other firefights I had been in that summer. What started out as light and medium indiscriminate fire quickly progressed to be high-volume and well-placed.

My unit was not full infantry. We were reconnaissance, surveillance, and target acquisition (RSTA). I don't think we ever went out on a fully-staffed mission, and that July, our trips outside the wire were composed of platoon-level forces of twenty soldiers. On the morning of the 27th, we fought more than a hundred insurgents who surrounded us on three sides from the mountain slopes above. The fight rolled out in two phases separated by a lull.

Within the first fifteen minutes of the fight, we suffered three American casualties and one Afghan casualty. I was the lone medic, and the casualties were not all grouped together. I moved between the two I could reach, working on their wounds for the first half of the fight, about three hours. I learned the locations of the third and fourth casualties: about thirty meters away and up the hill. The distance was complicated by the incredibly arduous terrain.

While making my way to the third casualty, I felt like I had been hit with a baseball bat. The force spun me around and dropped me to a knee. On the ground, I thought I had rolled into one of the thorny shrubs that covered the mountainside. The bushes looked like holly and had "ninja" leaves. We called them that because the leaves had thorns and could grab hold and stop you in your tracks.

The first shot had hit me in the right elbow—that's what felt like a bat striking me. Adrenaline immediately dumped into my blood. I realized I had not rolled into a bush like I thought. It was a second bullet. I felt absolutely zero pain because of the adrenaline. I took shelter in a small rock outcropping halfway to the casualty. I had made myself visible and been caught in the open. The enemy saw me and was targeting me. It was a lot of fire, and it was very effective.

I had time to think while I was stuck in that tiny little group of rocks being continually pelted by debris and ricochets. It was almost comical. More than once I thought, this is it. This is how I go out. I also thought about how this is what I had wanted. I chose this. I volunteered before September 11th. I was a medic with an infantry unit. I had literally put myself in this position, squeezed against rocks in Nuristan Province, taking fire. My thoughts turned to how to get out of my position and get over to the next casualty. I don't know how long I was stuck on the rocks, but it felt like an eternity.

AH-64 Apache helicopters arrived and banked overhead, providing the break in fire I needed to move to a better position with others from my unit. The enemy's attention was diverted to the circling Apaches and medevacs coming to take our wounded. A quick reaction force (QRF) had launched from OP Kamu and was entering the battlefield to evacuate casualties (CASEVAC) via Humvee. We had to get upriver where the medevac helicopters had space to land and effectively pick up the wounded. Word got around that I was shot twice, and when the QRF arrived, they instructed me to leave with them. I couldn't do that. There was no way I was leaving that area unless everyone was leaving. I couldn't tell you how long the helicopters were on station. Once the American and Afghan casualties were loaded up, and the Blackhawks left with the Apaches, the assault on our perimeter began anew.

The fighting intensified. Our positions received consistent, close-in effective fire. Then an RPG explosion. Through the smoke we heard calls for help. It was an incredibly surreal and scary moment when I realized they were calling for me, and I had to run through incoming fire to get to them. I don't know if I actually hesitated, but it felt like an eternity in my head.

I moved quickly down the hillside to the road where the survivors of the blast were gathered. They were pinned behind a small rock on the road, exposed on three sides with heavy fire dropping all around them. When I got to them, I knew my work was cut out for me. Our radio operator was hit with shrapnel in his face, eyes, and collarbone. He was hit bad in his clavicle and bleeding profusely. The Air Force fire specialist attached to our unit had also taken shrapnel. His Kevlar helmet was shredded, and he had multiple face and shoulder wounds. Facial wounds are the scariest because they bleed profusely and there is no way to work on the wounds without looking in the face of the person being treated. I had been able to hyperfocus on the task of treating the soldiers and removing the emotions I felt knowing them personally. It was an automatic compartmentalization. But this fight, on this day, with these casualties, destroyed that tactic. I was face to face with them. I felt their fear.

The RPG had hit our command post where I knew there should be four people, but I found three at the rock. I worked on the casualties in front of me and started figuring out the best way to move forward from the rock to the command post to find the last casualty. The effective fire on my position didn't let up, and it prevented me from moving, forcing me to stay prone while I worked on the injured. Once again, I had morbid thoughts of my own mortality. Again I thought, this is it. This is how I go out.

More explosions—this time up the hill directly above our main position. Man, they felt too close. From the north, the QRF from OP Kamu returned to the battlefield in Humvees mounted with heavy weapons. They took up positions on either side of our unit and gave the order to exfil to a defensible position. The balance of our unit moved to the road, and a part of the QRF went to get our last wounded at the command post. It was our commander, Major Thomas Bostick. He had been killed by the RPG explosion.

While carrying our commander to the QRF trucks, another soldier was hit in the leg. His femur shattered. I was called to the Humvee to assist with dressing the wound. It was bad. Femoral artery injuries are too often fatal. I let the platoon sergeant know that if we didn't leave then, the soldier would die.

I learned that the explosions up the hill had killed another soldier. The next part of the conversation with the platoon sergeant I don't remember well. I recommended that we exfil immediately to save the life of the man who was bleeding out from the leg. It was obvious he needed more medical attention than I, or anybody there, could give him. I will replay that conversation in my head every day until the day I die. I went back and continued treatment and heard via the radio that the entire unit was moving out.

The Humvees didn't have enough room inside to fit all of us, so some soldiers walked beside the trucks. I don't remember exactly how far we needed to go. We had to round a corner in the river valley to effectively end the fight. It was a distance of about 600 meters. That walk was brutal. On the road, the unit was exposed, and the enemy had started to maneuver down the hills and close distance with us.

As I sat in the Humvee in the front passenger seat and faced the rear, I worked on a tourniquet for the soldier's shattered leg. A round struck the window of the door. Trying to be funny, I said, "This must be our lucky day." Someone said, "I got shot in the face." Then I heard another soldier say, "I am hit." Those words came from outside, behind the vehicle. The soldier shot in the face was also outside, but in front us. Then another soldier behind the Humvee was hit. I jumped out of the vehicle to start treatment on them—just as an Afghan soldier was struck in the chest. He was standing less than two feet away. Not that I ever purposefully try to remember how it happened, the memory comes to me all on its own and replays in slow motion, the Afghan soldier getting hit in the chest.

Five more soldiers were wounded on the walk out, and the Afghan soldier died. We didn't have the capacity to get our last KIA. Staff Sergeant William Fritsche was left behind. (OP Wahrheit was renamed OP Fritsche in his memory.) We exfiltrated the battle to a more defensible position and evacuated the wounded and the remains of our commander.

That wasn't the last firefight I was in while in Afghanistan, and it also wasn't the last time we lost a teammate. But it was the most intense and personal fight I would ever be involved in. In July, 2008, my deployment ended. I wanted to stay with my unit, but it's normal for soldiers to move units after a deployment, and the Army found it fit to move me. Within the first three months after returning to Germany, nearly half the team with whom I served in Afghanistan was gone. Those of us who remained formed close personal bonds and professional friendships. My thoughts were more and more overrun by memories of the events of July 27th. Bitterness and anger replaced my normally happy emotions.

My Army career ended in April, 2011, and not a single day has passed where I have not thought of that day in some form or another.

Blissful Ignorance

One of the things that bothered me about returning to the civilian world was how naïve the general public was regarding the sacrifices, efforts, and personal losses that service members experience. I would sit in a coffee shop and overhear the most spirited complaints from otherwise aloof people about their idiotic behavior at the club over the weekend, or how rude the nail manicurist was. I even received backhanded compliments thanking me for my service. I felt a general disconnection between myself and everyone who had not served. I got the impression that they looked at me as if I was unable to do anything other than serve in the military. I also had the impression that they were unable to understand the military, our culture, and how it all fit together.

I might have fallen off the deep end and done something catastrophically stupid to myself or others had it not been for Vernon Valenzuela. Vernon was the dad of a childhood friend from the old neighborhood. I always knew him as the guy with the Corvette with the Purple Heart license plate. I didn't know how he was dedicated to helping Vietnam veterans work through their issues. He was the team leader of the VA Vet Center in Bakersfield, CA. I often stopped at his office and got a few minutes of his completely booked schedule. I didn't appreciate it fully at the time. Only now do I fully understand how much Vernon knew and what he did for me. He shared with me the concept of blissful ignorance. In many quasi-counseling sessions, he reframed my thoughts away from destruction and toward a productive outlook. Vernon eventually offered me a job as his outreach specialist targeting other Global War on Terror veterans. Less than a year later, Vernon died of cancer. His parting wisdom to me was not to waste my investments in education.

During active duty I had continued with my formal education. I finished the core curriculum and began to dabble in higher level courses. I had not completed the degree, but I had plenty of credits that would go to waste if I didn't finish the journey. Working at the vet center, I took night classes at the University of Phoenix and finished my degree soon after. It was absolute necessary for my transition—not only for the education but for how I felt now having a degree. Without a degree, people looked at me as though I could be nothing more than an infantry medic, as though my potential was capped out.

I kept dwelling on my experiences in Afghanistan. My thoughts would start quite innocently, but they inevitably took a dark turn toward the events of July 27 and the conversations I had about whether to stay on the battlefield or leave. I blamed myself for not getting our last KIA on the hill. I was awarded the Silver Star for my actions that day. That medal should have been a reminder of everything I did well. But it became a constant reminder of my failures. I keep the medal in its case closed in a box on the shelf of my closet.

I'm not the only person who struggles with what happens in war. But I did not lend my trust lightly, so I had difficulty finding the right network of people with whom to share my experiences. I am a bit stubborn when it comes to seeking help. Naturally, this led me to repress my emotions, and my emotions escalated to anger at everything I could not control. I found myself feeling more and more bitter at work. I came to realize that working for the VA and directly with veterans was not my path. Without Vernon at the vet center, I didn't feel a sense of belonging. There was absolutely nothing wrong with anybody I worked with, and I enjoyed working with them. The job wasn't right for me, and I left it a year after starting. I wish I had sought more counseling. As I said, my experiences were not unique. My stubbornness needlessly prolonged my negative feelings.

Working in the civilian world was about as big of a shock as the first days of basic training. The very simple structure of rank in the military is not present in the civilian world. There are many advantages to this; however, you really have to learn it. It was easy for me to manage up and down the military's rank structure, gaining approval and executing a task. To this day, I still find the civilian org charts to be convoluted. That doesn't mean they're wrong. They're better in many ways, as I can go straight to the right person without interacting with a chain of command in the middle.

In the civilian world, managing up requires a bit of politics. This was a skill I did not possess. Other glaring holes in my skill set appeared as well. A job's hard skills were never an issue for me, but navigating managerial bureaucracy was definitely an issue. The strategies and tactics used in the military work because of the general environment and culture of the military. What I mean is that having a captive audience whose priorities aligned with mine allowed me to be far more direct and blunt in my communication style. As a civilian, I didn't understand why my coworkers were so reluctant to choose my plans. When my plans were put in place, our teams performed better. We were more efficient. We were more productive. Overall we were just better. Yet I never found support from my colleagues. Often my ideas met with instant rejection rather than curious support. I was clearly missing something, some skill. I watched less experienced coworkers receive promotions while I stagnated. It was clear that I needed to change, but I didn't know how or what to change. I started researching master's programs in my area. I came across the expected results. I did not want to get a degree from a university that was considered a degree mill. I had a year left on my GI Bill. As I searched more, I found a new business degree program only for veterans at USC. It couldn't have been more perfect for me.

When I was a kid, before my parents split, my father took me to a USC football game. It was 1981 or '82. Like most young boys, I loved football. My father watched it on Saturday and Sunday. College football on Saturdays and NFL after church on Sundays was my family's fall and winter schedule. Because my older half-brother was a USC alum and season ticket holder, my father was a huge fan of the football team, so it was my favorite team as well. I was incredibly excited to go to the game. As we walked across campus, through McCarthy quad and into alumni park, my fascination turned to awe. Somehow, I knew I would attend USC. Life had prevented me from going there, or to any college, straight from high school. But my desire never waned.

I applied for the MBV program and attended one of the information sessions, where I met James Bogle. We had a fantastic talk. I didn't know it at the time, but I was being interviewed for the program right there. I asked James what the next steps would be. He told me there would be a phone interview and I would receive a decision afterward. A month later I received an envelope from USC. I hadn't done an interview. I assumed I didn't get in. I kept the envelope for a week and refused to open it. I couldn't handle reading the words that I had been rejected. Then I received a second communication from USC. It was a box, and I decided to open it. I remembered something James had told me. He said that even if I didn't get accepted, he would let me know what I needed to work on in order to reapply. In the box, I discovered some MBV swag. It was pretty odd to get so much merch and marketing materials, considering I wasn't accepted to the program. Maybe I should read the first letter, I thought. I opened it up. To my surprise and joy I had been accepted.

The USC MBV program is a year-long accelerated master's degree. Less time did not make that program easy; it made the program that much harder. On the business side of things, it did not introduce me to any new concepts. But what the MBV did for me was tie all of the disjointed lessons in my life into one comprehensive idea. It taught me many life lessons. Eighty percent of my success was not going to come from my knowledge of the job but from the success of my relationships with coworkers. I needed to identify my tribe, my network. I needed a goals partner, someone to hold me accountable to myself. The skills and tactics I had used to overcome cancer treatment made perfect sense when framed in these ideas. My unit was my tribe, and my friend was my goals partner. Vernon had been my goals partner when I left the service. I didn't find great success at that time because I did not belong to a tribe. I did not feel a sense of belonging in any job. Also, quite frankly, I ruined relationships with my coworkers because my bullish attitude caused others to focus on the way I delivered a message rather than the message itself. I succeeded in the Army because a bullish attitude was welcomed.

Managing up was a critical skill I needed to develop. It had nothing to do with my ability to operate the business. It was a gap in relationship building and communication. USC and the MBV program helped me bridge that gap. Now, I am far from perfect with this skill. And I'm forced to relearn it with every new relationship.

At USC, it was the first time I viewed myself as something other than a soldier—I was a Trojan. My personal outlook pivoted in ways I am still trying to grasp and appreciate. I was able to remove my emotions from complex problems and focus on garnering other opinions, so that as a team we could come to the best possible outcome. If I had been in a normal master's program, filled with civilians who did not share a common experience, this lesson would have been lost on me. I was able to accept their opinions as being normal rather than dismissing them entirely as coming from someone who had not experienced life the way I had. Changing the way I looked at myself catapulted me forward in regard to building relationships. I found other ways of addressing differences of opinion. I found constructive ways to disagree.

Now I work as an operations manager in an Amazon facility in DuPont, WA. I continue to experience all the struggles and failures I previously mentioned. The difference now is that I am quicker to recognize my failures and pivot my strategies and tactics. I run the Warriors Program for our building. The program is a network of veterans and some active duty and reserve servicemen and women. My goal for the group is to form a building-wide, or larger, tribe for every veteran in the warehouse. We volunteer and do community outreach, as well as share historical military success stories. I find my role very fulfilling. I no longer approach my managerial tasks in a formal manner but as an organic interaction.

As the leader, my goal is to impart to my employees the lessons I have learned, in the best way I can. If I mentored them formally, I would rob them of the personal journey they need to go through in order to realize why they are successful or not. That's the way I learn, after all. I need the tools and lessons that other people teach me in order to look back at myself and realize why I succeeded and why I failed.

Recently, I learned that my cancer may have come back. It took over a month of tests to know for sure. The tests came back negative. It was an emotional roller coaster, and it again reshaped my priorities. I am not done living. My kids are nine and eleven and they deserve more than what I grew up with. At this point they are still a bit too young to share all the lessons I have learned and the stories that accompany them. And up to this point I have lived an amazing life! I want the opportunity to share my life with my family. I still want to travel, and I still want to open my own business. Through the years I have learned who I am and what's really important to me. My time in the military shaped my skills, and USC and the MBV program helped me refine them and translate them into today's world. These are the life lessons I want to share.

"Scar Tissue"

Rob Fortner reflects on telling his story

I don't know how to explain the journey I went on while writing my story. The journey started years before I submitted the first draft. I was invited by Dr. Turrill to give my story when he was writing the first book. At the time, I didn't believe I was ready to reopen that chapter of my life. It was always easy to focus on a tiny aspect of that chapter, a smaller story within the story. I struggled for weeks and months to find the right way. When I finally got a few paragraphs on paper, I deleted everything and started again. This went on for several iterations. What I had not accepted was that it would never be easy to talk about painful moments. And it never will be. Rehashing past failures seemed more uncomfortable than it had been to experience them the first time. Given a second opportunity to participate in this kind of storytelling, I set to get my story on paper. Again I realized that there will never be a good time to write painful stories.

I struggled for more than a month to submit the first draft. When I finally broke through my own inhibition, I wrote everything in one sitting. I found my storytelling became more fluid as the story progressed. I had to come to terms with the fact that I had not dealt with the events, not correctly anyway. Like most veterans, I dabbled with VA services but slowly stopped using them. I learned ad hoc coping strategies, which were not effective and developed into bad habits. It's hard to replace old habits. In writing this story, everything I had done wrong was put down in obvious terms, and the writing ripped off scar tissue that had grown around the past. It gave me a fresh slate to lay proper foundations. I used the knowledge and experience I gained since leaving the Army to properly deal with what happened.

I wouldn't have done this on my own. I needed this book, and Drs. Arámbula and Turrill to come together, in order to sit down and write. They were my tribe and my goals partners.

Chapter 21

Picking Up What Others Put Down
by
Robert Goodwin

Robert Goodwin held senior positions at the US Agency for International Development, the US State Department, the White House, and the Pentagon. He supported humanitarian operations in Afghanistan, peace talks in Sudan, the rebuilding of the health ministry in Iraq, and he was the COO of International Aid.

A graduate of the US Air Force Academy, Robert served five years on active duty, achieving the rank of captain. He is a graduate of the Harvard Business School's General Management program and the University of Southern California's Master of Business for Veterans (MBV) program. He is the recipient of numerous military and civilian awards, including the Air Force Exceptional Civilian Service Medal.

Robert worked with the Mattel Children's Foundation as executive director and director of corporate affairs and philanthropy, partnering with Lady Gaga's Born This Way Foundation and Save the Children.

He is the cofounder of OceanCycle, which reimagines ocean plastic pollution. Recently, he helped launch a film and TV development company called Incline Studios.

Summers

I was born outside Chicago, IL, as the youngest of three children, and I was named Robert John after Senator Robert F. Kennedy and

President John F. Kennedy. My parents met on Cape Cod where my mother was a waitress; my father ran social events in Boston. Through his work with the local Rotary Club, my father met then Senator John F. Kennedy and was inspired by him. Our Irish Catholic family were true believers in the Kennedys and their ideals.

Much of my childhood was spent in Fort Myers, FL. My mother worked at an elementary school and had the summers off, and during those vacation months we spent time at my grandparents' house in White Horse Beach in Plymouth, MA, outside of Boston. Plymouth is home to the famous Plymouth Rock where the Mayflower pilgrims supposedly made their first landing.

White Horse Beach is famous for another rock known as Flag Rock, a huge boulder 200 yards from the beach and painted with a large American flag that can be seen for miles. My friends and I often braved the 60-degree water swimming to Flag Rock to look for lobsters and dive at high tide. White Horse Beach is affectionately known as the "Blue Collar Riviera" given the generations of construction workers, firefighters, and police officers who built homes there. Many of those fine people showed me the value of hard work and faith and taught me not to take myself too seriously. When our egos crept up, the constant ribbing from friends and relatives took everyone down a few notches. As the youngest of all the cousins, I took the brunt of their attention.

The community at White Horse Beach also showed me how to celebrate America with epic Fourth of July celebrations every year. Families built huge bonfires at each section of the beach (one year, someone burned a twenty-foot boat) and they fired off huge fireworks smuggled in from North Carolina.

When most kids would play war using cap guns, my friends and I shot bottle rockets and roman candles at each other. We were lucky to survive.

Those early summers were the best times of my life, filled with friends, cheering on the Red Sox with my grandfather, and visiting national historical sites. During those summers in Boston, I developed a deep love for this country, and it was during that time when I first started to learn about adversity.

Tragedies

My grandfather and I were extremely close. We watched Red Sox games daily and played golf at the municipal courses at Cape Cod. Since my father stayed in Florida to work, my grandfather was the male role model in my life. One summer, when I was about nine, my grandfather complained about terrible stomach pain. The doctors found a grapefruit size tumor. He had major surgery, and because the doctors thought they had removed all of the cancer, he didn't go through chemotherapy. I had two more years of great times with my grandfather until the cancer came back with a vengeance. It was inoperable, and additional surgeries took most of his intestines and colon, which required him to be fitted with a colostomy bag.

My grandfather fought like hell. During our golf games he didn't have much energy, and he would disappear into the bushes to empty out the colostomy bag. The smell was terrible, but nothing like what was to come.

My father was a brilliant and industrious man, but after multiple failed businesses and lost jobs, he became depressed and turned to drinking. He was mostly detached from the family and would sometimes become violent. I remember a few times getting the belt during his drunk rages. Thankfully, when I was eleven, my father had a minor heart attack, and it was a wakeup call for him. He stopped drinking, lost twenty-five pounds, and started exercising and eating healthy.

He opened a new business using his engineering talents installing heat pumps to provide air conditioning to homes in Florida. The innovative systems used the homeowner's pool as a heat exchange to draw heat from the house, providing cooling at a fraction of the price of traditional air conditioning systems, with the benefit of also heating the homeowner's pool. I got my first experience as an entrepreneur helping my dad assemble marketing materials and doing mailings to potential customers. I licked thousands of envelopes before I learned the trick of using a damp sponge instead.

As my grandfather's cancer worsened, he and my grandmother moved next door to us in Florida and my mother became my grandfather's primary caregiver. With my father trying to get his business off the ground and my mother caring for my ailing grandfather, I was left mostly on my own. My brother and sister were much older and already off at college and so consumed with their own lives they never really checked on how I was doing.

I coped by throwing myself into my studies and joining the Key Club, a community service organization that partnered with local Kiwanis Clubs on community service projects. At first, I joined the Key Club because all the cute girls from high school were in the Key Club, but eventually it became my lifesaver. The club's main advisor was a football coach named Dave LaRosa, and he become a stand-in male role model for me. I appreciated how he always treated everyone the same. From the janitor to the school principal, Dave was genuinely interested in their lives. He had a grounding effect on me that I needed during that time. In addition, after the trauma I had experienced, the Key Club projects that supported people with intellectual disabilities and those experiencing homelessness put my own troubles into perspective and helped me cope.

As my grandfather neared the end of his life, my grandmother became very confused. One day she kept saying there was a "tomato warning" on TV. While we found it funny at the time, it was a sign that my grandmother was in trouble. A few days later, she had a major stroke and fell into a coma in the hospital. While she was in the ICU, my grandfather succumbed to his long battle with cancer and passed away. My grandmother passed away two weeks later. These were my only grandparents, as my father's parents had died before I was born, so it was a huge loss for me. It was just the beginning.

While my father had done well recovering after his first heart attack, the stress from his new business wore on him, and he regained much of the weight he had lost. His work also kept him outside in the hot Florida sun, which added to a dangerous situation. One night he told my mother he wasn't feeling well and that it felt like his heart was disintegrating. She didn't think much of it. A few days later, while I was at school, a voice came over the classroom intercom and asked me to come to the office.

My brother, home from college, waited for me. He told me that my dad was in the hospital. As we raced there, I knew it was bad and prepared myself for what I would find. In the ICU, my father was full of tubes and wires, and though he reacted when I called his name, he wasn't conscious. He was slipping away.

When it became clear that he was becoming less and less responsive, the nurses encouraged us to say our final goodbyes. My mother and brother spent time with him, and then it was my turn. As a fifteen-year-old, I had my battles with my father, but we were still close. Prior to my grandfather passing, I wrote a long letter to him about what he meant to me, and I read it to him before he died. I never got to do the same for my grandmother, which I regretted. So I took the time to ask my father for his forgiveness and I gave him mine. I spent twenty minutes with him sharing all my thoughts and stumbling through a few prayers. I said a final goodbye, and as I left the room, my father coded, and the staff rushed in to intervene. He was gone, but to this day I thank God for that opportunity, and I believe he stayed alive long enough for us to have that moment.

My father had spent all of our savings on his new business and also hadn't kept up to date on his life insurance premiums. As a result, my mother only received $12,000 in life insurance, and her yearly salary was only $18,000 as an elementary school bookkeeper. This meant we both had to do whatever we could to make ends meet. My mom worked most nights moonlighting at a local pharmacy. I took jobs working at a grocery store during the week and at golf courses on the weekends, picking up golf balls on the driving range and cleaning golf carts. The good thing was that most golf courses let their employees play golf for free when they were not working, which allowed me to play at good courses despite our family not having disposable income.

I was a smart kid, and at the end of my junior year of high school, I was selected to be part of the Student Science Training Program (SSTP). Through the program, students from across the country took college courses and participated in graduate-level research at the University of Florida during the summer. Students selected for the program had the option to skip senior year of high school and start college early at UF. I decided to head back for my senior year.

I had always loved NASA and the space program, and watching Space Shuttle launches was a big deal in Florida schools. When Christa McAuliffe, a social studies teacher, launched with Space Shuttle *Challenger*, we all cheered at takeoff, only to be shattered a minute later when the shuttle exploded on live TV, and everyone aboard lost their lives. In the months that followed, we spent time remembering her and the other six astronauts. My interest in space was further cemented. With few male role models, I tended to look at public figures such as astronauts and political leaders for inspiration: people like astronaut John Glenn, who became the first American to orbit the Earth and later became a senator.

I got a small taste of politics when I unsuccessfully ran for class president of my middle school, and then in high school I ran to be elected as lieutenant governor of the Florida Key Club. I won and spent a lot of time traveling around the state meeting with different high school Key Clubs and doing service projects. I missed so much of my senior year with excused absences that I almost didn't graduate. My English teacher gave pop quizzes every week in an attempt to stop seniors from skipping school. I missed five of her quizzes, and even though I had permission, she demanded I get a failing grade, despite me being class salutatorian. Thankfully, the principal intervened.

The Academy

With no money to visit college campuses, I read as many books as I could to get some idea of where I should go to school. I learned about the Air Force Academy from a brochure in the guidance counselor's office, and it seemed like a perfect fit. No one from my town had ever gone there, and I had no connections to my district congressman or state senator. I was able to track down the Academy liaison officer in my region, and he ultimately became one of my strongest advocates. No one in my family was military other than my father, who did a short stint in the Korean War (which he never discussed), but I was convinced the military was where I needed to be.

My mother was so used to my absences as I drove around Florida that she didn't know I had applied to the Academy until I had a meeting with Senator Connie Mack's staff for a panel interview. The interview was a nerve-racking experience: being stared down by seven people, including former general officers who assisted with the vetting process. Luckily, or providentially, I received a nomination from both Senator Mack and my local congressman, Porter Goss.

Months later, after my appointment to the Academy, we were invited to meet with Senator Mack. I remembered to take my suit, tie, and dress shirt, but, to my mother's horror, I left my dress shoes behind at Fort Meyer. I only had the sneakers I was wearing, which barely held together. We found a store that was still open before the reception, and all seemed right in the world. That was until, to my horror, my mother told the entire story in embarrassing detail to Senator Mack.

I couldn't visit the Academy prior to reporting for basic training. To prepare me for basic, I only had what I had read in books or watched in promotional videos to guide me. When I reported to Colorado Springs for duty, I wasn't prepared for how quickly I was relieved of my freedom, my civilian possessions, and all my hair. This Florida kid had a hard time adjusting to the Academy's altitude and temperature changes. I ended up getting pneumonia during basic training. It got so bad that I was hospitalized and missed the issuing of all my books and supplies for freshman year.

The upperclassmen were trained to weed out any weakness in the younger students, so I became their special project. They did everything they could to make my life difficult. At one point I was fed up, and I spoke to a Catholic priest about leaving. However, I had never really quit anything in my life, and I decided to stay. The four years at the Academy weren't a fun experience for me, since I had been on my own since I was twelve and didn't like to conform. But I am glad I stuck it out. The Academy became "finishing school" for me, and it's where I learned to push my mental and physical limits—something that would serve me well in the years to follow.

The first Gulf War started during freshman year, and we spent hours in dorm rooms glued to the TV and watching CNN as bombs fell on Iraq. By the time we graduated three years later, the US Air Force was going through a major drawdown as part of President Clinton's reduction of Department of Defense spending. In previous years, almost every graduate went on to pilot training, and if they didn't want to fly, they had to talk to the three-star superintendent about their reasons why not. But during the drawdown, only twenty percent of the class got to fly planes, and people did everything they could, including undercutting other classmates, to get a pilot slot. What was normally a supportive environment took on a cutthroat and toxic atmosphere.

Saudis and Cartels

Many of us were sent to bases and into career fields that were different than what we expected. My first assignment was to Wright-Patterson Air Force Base in Dayton, OH. I was in charge of test and support equipment for components of F-15 and F-16 jet engines. To say this was not a choice assignment is an understatement. Most of the officers I worked with were passed over for promotion or being forced to retire. I was recognized as one of the top young officers on the base, but I was not able to move to another program office. As I investigated my options, I decided to use a little-known loophole that allowed lieutenants to transfer after only a year of duty. It was primarily used for officers graduating from pilot training, but I used it to transfer to Hanscom Air Force Base outside of Boston. There was a captain's position there that hadn't been filled for two years because it required extensive travel to Saudi Arabia. When I applied, they had to evaluate me.

After they realized I wasn't a slacker or a psychopath, I was on my way to the Middle East to work on a $6 billion air defense program called Peace Shield. My duties included leading the construction and activation of large fiber optic networks connecting radar sites and command centers as part of a warning system for Saudi Arabia. I was based in Boston, technically, but I hardly spent any time there and lived mostly on a compound called The Rock in Riyadh.

My first trips gave me quite a culture shock. Growing up, my family didn't have any money to travel, and my sole childhood experience was visiting Canada as part of a visit to Niagara Falls. I shadowed an officer who had been working on the Peace Shield project for the past few years. I was interested in learning the culture, and I paid attention to cultural nuances like not showing the sole of your shoe. I also learned some basic Arabic. My colleague didn't care at all.

Several months into our work, despite being junior in rank, I carried more influence with my Saudi counterparts because of my knowledge of the culture and language. I was able to get us into facilities previously inaccessible, which helped make major progress. I also made friends with several of my Saudi colleagues. They took me into the desert to ride camels and participate in a traditional goat grab, where a goat was slaughtered and cooked in a pit under the sand. The guest of honor was given the eyeballs and brain as a show of respect.

Eventually, the Saudis chose me for some engagements with the royal family, and I became the executive officer to the program director. I normally attended the senior meetings, but one day the top Saudi general asked to see me as he needed my help to get some things done. He spoke in Arabic to his staff and when he turned to me to explain in English, I said, "I understand," and I proceeded to repeat back to him what he had said to his staff and what he wanted me to do. While he appreciated my proficiency in Arabic, from that moment on I was no longer allowed to attend senior meetings between US and Saudi officials. The Saudis liked having sidebar conversations without the American officials understanding what was going on. In their eyes I had become a liability.

After my experience in Saudi Arabia, I wanted to continue working on international efforts. I was selected to run radar and communications programs supporting counter-drug operations in Central and South America. I again took advantage of the opportunity to learn the language and culture of the region. This time my life depended on it.

I was often the only US military in areas such as San José del Guaviare, Colombia, where about eighty percent of all the cocaine in the world was processed. My job was to set up radar and communications sites to track aircraft carrying drugs into the United States. The range of the new radar was 250 miles, and once it was operational, it would complicate cartel business.

I was a big *Star Wars* fan and remembered how, in *Return of the Jedi*, the Empire made the Death Star look like it was under construction, even though it was fully operational. I did the same thing and made the radar site look as if it wasn't working even though it was. This helped the Colombian Air Force shoot down a bunch of drug planes in just a few days. When the cartels learned of our deception, they weren't happy with me, and I became somewhat of a focus of their attention.

Making a hit list is a double-edged sword. On one hand it is scary. On the other, it meant I was having an impact. Two assassination plots against my life were discovered. The first entailed an attack to be carried out as I was being transported from the San José del Guaviare airfield to the Colombian Army base where the radar site was located. What saved me was that some of my colleagues were former NSA, and a listening post picked up the cartel's communications, allowing us to change our plans. The second plot proposed to infiltrate the construction workers and attack our living quarters. It was discovered by chance when the Colombian Army overran a FARC encampment and found the documents.

I couldn't help but feel compassion for the people living in the remote jungle towns of Colombia. Many of them worked with the cartels out of fear for their lives or because they had no other way to feed their families. I came to understand that poverty made people vulnerable to influence and was in many ways a root enabler of the drug trade. If I wanted to reduce drug trafficking I needed to get at the root of poverty and strengthen communities and families to be able to resist. It's an insight that continues to shape my career today.

USAID

When I left active duty, I ended up going to work for a defense company near Washington, D.C. My plan was to work there for a year until I entered business school. I was still interested in politics. I started working on grassroots campaigning and get-out-the-vote efforts in the Washington, D.C., area, and after a short period of time became known as someone that could lead teams and get things done.

Though I worked on local races, I gained the attention of leaders within the Republican Party, and when President Bush was elected in 2000, I had the opportunity to volunteer on his inaugural staff. I worked for months for no pay. We had limited time to plan the inauguration, given how long it took the Supreme Court to finalize the election results. After months of working for free, the day came when President Bush would take office, and I still didn't have a job. With money running out, I wondered if all my effort had been in vain. A week later I got a call from the White House, and on February 7, 2001, I reported to the Pentagon and became one of the first several hundred appointees in the Bush administration. Business school would have to wait.

My first job as a presidential appointee was on the Department of Defense transition team, staffing up senior positions to run different departments. But I didn't want to be at the Pentagon so soon after leaving the military. One of the senior people working on the transition told me to look at the Plum Book, which listed all political appointments in government. I petitioned to join the international disaster response team at the US Agency for International Development's (USAID) Office of Foreign Disaster Assistance, and the

White House approved my request. No political appointee had ever held a position in the Office of Foreign Disaster other than as the director, so the USAID administrator Andrew Natsios had to personally approve my appointment.

At USAID, my job was to reset the relationship with the government's domestic disaster response arm, the Federal Emergency Management Agency (FEMA), which, during President Clinton's time, had taken on a more prominent role under Administrator James Lee Witt and been elevated to a cabinet position. Even though FEMA had a domestic mandate, Witt was called upon to engage in international disasters and had expanded the agency's purview beyond its constitutional authorities. I was tasked with helping rein in those responsibilities for USAID and the new FEMA administrator, Joe Allbaugh, one of the architects of President Bush's election victory, who had no desire to play an international role.

As that work progressed, I was asked to be part of a special task force establishing a new initiative called the Global Development Alliance (GDA). The GDA leveraged engagement with the private sector and pooled resources so that US companies could achieve business objectives while the government improved sustainable development programs, essentially finding ways for the government and private sector to collaborate.

Terrorism

The morning of September 11th seemed like a normal day. I was working on the GDA task force at USAID in the Ronald Reagan Federal Building a few blocks from the White House when I learned that a plane had hit the World Trade Center in New York City. Several colleagues were sitting around a TV watching smoke rise from the building. I recalled that many years ago, a plane had flown into the Empire State Building, so I thought it was another accident and went back to my desk. Then I heard screams from my colleagues who were watching the TV as a second plane hit the towers. I spent the next few hours getting people to evacuate the building. We heard the Pentagon had been attacked, and there was a rumor that a bomb went off at the State Department. There was chaos with communications.

I lived in Capitol Hill, and as I walked home, I passed a makeshift operations center that a congressman had set up, and I pitched in to help. Later, I learned that one of my former colleagues at the Pentagon, US Army Sergeant Tamara Thurman, was killed in the attack. She was twenty-five years old. I had left the Pentagon only a few months before. I tried to go there to help with the cleanup and recovery, but the area was locked down, so I organized a group of fellow appointees to deliver sandwiches to the Capitol police.

With not much to do in D.C., some of my Pentagon colleagues and I decided to go to NYC to help. It was an eerie drive. There were absolutely no aircraft in the sky and very few cars on the highway. While staying at a friend's apartment, we made our way to Ground Zero. Our government IDs got us through the checkpoints. We saw the scope of the destruction for ourselves. It was hard to believe. Active fires raged and layers of dust blanketed the streets. There was incredible sadness all around, with makeshift bulletin boards everywhere posted with pictures of missing people and pleas for information on their whereabouts. We spent much of our time consoling the firefighters who worked at Ground Zero night and day who had lost dozens of friends, colleagues, and family members and desperately searched through the rubble for signs of life. I had always loved New York, and the experience at Ground Zero shaped the next few years of my life.

I was recalled from the GDA task force and became the logistics director for USAID disaster response. As the bombs began to fall on Afghanistan, I was responsible for providing food and relief supplies to Afghan civilians displaced by the conflict. My military and Pentagon experiences came in handy when coordinating supply pickups on C-17 aircraft and choosing the color of packaged food that was air-dropped into the region. We switched from a yellow to an orange color because we didn't want the food packages to be confused with unexploded cluster munitions.

As the first phase of the war ended, I was asked to work on a special task force at the State Department that was focused on bringing an end to the twenty year-long Second Sudanese Civil War. Sudan was a hotbed for terrorism and where Osama bin Laden lived in the 1990s, and it was a key priority for President Bush. I was led to believe that the war in Sudan was about religious and ethnic conflicts, but I learned that it was really about control of oil, land, and power.

I became the main contact and travel companion for the special envoy to Sudan, former Senator John Danforth, a man of true character whom I admired, and we became constant companions traveling around Africa and meeting with foreign leaders to gain support for the peace process. On one occasion, we had breakfast in Egypt with senior staff from the State Department, including the ambassador to Kenya and the assistant secretary of state for Near Eastern affairs. Senator Danforth was eating a muffin, and a big piece of it stuck to his chin while he was speaking. I brought it to his attention, and in his booming voice he said, "What's that, Bob? I have something on my face?" The senior staff looked disapprovingly at me. Senator Danforth went on to tell a story about the time he was the graduation commencement speaker at the University of Missouri. Forty-five minutes before he had to go up on stage, he greeted parents and students, and he noticed that they all seemed a little weird as he spoke with them. When it was time to take the stage, he realized he didn't have his speech. He had put the large folder that held his speech on top of his cap while fixing his gown, and no one had said anything to him. He then turned to me at the table and with a wink said, "Thanks for telling me."

Iraq

On the verge of a peace deal in Sudan, I was approached by the team I worked with at the Pentagon to discuss the situation in Iraq. They were putting together a group of advisors to support Iraqi government ministries after the fall of the Saddam regime. I had just returned from the Sudan peace talks in Kenya and met with the man selected to advise the Iraqi Ministry of Health. His name was Jim Haveman, and he was the former health secretary for Michigan.

I wasn't interested, but Haveman was determined to have me as his chief of staff, and after much prodding I agreed. Two weeks later, Jim and I, along with Iraqi advisors, landed in Baghdad.

Haveman had incredible talent and understood healthcare, but he had never been anywhere near a war zone. I went into warp speed procuring everything we needed, from lodging to satellite phones. When we arrived at the Ministry of Health, it was looted down to the electrical sockets. In the parking lot, some fifteen hundred employees waited for someone to tell them what to do. Their first request: they wanted chairs so they could wait in the parking lot more comfortably.

We hired local contractors and artisans to rebuild the ministry, and we spent the next eleven months getting Iraq's 200 hospitals and 1,200 clinics back up and running. We had to plot out a new strategy for the country's health system. Under the UN resolutions and authorities of an occupying power, Haveman and I had the responsibility of running the Ministry of Health. However, we shifted more to a support role once an interim Iraqi minister was appointed. We worked out of the ministry building every day.

Our team handled security and travel logistics, as the Ministry of Health was a target for snipers and IED attacks, and we had to travel between there and the offices of the Coalition Provisional Authority (CPA), housed in one of Saddam's former palaces. Iraq was seven hours ahead of the US East Coast, so when it was afternoon our time, Washington was just waking up. Most days we ended up working late into the evening to "feed the beast" in Washington. Many nights we were bombarded by mortars and missile attacks. The "giant voice" warning system would tell us to take shelter in the basement. After a few months, we were so accustomed to it that we just stayed at our desks working.

Haveman and I made a point of engaging with the Iraqis in the local community. It was important, especially in forming relationships with the doctors and medical professionals we needed to rebuild the ministry. One night we went out for dinner at a local Iraqi restaurant. We were the only patrons, and when the bill came, we saw that the waiter had overcharged us by more than three times for a bottle of cheap Lebanese wine. It turned out the waiter knew there were snipers outside waiting to kill us as we left the restaurant, and he was looking to maximize his tip from customers who would not be returning. The security staff was also onto the plot after hearing someone ready their weapon from the window of a nearby building, and we were able to escape through the side door of the kitchen and run to our vehicles. To this day, I am more upset at the waiter than the snipers.

My time in Iraq consisted of periods of normalcy interrupted by extreme chaos. One week I started out having to fire and relocate an Iraqi-American on our team who was suspected of working with insurgent forces. Then at night two of our military staff were wounded by gunfire. The next day I ran a three-day conference to discuss the future of healthcare in Iraq that involved ministry staff, the United Nations, World Bank, European Union, and many humanitarian organizations.

The meeting was a huge success and finished up on the afternoon of August 19, 2003. The team was supposed to reconvene that night for a celebratory dinner. An hour before we met for dinner, the United Nations headquarters in the Canal Hotel was attacked by a suicide truck bombing that killed twenty-three people, including United Nations High Commissioner for Human Rights Sérgio Vieira de Mello. More than a hundred others were wounded, including several from our conference. Christopher Klein-Beekman of UNICEF also died in the attack. I spent the next forty-eight hours working frantically to organize military and ministry resources to care for the wounded as well as evacuate UN staff for treatment.

Even though I was a civilian in Iraq, I ended up seeing more danger than when I had been in uniform. Members of our team were wounded during the rocket attacks on the Al Rasheed Hotel where we all stayed in the Green Zone.

Though our team was at great risk, our Iraqi colleagues took even greater risks coming to work with the Coalition Provisional Authority (CPA). Many of our key staff were killed as they left their homes or traveled to the ministry. My closest Iraqi colleague and friend, with whom I worked daily, was Deputy Health Minister Ammar al Saffar. He was kidnapped and assassinated for his work to reduce corruption in the ministry.

Despite all these challenges, the Ministry of Health was seen as one of the success stories of the occupation and became the first ministry to hand over sovereignty to the Iraqis prior to the formal handover of power by the CPA. It was also a great learning experience for me and gave me an understanding of the role of health in a functioning society. Sick children cannot learn, and sick employees cannot work. Without security, health isn't possible, but once security is established, health is the fundamental next step for a society recovering from war.

Back to Washington

After Iraq, I was asked back to the Pentagon to work on special operations and low-intensity conflict matters. After the UN headquarters attack, most of the NGOs and civilian institutions left the country and the military filled the void by building schools and performing other civilian functions—something they are not trained to do nor are good at. So, I helped think up possible ways to build new civilian institutions that could better handle such insecure environments.

My next assignment took me to the White House Presidential Personnel Office where I helped with staffing senior national security leadership positions. I was then selected to be the deputy assistant secretary for the Air Force with oversight responsibility of the US Air Force Academy and personnel policy for active duty and civilian personnel. I spent months helping the academy recover from sexual assault scandals and reports of religious intolerance. After a year and a half, I was asked to take over the duties of the assistant secretary of the Air Force for manpower and reserve affairs—a four-star general-equivalent position in charge of all personnel issues for the US Air Force.

I was the youngest to have ever served in that position but thankfully I had great mentors who helped me navigate the responsibility. One such mentor was Deputy Defense Secretary Gordon England, who shared with me that "in these senior positions we have certain duties and responsibilities that we need to uphold, but when our time is done, we also leave these titles and responsibilities behind." It was advice I took to heart, and many days I wondered if members of Congress were afforded the same wisdom.

After seven years of the Bush administration, I found myself a little burned out. While I appreciated the resources I commanded at the Pentagon, I also realized I wasn't fulfilling the vision I had many years before, during my time in Colombia—to improve global security through reducing poverty. It had become even more important to me after 9/11 and what I learned in Sudan and Iraq. Despite the shock of my colleagues and staff in Washington, I left behind the trappings of power and moved to a small town in Michigan to become chief operating officer of a Christian nonprofit called International Aid.

Disaster Relief

International Aid was my first time working for a nonprofit. I was originally charged with growing the organization based on a projected $12 million budget. However, after six months on the job, I realized the CEO had calculated the budget by adding twenty percent to the previous budgets, not considering that much of the money had been raised from an outpouring of support in response to the 2004 Indian Ocean tsunami and Hurricane Katrina in 2005. When no major disasters happened on my watch, it turned out we were projected to receive only about $8 million in donations. So, instead of growing the organization, I was forced to lay off half.

Also during this time, I completed an executive education program at Harvard Business School. With some classmates, we founded Executives Without Borders (ExecWB), which recognized that many nonprofit leaders had great skills in healing and supporting people, but many lacked the business skills and expertise to implement their goals effectively. A prime example was the budgeting fiasco at International Aid.

I had planned to use the capabilities of ExecWB to strengthen International Aid, but the CEO wasn't very receptive. I think he felt threatened and didn't want any more scrutiny placed on their operations. When I was asked to be the first CEO of Executives Without Borders, I accepted. From there I focused my energies on supporting nonprofits in India, Rwanda, and Honduras, among others. In Brazil, I worked with the US Chamber of Commerce and multinational companies to create a business version of the Peace Corps.

My life took a turn when a major earthquake hit Haiti on January 12, 2010. From my experience with international disaster response teams, I knew from the initial reports that the damage would be extensive, but I never envisioned the extent of the destruction and loss of life. Like many others, I was moved by the stories and images to do something. After first raising money in support of safe water and medical services, I then took a delegation of senior leaders from FedEx, Dow Chemical, Harvard Business School, and other organizations to see what we could do after the initial response. The companies donated more money and supplies to support the Haitians, but donations were limited as Haiti wasn't seen as a future market for their companies.

When the delegation left, I decided to stay in Haiti to see what could be done from the standpoint of Executives Without Borders. People were infected with cholera, which was spreading from exposure to bacteria in bad water. I learned that much of the cholera outbreak was due to flooding from the clogged drainage systems, and the culprit was plastic garbage damming up the canals.

Ramase Lajan

I investigated the status of plastic collection and recycling and found a robust industry for metals recycling but not for plastics. Over the course of the next six months, I brought together a partnership between GS Industries (a Haitian company focused on metal recycling) and Samaritan's Purse (a Christian NGO treating many of the cholera cases) to establish local collection centers. A company I worked with in Iraq, called CSS Global, happened to be helping with construction projects in Haiti. CSS Global staff helped me with a model of recycling in Haiti. They modified ocean containers into storefronts which would be deployed across Haiti.

Once we found a location and built a collection center, we found a Haitian entrepreneur to run the local operation, paying cash to locals for the plastics and aluminum cans they collected. The first time we opened a center in Leogane, a man, dressed in rags, brought a small bag of plastic to the modified container storefront. We showed him how to sort the plastic and wash it, removing dirt and sand, then we weighed it. We paid him cash, and he looked at the money in his hand in disbelief thinking it was all a big joke. My Haitian staff asked the man, "What are you going to do with the money?" The man said, "First, I am going to buy some food because I am hungry. Then, I am going to buy some food for my family because they are hungry." How about after that? "I am going to collect more plastic," he said.

The program expanded across the country, and the Haitians nicknamed the project Ramase Lajan which literally means "picking up money" in Creole. The program created 1,500 full-time and part-time jobs, recycling more than 200 million bottles and generating $600,000. Cleaning up the environment was an essential part of the program, but I found the additional income it provided families to be the most critical part, as it meant more food on their tables and more children in schools, in addition to the reduced flooding that helped stop the spread of cholera.

Mattel

After years of running Executives Without Borders, I found myself advising companies regularly on how to make money while helping better the world. Most of the companies liked the idea, but they didn't really know how to execute the concept. It led me to conclude that I needed to join a multinational corporation and show them how to integrate business and purpose from the inside. When I was recruited by Mattel to run their foundation and corporate philanthropy programs, I jumped at the opportunity.

At Mattel, I had a $12 million cash budget and $20 million-worth of toys to give away. In the past, the primary strategy had been to give cash and toys to organizations in Los Angeles and where Mattel had operations, but I looked for opportunities to drive a larger impact.

One of my first initiatives was to launch a nationwide program called Speedometry that used Hot Wheels to teach math and science for students in kindergarten through fourth grade. From my mother's experience in elementary schools, I knew how under-resourced classrooms were. The strategy entailed providing classrooms with kits that included cars, tracks, and education videos with the curriculum. Teachers and students loved the program.

In addition, I established a global partnership with Save the Children. The combined efforts of Mattel and Save the Children caught the attention of the president of Indonesia and helped create conditions for the rapid growth of Mattel's business in the region. I then established a partnership between Lady Gaga's Born This Way Foundation and Mattel's Monster High brand, with the goal of finding brand synergies to reduce bullying and inspire bravery among children, which eventually led to the leveraging of Mattel's talents to design the Monster High Zomby Gaga doll that raised over $1 million for her foundation.

Unfortunately, Mattel's business was suffering overall, and its philanthropy budget was based on a set percent of Mattel's profits. My budget was reducing each year and would be further constrained as

Mattel earmarked $50 million over twelve years to build a new Mattel Children's Hospital at UCLA.

OceanCycle

At the same time, I was a strategic advisor for the newly established Master of Business for Veterans program at the University of Southern California, and after a year of conversations with the program director, James Bogle, I decided to go through the program myself. I was impressed with the diversity of the class. I was one of the "old guys" with less current warfighting experience but more current civilian workforce experience I could share.

The most impactful part of the program turned out to be the entrepreneurship portion taught by Professor Tommy Knapp. I was always entrepreneurial in my work, but I was more focused on purpose than profit, and I wondered if there was a way I could bring the two together.

When I ran Executives Without Borders, I found it a somewhat self-defeating exercise. The more successful my programs became, the more money I had to raise to keep the nonprofit going. Instead of just being able to do the work, more and more time was required for fundraising and communications. I often felt that I had three jobs: telling the organization's story, raising the money, then doing the work. I found it unsustainable, and it was a common challenge within the nonprofit community.

During my time at USC, I learned more and more about how plastic was impacting the planet in negative ways. I learned in Haiti that if we didn't collect plastic and it rained, the plastic washed out into the ocean, but I didn't know the extent of the ocean plastic problem. Not until I connected with people who were making consumer products and sourcing plastic from my program in Haiti did I understand the scope of the problem.

When I learned that ninety percent of ocean plastic starts on land and once in the water it can't be recycled, I wanted to see if I could replicate my Haiti project in other parts of the world. At first, the idea was to make products such as backpacks and beach chairs out of collected material, but I concluded that while that effort might make money, it wouldn't make a dent in the global problem.

I also learned in Haiti that if there was no demand for plastic, then a clean beach would soon get dirty again. However, when we saw consistent demand and purchasing of plastics, the beaches stayed clean. The strategy shifted to building demand for ocean-bound plastics by helping companies make millions and millions of items out of collected plastics. That was how OceanCycle was born.

My business partner and I had first envisioned OceanCycle as a nonprofit. But after the MBV course, we decided to set up OceanCycle as a social business with the idea of funding the organization from our operating income instead of from donations. We developed new standards and instituted a system of auditing so we could be completely transparent and honest about our claims. We implemented an ability to track collected plastic "from shore to shore" and emphasized the livelihoods of the waste-picker communities responsible for collecting the plastic. After my work running the Mattel Children's Foundation and my time with Save the Children, I put a special emphasis on making sure we prevented child labor and forced labor from entering in the supply chain of our recycling partners.

The trust and transparency we established resulted in OceanCycle becoming the largest global certifier of ocean-bound plastics. We worked with top brands to make products out of material sourced from countries like Thailand, the Philippines, and Indonesia. With OceanCycle's help, a leading startup called ZenWTR made the first bottled water out of 100-percent ocean-bound material. Dell made soft goods such as backpacks and laptop bags, and Patagonia and Helly Hanson made outerwear, from OceanCycle certified material. Even Herman Miller made their flagship Aeron chair out of recycled ocean plastic.

In total, our efforts resulted in the collection of more than one billion bottles and generated approximately $5 million for coastal communities vulnerable to plastic pollution. I am proud of our success in strengthening recycling around the world, but we are a long way from achieving our objective of stopping ocean plastic pollution by 2030.

As a social enterprise, we were able to maintain OceanCycle through operating revenues, but we grew slowly because of our investments in marketing and storytelling were limited. We were the first and most robust certification program for ocean-bound materials, but our lack of investment gave other companies with inferior methodology and standards the opportunity to gain traction in the market. In addition, we found it difficult to pay key staff enough to invest in things like their retirement or a home.

That wasn't as hard for me because I don't have a wife or children, but it was more difficult for my business partner, who had a young family. We tried to raise money to support OceanCycle, but soon we realized that people knew how to give money only one of two ways: either to nonprofits or to highly-profitable enterprises. They didn't know what to do with a social business. Capital doesn't flow to social enterprises, even if they are more efficient than nonprofits, simply because they don't offer tax deductions. OceanCycle continues to this day, but it will never compete financially with better opportunities for investors.

Building a Better Future

Our approach at OceanCycle was similar to my strategy in Haiti: build recycling systems for high-value plastics (e.g., bottles and rigid containers) and leverage that infrastructure of people, trucks, and processing centers in order to get at low-value plastics (the flexible plastics, such as plastic bags and Styrofoam). Even after years of collection of high-value plastics, the low-value plastics are still being left behind and are the majority of plastics polluting the ocean.

Part of the reason this continues to happen is simple economics. A metric ton of clear PET plastic (e.g., water and soda bottles) is worth about $1,200 on the world market. The same weight of plastic bags takes much longer to collect and is only worth about $200. The waste pickers combing the beaches to pick up bottles tend to leave the other plastic behind because it isn't the best use of their time and won't earn them enough income to support their families. With this reality in mind, we decided to tackle the problem in two ways. First, find ways to further incentivize collection of low-value plastics; and second, find alternatives to replace single-use plastics in areas where recycling is virtually impossible.

So far, we have used philanthropic dollars to do beach cleanups, but that doesn't solve the problem in the long term. We need to find ways to increase the value of the material, so we are doing pilot projects to generate plastic credits. As those credits are validated and purchased, the income can be used to increase the prices of the low-value plastics. For example, a metric ton of film that might have been worth $200 would then be worth $800 and pay waste pickers more.

Many brands criticized for their plastic usage have begun experimenting with buying credits to offset their plastic usage. Similar to how private jet owners try to reduce criticism by buying carbon credits that offset the impacts of their travel. Some brands like ZenWTR are now buying credits, so that if they use twenty tons of plastic in their packaging, they can fund the cleanup of twenty tons of plastic from coastal communities and become plastic-neutral.

Similar to the early days of the carbon credits market, there are a few players in the plastic credit space but there is little reporting or traceability behind their claims. We plan to use our systems and methodology in tracking material "from shore to store" at OceanCycle to prove that material was recycled or properly disposed of.

Reflections on Work and Service

Much of my life has focused on trying to make the world better. I have been driven by people I lost and the sense of my own mortality. I also feel called by my faith to serve others and be a good steward of the planet.

In government, there were many resources available which weren't implemented well or efficiently. Working for NGOs, there was great passion and ability but a lack of resources and expertise to scale solutions. Private companies had the money and expertise to drive scale but were primarily focused on making more money and, in many cases, didn't think about the harm they caused. I have always wondered—what if we had purpose-driven companies that integrated purpose and profits? I have made it my life's goal to build and support such enterprises.

Veterans can be a great resource for private companies and nonprofit organizations, but veterans individually and as a group are often misunderstood. Many people working in private companies carry all kinds of stereotypes about them, and in some places, they have had zero exposure to those who have served. Before accepting the job at Mattel in 2014, one of the leaders I interviewed said, "You know you can't yell at people here like in the military, right?" I told them I hadn't been in the military since 1999, and I had been running global nonprofits since then.

When I was leaving Mattel, one of the last things I did was go to the annual Toy Fair in NYC, and I had drinks with some colleagues afterwards. The lead designer for Barbie said to me, "Wow, you are kind of a nice person." I said, "Thank you, I try to be." She proceeded to tell me that she had hated me for three years. "You hated me?" I exclaimed. "We've never had a conversation." She had assumed that because I was former military, I was conservative and didn't like gay people.

I also saw this bias during my time at USAID and while working for NGOs, where people from more liberal or social service backgrounds didn't understand how someone could choose to join the military. They assumed that people who served were unintelligent and misinformed and willing to see every problem as a nail in need of a hammer. Unfortunately, that point of view neglects the fact that those people who have been in the thick of it and seen the consequences of disaster and conflict tend to have more compassion.

Veterans also tend not to think about themselves as much as others do in the corporate world. They value teamwork and impact. They are also willing to make sacrifices, especially of comfort and personal time, to work on projects and meet important deadlines. In my personal experience, veterans have more integrity than average and are willing to speak truth to power, which are qualities that are not always valued in the "yes man" culture that surrounds some corporate leaders.

This perspective drives my work and my desire to integrate purpose and profit. I like money and the freedom and opportunity it brings, but I have met too many wealthy people who are unhappy despite their wealth. And I have met many people living in poverty who have a richness beyond measure in terms of their faith, outlook, and community.

I don't have all the answers, and I look for opportunities every day to serve. I don't have any guarantee that I will see the fruits of my labor. I think about the perspective a good friend of mine has. He is one of the top bluegrass singers in the country. He is always so complimentary of those who came before him, musicians that mostly ended up penniless but created the foundations of bluegrass music that now allow him to sell out 15,000 seat arenas. Ultimately, it is up to us to till the soil and plant the seeds, but only God can make those seeds grow. I hope to see a world where we don't need to make trade-offs between profits and purpose, but if that's not possible, I hope I can help create the conditions for such a world to happen in the future.

"A Desire to Make the Most of It"

Robert Goodwin reflects on telling his story

This chapter was difficult for me to write. First, I don't like writing about myself. I learned how much work it takes to make an actually interesting story. I always admired my uncle Jerry and my aunt Donna who could keep the attention of the entire family with their vivid storytelling about global adventures. I am not accustomed to telling my own story, even though a big part of my life has been advocating for causes I believe in. I find it much easier to advocate for worthwhile ventures and to hype up others. It's something I need to work on for myself but hard to find the energy to do so.

While writing, I kept asking myself, "What did it feel like?" and "What was I feeling?" Dealing with trauma and difficult situations in life, I tend to default to the factual account of what happened and not dwell on feelings as much. I think it is evident in my writing. Feelings are surely there when I reflect on them, but I have likely repressed many as a coping mechanism.

I bet that other veterans and people who have gone through similar situations feel the same. It's likely why we can wade into difficult and chaotic situations and perform well, but it also might rob us of a bit of the flavor and joy of life.

I am fifty-one years old as a write this, and I realize that I am at the tail end of my life and career. My father died at fifty-nine and both my grandparents died in their early seventies. I am struck with how little time may be left as well as a desire to make the most of it. The one blessing I have from all the death I have experienced is that my mortality is very real to me, so I try to do the things I want and feel I should today and not wait until tomorrow to do them. I have always been cognizant of the fact that we never know how much time we have.

Chapter 22

You Have to Happen to the World
by
Erdavria Rose Simpson

Erdavria Rose Simpson was a second-class petty officer in the US Navy Reserves where she served as an intelligence specialist, supporting multiple international joint intelligence missions.

Erdavria graduated from the University of Southern California with a Master of Business for Veterans (MBV) degree and a master's in science in entrepreneurship and innovation. She earned a BA from California State University, Fullerton.

She is a cofounder of It's Gameday, a corporate transition and soft-skill eLearning application that helps athletes, veterans, and minorities. She currently serves as a project manager for Millennium Space Systems, a subsidiary of Boeing.

Erdavria is an active member of the MBV Foundation and USC Black Alumni Association. She volunteers for the Girl Scouts of Greater Los Angeles and leads corporate diversity, equity, and inclusion initiatives. In 2022, she cofounded The Lazy Rose Café in Mid City Los Angeles.

Keep Moving Forward

When I was finally accepted to the University of Southern California, I was caught off guard by the rush of emotions I felt. I had lived my life by the motto "keep moving forward," and in order to arrive at that moment I had to pick myself up from several past failures and—quite literally—keep moving forward. I got the motto from the Disney movie *Meet the Robinsons*, which emphasizes that you learn your biggest lessons from the moments that feel like your greatest failures. You have to face these failures and keep moving forward in order to achieve a greater future.

I had to face what I saw as my biggest failure—the failure to be accepted into USC when I first applied as an undergraduate. This failure set me back from the golden path I had imagined as a young girl. As a Los Angeles native, I grew up three miles from USC. I even went to a magnet middle school on its campus. USC played a significant part in my childhood, and I very much viewed it as the place I had to be in order to be successful. To get there, I worked tirelessly to check all the boxes that were required of me, but I still didn't get in. So many emotions rushed back when I first sat in the classroom.

I quickly found a new motto from a maxim that Professor Turrill shared with my class on day one: "You can let the world happen to you, or you can happen to the world." I held to those words in order to conquer the feelings of self-doubt and imposter syndrome. After hearing that maxim, I began to understand that I have always been my own worst critic and my greatest obstacle. Feeling as if I'm always getting hit hard by the world, I decided in that moment to fully accept that I belong in the room and that I, Erdavria Simpson, am ready to make an impact.

This raised an important question: Why am I my own worst critic, and why do I always seem to be in my own way? Everyone in my cohort had to walk their own path to arrive at this place, and I was no different than the person sitting to my left or right—but it's easy to lose sight of that.

The Difficult Path

My childhood encircled the USC campus. More childhood dreams than I can count included walking by Tommy Trojan between classes while discussing with my classmates the most recent lecture.

It was this dream that led me to always take the more difficult path. I studied harder than my peers in order to get the highest test scores and the highest GPA. Even if my GPA was not the highest, I rested easy knowing I worked harder than anyone else. I focused on community service by working through the Girl Scouts and achieving a Gold Award. As with school, I overachieved as a scout at every step. I was even a leader in the dance programs and on the cheer squad at my high school, Hamilton High.

Writing down these achievements even now stings me knowing the setback I would suffer. But even more pervasive among my memory of childhood achievements is the thankfulness I feel for my family, who went above and beyond to ensure I could succeed as a child. I have always believed strongly in the importance of family, especially when it comes to my mom, who raised me by herself. She worked tirelessly to ensure I had the best of everything she could provide. A lot of my work ethic comes from seeing her work to put me through the best schools, ensuring I could achieve my goals. It was not easy for her, and it makes me happy when I can provide her some experience she missed out on due to the sacrifices she made to ensure my sister and I had the best of everything.

Cal State Fullerton

After being rejected from USC, I was accepted into Cal State University, Fullerton, with the intention of studying broadcast journalism. I don't know that I fully engaged with my college experience initially. I was the manager at the Gastronome, the student café on campus, and I worked at a Ross thirty miles from campus. I split my time living with my mom and in an apartment. The work-life balance wasn't easy, as I spent many nights working until so late that my best option was to take a nap in my car on campus before morning classes. I don't even really remember many of the classes I took during that time, as my work-life-school balance was so heavily tilted.

Navy Reserve

With the goal to improve the trajectory I was on, I enlisted in the United States Navy Reserve as an intelligence specialist. I believed working in this capacity in the Navy would bolster my dream career as a broadcast journalist, as the recruiter told me about all of the experiences I would have briefing leadership on the topics of the day. In the cold of January, 2014, I left Los Angeles for Recruit Training Command near Chicago.

Though I did not yet know the phrase "you have to happen to the world," I definitely happened to the Navy. At every phase of training I excelled and received honors. Day one, I raised my hand to volunteer to be in the music and flags 900 Division. I was intent on being the greatest sailor in the world's greatest navy. The work ethic I brought with me really paid off in the structured environment of boot camp, and I graduated with honors, even receiving a special cake-cutting with the then commanding officer of boot camp, as well as a commendation letter to carry forward.

From boot camp, I traveled to the Navy and Marine Corps Intelligence Training Center in Dam Neck, VA. Once again, I overachieved. I was a class leader, a class mentor, and I had a much better GPA there than my GPA in undergrad. As the class leader, I often led us when we marched from the barracks to the school building, and I wanted to make sure we looked our best when we did.

I also had a great awakening at this time. I was bitten by the travel bug. Northern Virginia provided an excellent base for weekend getaways to many historical sites from American history. One of my first getaways was to Williamsburg, VA, around Memorial Day. I felt patriotic pride as I watched a fife and drum corps march down the streets. Later, at Monticello and Mount Vernon, I visited the homesteads of two founding fathers. I attended my first real concert. But what impressed me most was seeing Washington, D.C., for the first time, standing in front of the Lincoln Memorial, and reading the transcription of the Gettysburg Address on the wall. Traveling through that part of the nation and its history really filled me with patriotism and reminded me of the impact I could have on the world.

First Assignments

As a new sailor in the Navy Reserve, all of your training is given to you initially when you enlist, and then you... return home to your regular life. In August, 2014, that is what I did. I went back to reality and left the structured environment I was finding so much success in. But I had a renewed zeal to succeed. I took more pride in my undergraduate work, and my GPA recovered significantly. I took remote and in-person classes, so that my work-life balance was not thrown askew like it was before. My time management was significantly improved, and I was able to really start to succeed.

I also continued to travel. I went on a cross-country trip from Richmond, VA, to Whidbey Island, WA, stopping at Nashville, Kansas

City, Mount Rushmore, the Badlands, Yellowstone, Boise, and Seattle. I went to New York City to see the Statue of Liberty, and I went to Walt Disney World.

Between school and travel, I still spent time with the Navy, and I continued to succeed. I was assigned to the Joint Forces Training Base in Los Alamitos, CA, and quickly advanced to the rank of petty officer second class. I forged strong bonds with several sailors in the unit, and I was able to leverage their expertise to grow as an intelligence analyst. I gained many of the skills which I now credit with being integral to my professional success in the time since.

In fact, an officer in my unit was a deputy mayor of public safety for the city of Los Angeles. My strong work ethic and dedication impressed him, and he suggested I send him my resume once I completed my undergraduate studies.

By the time I finished undergrad, I had cast aside many of the fears and much of the self-doubt I had harbored since being rejected from USC. As my undergraduate degree was in communications, I had hoped to be hired as a broadcast journalist. However, the opportunity I had through my Navy connection was too good to pass up, and I was hired to work in the mayor's office.

Mayor's Office

From day one at the mayor's office, I beamed with pride for my job. There I was, an LA native, walking through the doors of City Hall with an employee badge and serving the community I loved. Normally a fresh college graduate is happy just to have a job, but in my case, I not only had a job, but I also had a mission I believed in and—initially, at least—I felt fulfilled. I continued to overachieve, volunteering for the Mayor's Crisis Response Team. The experience gave me a holistic view of the town I grew up in and provided an education I could not get in a classroom.

Another important change happened at this time: I was transferred from Los Alamitos to Naval Base Coronado. This presented my greatest struggle with the Navy Reserve. I left a group of people I cared about and moved to a place that seemed much less personal. All of the positive experiences I had with the Navy came undone with this one transfer. The sailors there seemed more interested in looking out for themselves than helping others, meaning their success would come at the cost of the success of those around them. Nothing highlighted this more than the story of a senior chief who confiscated an item from me. For one drill, we were assigned to wear our khaki uniform, and I put on the pin from my previous command as a way to show my pride in that command. Within the first hour, rather than telling me it was out of regulations, the senior chief confiscated the pin and never returned it. This symbolized the end of the line for me, as a senior leader had essentially robbed me of something I felt great pride in.

My work at the mayor's office was proving to be a challenge I was more than up to, and I was gaining the necessary confidence to overcome my worst enemy—myself. The strength the job gave me made me feel like I mattered and was making a name for myself. It led me to reapproach the source of my greatest personal defeat to date: I reapplied to USC for the Master of Business for Veterans program.

I was overjoyed to be admitted into the program. It meant I was finally achieving my dream of walking onto the USC campus as a student. However, in spite of the program recognizing my success and letting me through the door, I still had some lingering self-doubt.

I was certainly intimidated by the eighty people in my cohort, but I knew I was their peer, and we were USC graduate students together. I shared many experiences with them, and the encouragement I received from the cohort is something I would not exchange for the world.

Once I recognized that I was just like everyone else in the room and equally deserving of being there, I realized that everyone was facing many of the same challenges though they may have looked different than my own. In the leadership course I realized that I wanted to grow as a leader and inspire those around me.

Spirit Leader

Shortly after my arrival at USC, I saw a flyer for tryouts for the Spirit Leaders, a co-ed cheer team. They carried school spirit to many of the sporting events on campus as well as represented the school at community events. I had been a captain on my high school cheer team, and when I saw the flyer, I knew I had to try out.

Graduate students weren't typically accepted to the team. I took a major leap of faith and applied to be a Spirit Leader despite being a graduate student. I was accepted—the first graduate student ever to be a Spirit Leader. It was a full circle moment for me that cemented the sense that I belonged at the university.

Meanwhile, I was afforded an opportunity to transition away from the mayor's office. I began a new career in the space industry at Northrop Grumman working as a program schedule analyst. The leadership lessons I learned inspired me to take full ownership of my career and gave me the confidence to operate in the business world, a world dominated by men, with confidence and poise.

The potential for growth at Northrop was something I could not overlook when weighing the appropriate next move away from public service in the mayor's office. At Northrop, I learned new skills that drew heavily from the attention to detail I had developed in the Navy, from lessons I learned in the MBV program, and from my experience at the mayor's office. The confidence I had gained I applied to my new position, and I found myself at the center of some unique projects.

Starting Businesses

Shortly after moving to Northrop, I fulfilled one of my biggest goals and purchased property in the city of Los Angeles. The duplex needed a lot of work to make it livable. I learned more about home building than I expected. The house needed everything from a new roof to new ceilings and floors. But for the first time I owned a piece of the backyard I grew up in. I had studied the real estate market carefully, and early in my career, I had passed the California real estate board. I had also worked briefly in the field, and my professional skillset was a valuable part of my effort to purchase property.

Meanwhile, I also started a business, inspired by my time as a Spirit Leader. I cofounded It's Gameday with a fellow Spirit Leader as business partner. Our mission was to inspire young athletes to translate their success on the field into the realm of professional life. We recognized the rough transitions many college athletes make after college, or after sport, and we hoped to motivate them to take control of their careers.

I took another, even greater, leap of faith and opened The Lazy Rose Café. Opening a brick-and-mortar store had long been a personal dream of mine, and I wanted to give back to my community by giving people the opportunity to seize their own lazy day. I took the café from vision to reality using the lessons from USC's MBV and Master of Science in Entrepreneurship and Innovation (MSEI) programs. Within three years, my partners and I went from having a fledgling idea to opening our doors in Mid City. With the café, I operated with a different mantra: Everyone deserves a lazy day. I regularly hosted yoga sessions, book clubs, and markets for local vendors. As a Black woman veteran-owned business, the goal was to serve as an inspiration for the community.

The concept of a lazy day was something I found foreign, because at any given period of time I was starting businesses, dealing with contractors, doing homework, rushing to Spirit Leader practice, and making sure I was progressing in my nine-to-five corporate career. I firmly believe that it is not what you do during your nine-to-five that counts, but how you make the hours after work matter.

Millennium Space Systems

Upon completing the MBV program, I enrolled in the MSEI program at USC. The program carried through the 2020 COVID-19 pandemic and much of the coursework had to be recreated. It taught me ways to be flexible in times of hardship. Scholastically, I struggled with being spread very thin. I completed much of the coursework as I went through many major life changes, including transitioning out of the Navy Reserves.

Between rebuilding the house and opening the café, I began to flourish. I received an offer to transition to Millennium Space Systems, which turned out to be a great blessing. I believe in the company's mission, and I go to work every day feeling as though I am making a difference for my team. In my current position, I lead by example developing junior scheduling analysts into successful employees and ensuring their careers are successful. In addition, each time our company succeeds, I find that I am right in the thick of that success, and I can take pride in what we have done.

"So Much Has Changed for Me"

Erdavria Rose Simpson reflects on telling her story

I was incredibly excited to share my journey, because I hope elements of my story can be useful for others considering how the MBV program may impact them. I believe my path to the program was not necessarily straightforward, and I hope it can serve as inspiration for others. Since finishing the program, so much has changed for me: I opened a brick-and-mortar café then closed the location to pivot into the e-commerce space. My company promoted me to management. I got married. All of this in the space of just five short years.

Sharing my journey through this effort has allowed me to reflect on my path and helped me decide the narrative I want to carry into the future. While it may appear only reflective, writing my story has been transformative, as it allowed me to sit with my experiences—the good, the bad, and the in-between. I have come to appreciate even more how valuable the MBV program has been to me and my growth as a person.

The biggest obstacle to writing this was homing in on which parts of my struggle I wanted to share, as well as selecting the moments I am not as proud of. Finding the balance between telling the struggle, avoiding certain moments, yet still maintaining a raw and good story. While this is hard, it allows me to share the impact of my successes in greater vivid detail.

Chapter 23

When I Found Peace
by
Sonny Tosco

Sonny Tosco is a father, entrepreneur, and veteran of the US Army. After graduating from the United States Military Academy at West Point, he commissioned as an air defense artillery officer. Through three deployments he earned an Army Commendation Medal. After leaving the army, he cofounded three companies (Limelight, Anvyl, and Ronin Dev) in Silicon Valley while raising two children.

After graduating from the University of Southern California's Master of Business for Veterans (MBV) program and developing more than twenty apps, he worked at Meta as a technical program manager. In 2023, he joined Google as chief of staff for data center operations. He currently resides in Charlotte, NC, and is expecting a daughter.

Eleven Important Dates

I want to write about eleven key moments that shaped where I am today. Each of these dates initiated a series of events that significantly changed my life. Like in a football game, had the ball bounced another way the outcome would have been significantly altered.

Some things are obvious to success, like working hard and having good character. But there are other insights specific to my life experiences that I want to share. More important, as my babies get older, they ask stories about my past. I have to pick the tame PG stories to tell them, as they are not old enough to hear about the crossroads I have stood at.

August 11, 2000

I had just turned sixteen years old and wrapped up the summer swimming season. Growing up in a strict household, I had to forge swim meet permission slips in order to gallivant around Northern California with my friends. On one of those nights, we went to the Santa Cruz beach boardwalk. The plan was simple—the crew would pick me up for the "meet" and we would hangout in Santa Cruz. This was the first time we did ecstasy.

There we were, three eastside kids trying to escape our broken homes for a night, when I felt a burst of serotonin enter my system that I had never felt before. The new sensitivity let me perceive the world in a new way, but it also resurfaced some suppressed traumas.

As we watched the sun set against the Pacific, I told my friends about the verbal and physical abuse my parents inflicted on me. My mother told me that I would never amount to anything if I came home with a B, and my father, after a bad day at work, would hike up his pants every time he smacked me with the belt. The abuse was the lighter fluid that lit a blaze inside me to succeed in life and never have to depend on them for anything.

My friends listened to me, and it was a formative moment for all of us. We remained close and showed up for each other as we continued to grow. All of them were in my groom's party and I was in all of theirs. As kids entered the picture, we became uncles to each other's families. Although we have taken different paths in life and distance has come between us numerous times, we still find ways to maintain our friendships.

Our relationships taught me about loyalty and the importance of family. We are each other's sounding boards, and we keep each other grounded.

June 6, 2002

I was a week away from graduating high school and was ready to start firefighting school in a few weeks. I was accepted to San Jose State, but I knew there was more I wanted to do in life than party for four years.

I had also been rejected twice from West Point. One time for having asthma, and the second time for being born with a heart too big for my body. No joke, it's a condition I have. My heart doesn't have to work as hard as the average heart because it pumps blood more efficiently, and even more so because I'm in shape. The downside is that when I go to sleep, my heart rate falls to eighteen BPM. The BPM anomaly raised concerns among medical staff because you're declared dead at zero BPM.

I was devastated. I had worked hard since middle school to get accepted to West Point. I needed closure around the rejections, so I wrote a letter to the superintendent. I told him how hard I worked to be a competitive admission and what I planned to do after joining the Long Grey line. I sealed it, mailed it off and started moving on with my life.

Not too long after, I came home, and my dad motioned to a package addressed to me. Inside was a green dossier with a formal offer of admission to West Point, my dream school. The greatest adventure of my life was about to begin.

May 27, 2006

Hungover on the morning of West Point graduation, I almost missed my last breakfast formation. The night before, my mother had told me that everything I had ever accomplished in life was because of her. That hurt and enraged me, and I took to the bottle to suppress the pain. I smelled like old potatoes as I sat through President Bush's commencement speech.

I continued to think about her words and how she was taking credit for my accomplishments of the last four years. There was a part of me filled with self-doubt, and I wondered if maybe she was right. But I wanted to own the fact that my actions, my grit, and my perseverance got me to graduation day.

I barely remember walking the stage. There was a problem finding my diploma, and I stood there awkwardly for what felt like an eternity before they found it. I shook the President's hand, the hands of a few Joint Chiefs of Staff, and smiled for what felt like one of the longest times of my life.

My parents didn't make it to my commissioning ceremony, and I didn't speak to them for the next four years. What should have been one of the most special days of my life was lonesome because of ego and our inability to listen.

After graduating, I was stationed in Fort Bliss, TX, from 2006 to 2012. The area wasn't bad, but not the right environment for me in my twenties. Of course, I have to own up to my bad decisions, such as drinking in excess most weekends, moonlighting as a bouncer, and flying back from Jeddah dressed as a Saudi.

Over the course of three air defense deployments and some side missions, I grew up.

On my first deployment, I served as the executive officer and lead tactical control officer for the first Patriot missile battery in Bahrain. The role positioned me to consult with our host nation and our allies on how to optimally use the air defense system and how to integrate it into our overall air defense posture. It sounded like a bunch of military corp-speak if you ask me, but our mission was noble and just. Learning how to work with a host nation and allied stakeholders gave me my first exposure to real-world program management, which I later pursued professionally in the tech industry.

My second deployment, to the United Arab Emirates, was many things, but it was especially the time when I started becoming the leader I am today. I was a tactical director for our battalion, which meant that I was responsible for four Patriot missile sites while on shift. Outside of the Groundhog Day-nature of shift work, my constitution was really tested by the senior leaders of our battalion, who made unethical decisions and eroded the confidence of our soldiers. I could have turned a blind eye to it and complied with their rank, but I called it out, and I even locked up the senior noncommissioned officers when it came to that point.

Because of my antics, I knew if I left the Army at that point, I would have nothing good to say about the experience. I was a senior captain, and we were still made to sit at a desk for twenty-four hours of staff duty so the lieutenants could study for their air defense artillery Table VIII exams.

The next morning, delirious from lack of sleep, I went to the lieutenant colonel's office. He was the one battalion commander everyone said was the best officer they ever served with. I volunteered for a third deployment with his unit. I joined their team, did the predeployment field training exercises, and the team and I both realized how underutilized I had been in my last unit. They made me the battalion operations officer.

On my third deployment, I became a man. In the months leading up to it, my wife had a miscarriage. Despite it, I had to keep focused on the mission. Things were coming full circle. The first country I deployed to was the last country I deployed to. In 2008, I had deployed as a boy, and in 2012, I left as a man. My father passed during that deployment, and while I was stateside for two weeks, my unit carried the rock, which was a testament to the team I had built.

At twenty-eight years old, I didn't think I would make it to that point. Despite my positive experience on my third deployment, I knew that chemistry couldn't be replicated again. I decided to leave the Army. Most of my time in the service had been marred by toxic leadership and antiquated standards, which eighteen memorable months with 2-43 ADA couldn't make up for.

Back home in California, I didn't realize how much I had changed as a person. Home felt just as foreign as the sands I had been deployed to.

August 2, 2008

The day I got married. My wife and I had known each other since high school, became serious in college, did long distance for a few years, then finally tied the knot. My emotions about the marriage were complicated, to put it lightly. I didn't know if it was the right path for me. In the vestibule I sat in a tuxedo on what was supposed to be one of the happiest days of my life, unable to feel the special gravity of the moment.

My parents and siblings didn't attend the wedding. Only a few family members did. The absences communicated a lack of support but subconsciously validated the marriage. Not until recently have I started sorting out those feelings.

With the decision to get married, I was jumping into water and letting the current take me. In your own heart, you know what doing the right thing feels like. Going against that innate feeling creates anxiety and abnormalities in your behavior that can only be resolved once you get honest with yourself. I have two beautiful kids from my marriage, and I feel no ill will towards my ex-spouse. It has been a lesson about the importance of always aligning your actions with your inner voice.

May 8, 2013

I woke up in our one-bedroom in Walnut Creek to find out my wife's water had broken. Off to the hospital we went, and at 5:29 p.m. I became a father.

Honestly, I didn't know how to handle the situation. All day I went through emotions. I was happy but unable again to process the gravity of the moment. Shouldn't I be happier? I held my son for the first time and didn't know how to feel. My mind raced with thoughts of all the things I wanted to do with him when he got older. Playing catch, teaching him how to swim, seeing him become a man. But I also knew my life was going to be different from that point forward.

I texted friends and family to tell them they were now uncles and aunts. One of those people was my good friend Tony Fusco. Tony was a West Point grad I met on my last deployment. He had become my protégé in the unit. When my father passed, Tony oversaw the operation and executed my playbook to a T. Before I left the Army, Tony threw a party in my honor, which was one of the nicest things anyone has ever done for me. There was a lot of friendship between us. He was one of my closest friends from that chapter of my life.

When I texted Tony, he responded with a reference to the Kanye and Jay-Z song "Otis." We used to blast the song in the tactical operations center before we had to brief a distinguished guest. I must have already compartmentalized that part of my life because I didn't immediately recall the reference. Through the phone, I could feel that he was upset. I didn't pay it much attention. I just had a baby, bro, your feelings can wait.

Five days later, I received a call from a mutual friend who told me that Tony had taken his own life.

When I attended his funeral, I saw agony on his parents' faces which I wouldn't wish on my worst enemy. It's a pain I never want to feel. I made a promise to myself. I would build a relationship with my kids so that whatever they were going through, they would always feel comfortable reaching out to me. No parent should have to bury their child.

That's what shaped my parenting style. Plus, making good on the promises I made as a kid that I would do better for my own children than my parents did for me.

April 8, 2014

I was two years out of the military and two years into working at Jostens as a yearbook salesman. I took the job because it was pitched to me as an entrepreneurial role, which spoke to my ambitions of wanting to start my own business one day. What I didn't understand was how to evaluate an opportunity. After a year, the company reneged on the salary they promised. The pride in me made me stay to keep grinding.

This was a strategic blunder that led me to work longer hours for a $75,000 base salary to support a family of three in the Bay Area. We burned through our savings and faced a dire position. I did the only thing I knew how to do—worked longer hours and spent more face time with prospective clients. This approach only hemorrhaged my sanity.

April 8, 2014, could have been the last day of my life. My ex wasn't working and hadn't worked for the last year because any income she brought in went immediately towards childcare. This placed the burden of providing solely on me. Our relationship wasn't in a good place, and I was already thinking about divorce, but I held on to be present for my son.

That evening, she was in a panic because we only had $10,000 left in our savings. I was applying to jobs in tech, but I had been unable to gain any traction. I didn't know how else to generate income, so I proposed that I could drive for Uber on the way home from work. She didn't like the idea. That's when I flipped.

I yelled, asking why she was against the idea. She said it was because I would be out of the house more often, which didn't make sense because I definitely wasn't solving our problems at home. I lost it and broke down and started crying. I yelled at myself and said I was a failure.

"I left the army as an S-3 to sell yearbooks," I sobbed. "And now I'm a fucking Uber driver."

She told me it would be alright, to which I replied it meant nothing. There was no plan. I knew what false hope looked like. I thought about Tony and how if I were still in the Army then maybe he would still be alive.

I went to the bedroom, grabbed my father's .357 Magnum, and jumped in the tub ready to make a scene.

My son crawling towards the door stopped me.

Holy fuck. I sat there for a little longer and came to my senses. I left the house that night, got high in my car, and fell asleep.

I woke up the next morning with a newfound clarity. I realized that I had touched the bottom. I truly felt like I had nothing left to lose at that point.

There had been numerous times in my life when I felt that something held me back. When I was a cadet, I ran a nightclub in my West Point barracks. I DJed and sold liquor out of my room, which was frowned upon. When I was a lieutenant at Fort Bliss, I wanted to open a nightclub. My battalion commander didn't think moonlighting as a bouncer was the right way to network.

But now, for the first time in my life, I was free. Truly free to make unbridled and unhinged choices for myself.

I had an idea for an app called Limelight. It gnawed at my insides, but I had been hesitating to take the risk of making it.

But bro, you almost took your top off, I told myself.

From that moment on, I embraced my entrepreneurial nature and made Limelight a reality.

I went to my first appointment of the day at a high school. My client, the yearbook advisor there, didn't address me, so for an hour and a half I sat in the corner of the classroom wasting time, which the client didn't have a problem with. That made me contemptuous of the conversation that was about to transpire.

Because my commission was tied to the school's timely delivery of their yearbook materials, it was critical that they made the deadlines on time. This yearbook advisor had made zero out of five deadlines, so at that point I wasn't getting paid on the account.

Finally, she said, "I'm so sorry it's been taking me awhile. The kids aren't listening to me."

"I'm gonna stop you right there," I said. She looked shocked as if she had never been interrupted before. I continued. "Respectfully, I gave you the playbook to follow at the beginning of the year and you didn't even try to follow it. I'm actually quitting today, so this is your mess to figure out moving forward."

She laughed nervously, expecting me to say I was kidding.

"What are you going to do next," she asked.

"I'm going to start a company. Wish you all the best."

May 15, 2015

After that, I was on the clock. I lived off four months of severance and my credit card to finance my company. I listened to podcasts to quickly level up my Silicon Valley acumen. When Gil Penchina, a prominent angel investor, first immigrated to this country, he went to a networking event every day for a year to build his network. So I did exactly that.

After a while, I started seeing regular faces at the events. I went through three cofounders who had varying levels of commitment until I found my CTO. When I say things happen for a reason, this was a prime example of that.

That afternoon, my wife was in a fender bender with my son in the car. It wasn't bad, but I could have easily skipped the event and gone home. I'm glad I stayed, because that's when I first met my business partner. He was an engineering manager at eBay looking to build something else. His joining the team gave me confidence to move forward.

I recruited our first front-end engineer on a networking app called Weave, which was a combination of Tinder and LinkedIn—swiping left or right on professional profiles and connecting with the people behind them. The engineer I found had immigrated to the US from Portugal and wanted to chase the American dream.

Our backend engineers I found on AngelList. I didn't have the liquidity to pay them, but we worked out a deal where I covered their winter break airfare to China, and they built the backend. My cousin donated his frequent flyer points to cover their trip. A couple months later, we had an app.

I learned countless personal and professional lessons while riding the rollercoaster of starting my own company and creating an app. I don't regret getting on the ride at all. I turned down a six-million-dollar acquisition offer because entrepreneur and angel investor Jason Calacanis called Limelight his next billion-dollar company. I lost close friendships in the process. We got hacked by the Chinese Communist Party. Had to pivot our business model a few times.

Then, after the business went on life support and I moved in with my in-laws, reaching critical burnout and huge debt, I decided to step away. I felt like I was again at the bottom, but I was there knowing I had grown from the experience.

May 23, 2016

We were a week away from sharing Limelight at TechCrunch Disrupt, a marquee conference in Silicon Valley, when we found out our app had been hacked by the CCP. After user data had been compromised, most of the user base left our app.

My wife and I found out that we were expecting a second child. The timing wasn't the best. It wasn't opportune given that my source of income was a fledgling startup that had just been hacked. I made the adult decision to take a job at Yelp in order to provide income and insurance for our developing child.

We wanted the child's gender to be a surprise. The pregnancy was different than the first in the sense that I noticed my wife was more difficult to deal with. A close friend said it was because I was having a girl. We found out he was right.

With the impending decision about what to do with the company moving forward, I was too busy to fully appreciate the moment. But the right thing to do was move on. With a second child in my life, I needed to find a new way to provide for the family and level up.

On paternity leave, I reflected on my startup experience. I loved the fast pace of the sector. I loved building a team and showing up to work with the adrenaline rush of not knowing if a deal would make or break for the company. And I knew I needed to evolve in the space.

The week my daughter was born, I took my son to a BBQ spot on a Sunday night. Delivery drivers waited by the counter for DoorDash orders. Most of them looked fifty-five years old or older. That's how they were spending their Sunday night. Likely not because they enjoyed driving deliveries, but because they needed to make ends meet. The moment connected to the time I found out my father had been laid off during the dot-com crash, which led to a rough time for our family before I took off for West Point. It was the last job my father ever held.

The realization hit me that if I didn't find a way to expand my skillset, making food deliveries was eventually what I would end up doing at that age.

In the valley, I noticed two types of people, those who considered themselves tech, and those who were non-tech. The non-tech people got MBAs but couldn't speak engineering. On the other hand, if you asked an engineer what time it was, they told you how to build a watch. I knew that I needed to understand both sides to climb in my career.

Financially I was fucked. If I didn't do anything, in a year I would be exactly where I was, pushing Yelp's product. I mitigated my risk by completing Salesforce's Vetforce program. This gave me the confidence to enroll in a full-stack coding boot camp. To help with funding coding school, a friend pointed me to the Commit Foundation, a nonprofit that helps veterans transition to the private sector.

April 28, 2017

Two months into coding school, things started to come together in terms of how I understood software. I realized where I had had unrealistic expectations for Limelight, but I also recognized where my team had lacked experience.

An opportunity came from another West Point grad who wanted to pick my brain (like they all do) about a startup idea. After a year of working on it, the only deliverable was a ten-page business plan. I told him to keep working on it and best of luck.

He persisted to reach out, and shortly after the birth of my daughter I agreed to work with him on his startup. The company was looking to disrupt the manufacturing space by providing transparency to the sourcing, procurement, and manufacturing processes.

I made a mistake. Because we were both West Point grads, I agreed to terms with a handshake, naively thinking that we could move forward and instead put any lawyer fees into the product. I'm shaking my head typing this, just as you are reading it.

I was confident that we could build a minimum viable product (MVP) with talent I knew from coding school. We built an MVP and started putting it in front of potential customers for feedback. Leveraging my experience and relationships from the first startup, it was significantly easier the second time around.

I was able to land our first significant investor who committed $750,000 to the company's seed round. Other investors followed suit. We raised $1.5 million pre-revenue. If you are shaking your head thinking it was too good to be true, you're right, it was.

That spring my business partner flew out to San Francisco to meet with me. The energy felt off, and I had an intuition that I was going to be pushed out of the company. I confronted him about it, and he denied and deflected.

When it was time to formally incorporate the company, the other shoe dropped. In the contract language at signing, my partner had vested over a million shares in stock, whereas I had a six-month vesting cliff on my shares. The closest thing I can compare to reading that contract is realizing your friend is wearing chain mail at the Red Wedding.

I confronted my business partner and our lawyer, and neither would give me a straight answer. They gaslighted me, and I was terminated for allegedly trying to sabotage the round.

I didn't care about the money, but it took me a while to get over being betrayed by a brotherhood. What I can't get back is the time I spent working on the startup. I could have spent it with my newborn daughter.

In hindsight, I'm glad it happened. I avoided moving to New York City and spending the 2020 pandemic and quarantine there. It made me recognize that I needed to be extremely selective with my business partners.

The experience was also a sign that I needed to level up in other areas. I understood the technical language, but I needed to better understand business at the highest levels. I linked with an alum of USC's MBV program, and the alum connected me with James Bogle, the program director. A month later, I was accepted into Cohort V.

Grad school taught me how to speak the language of business. Having a conversational understanding of what was important in a negotiation made me more effective in app development. I was able to close more deals because I could convey my value to business and technical leaders both.

What made the MBV program especially valuable was the common experience we all shared as veterans. Sure, I learned about balance sheets and marketing, which I had gone there to learn. But the feeling of being back in the barracks felt special.

In a progressive industry like tech, and while living in Silicon Valley, I often felt alienated and out of place. My demeanor came off as intimidating and abrasive. The MBV program taught me how to soften my approach while maintaining an intense pursuit of my goals.

By the time I graduated, I had learned so much professionally and personally that I was ready for the next steps in my career. I learned to forgive and move on from my second startup experience as a direct result of learning from mentors who I can only aspire to emulate.

The app startup just raised a valuation of more than $100 million. Friends send me updates about the company. It used to affect me, but I am satisfied knowing that the company's story can't be written without me.

July 3, 2023

After graduating from business school, I ended up at Meta. I performed several roles there and provided my family with a better life. My first year, I had a five-hour commute to work. A year later, I had a three-hour commute. When COVID-19 hit in 2020, I had no commute.

It was during the pandemic that my wife and I made the difficult decision to get a divorce and co-parent. This was a result of years of unaddressed issues. In the military as a cadet and then on active duty, we never had the chance to develop a relationship. We were long-distance until we were married. Early on when we lived together and had problems, afterward I would deploy for a year. We missed each other, then we would start fighting again, and I would redeploy.

When I transitioned to civilian life, we had no choice but to work out our issues. I wanted to confront them, but she wouldn't, so we swept the problems under the rug. She did a better job of compartmentalizing than I did. I was always noticing that the pile under the rug was getting larger and larger, becoming a tripping hazard. All I needed was for someone to say I mattered.

I had watched the cancer slowly degrade my father for years, and when I spent his last days with him, he told me that if it had not been for my siblings and me, he would have left my mom years ago. I never forgot what he looked like when he said that, worn down and defeated choosing to stay in a situation that was detrimental to his well-being.

Growing up, I saw that their relationship was characterized by verbal and physical fighting and no displays of affection, and they still had the gall to tell us they loved each other. I'm grateful for what they did for me, but I have to call their relationship for what it was—two immigrants trying to make it in a new world.

Seeing the same dynamics play out in my own relationship, I couldn't blame my ex-wife for what she had also been indirectly taught. But I also couldn't see myself ending up in my dad's position with more education to choose differently.

Co-parenting was as big of a transition as leaving the military, if not more so. Through it all, my kids' welfares were my top priority, which is what motivated us to move to North Carolina.

Shortly after living independently, I was hired at Google. Google and my team there were a better fit for me, and I'm excited now to show up to work every day.

I took my daughter to the beach for a father-daughter trip. We ate at a restaurant on the pier that overlooked the ocean. The setting sun made a memorable scene. For the most part, I have spoken to my children like adults and never avoided difficult conversations.

"People say Daddy works a lot, but if it wasn't for me putting in that work, we probably wouldn't be able to enjoy a life like this," I said.

"Don't listen to them, Daddy. They're crazy," she replied.

That's when I found peace.

For the longest time, I felt like a distant father. Working all of the time made me feel like I was inconsistent as a parent. I felt guilty having to take calls at odd hours and on weekends. My only hope was that it would all pay off some day. Despite the failures and rough transitions, the way my daughter understood it made me feel at peace with my journey.

I hugged her, kissed her forehead, smiled, and thanked her.

December 6, 2023

I met a woman on the dating app Hinge, and we went on our first date on December 6, 2023. We had main character energy. The world around us felt like it was nonexistent. We fell in love a few weeks later and conceived a child together.

When my mother and I reconnected during my divorce, she told me the main reason she had opposed my first marriage was because by marrying early I would miss out on dating in my twenties and thirties and the personal growth that came along with that. My mother was right.

And I was ready to settle down. I realized there were things I could improve on, and I understood now what I needed in a partner.

I heard people say that the most important decision you will make in your life is choosing a partner. I didn't understand that until my current relationship. Half of my understanding was doing the work to realize that as a child I hadn't been given the best examples of love, and the other half was becoming a person my partner would fall in love with.

"My Truth in Its Raw Form"

Sonny Tosco reflects on telling his story

For awhile now I have been trying to write a book. I update my circle periodically about what I'm up to, and every time they sit back aghast at what has been going on.

I have arrived at a point in my life where I can sit back, synthesize, and attach meaning to the decades. Last summer, my daughter told me that everyone else is crazy. I sent my exes texts saying I found peace and I wished them well. Not until we are ready to heal can we afford ourselves the time to heal.

I approached my story as a long-form continuation of my usual Instagram posts. I'm at a place in life now where my priorities are clearly defined. I have my family, I have my career, and I have my health. I rarely entertain anything that deviates from these.

My story may be familiar, but it's unique to me. I work in the tech space and wonder about the impact of AI on society. I'm conflicted about it, as I have a career, and one of my duties is to my organization. But I cherish the things that make the human experience. Candlelit smoke sessions in papasan chairs are something that Gemini will need more than a data center to train.

Writing this has been therapeutic for me. I can comfortably share my truth in its raw form. I hope my truth inspires you to live authentically.

Chapter 24

Blessings in Disguise
by
Troy Baisch

Troy Baisch served for thirty years in the US Navy and retired as a chief warrant officer four. He served as a fire controlman for twenty years and supervised the operation and maintenance of shipboard weapons systems.

Troy is an alumnus of the Master of Business for Veterans (MBV) program at the University of Southern California's Marshall School of Business. He received a BS in management and computer information systems from Park University in Missouri, graduating magna cum laude. He holds an AS degree from MiraCosta College in San Diego.

After graduating from USC, Troy founded Charlie-Six Partners, a business-to-business promotional services firm. With the US Census Bureau, he managed the 2020 Decennial Census operations for North San Diego County.

Currently, Troy is accumulating flight hours for his flight instructor certification. He volunteers at the local animal shelter where he lives in Oceanside, CA.

Acacia Confusa and Okinawa

My story begins with a tree, *acacia confusa*, the small Philippine acacia. The park in my neighborhood was bestowed with a good number and variety of trees, including the acacia. There are also some picnic tables, a small playground, and a walking path paved around the perimeter.

Every time I walk my dogs to the park, I make sure to visit my special friend who lives there. It may seem a bit strange that my friend is a small Philippine acacia. Seen from a distance, my friend is unremarkable, even plain and ordinary-looking. It's not the tallest tree. That honor goes to the sycamores. And it doesn't have the most beautiful blooms. The jacarandas take that prize. Unlike the African sumacs with their lemony scent, the acacia doesn't stand out on the nose. There's not a whole lot to recommend about it.

But if you look closely, you can tell that the tree has been through some things. It's scarred. It has been climbed on and hacked at, broken and blown by storms. Lightning struck it. It's damaged goods. I thought the park's arborist might have to take it down, but it's still standing. There's no doubt in my mind that it is the strongest tree. The scars define its beauty. Despite what it has been through, thanks to its strong roots, a strength that is hard to comprehend, the tree still stands.

I can relate. I went through some things. It's comforting to spend time with others who have shared similar experiences, even if they are a tree.

My dogs like the park. It's a place to play, meet old and new friends and bark their disapproval at interlopers. When Gojira, my Shiba Inu, isn't terrorizing his sister, he likes to plop down in a prone position on the grass and keep watch on the world around him, alert and ready should he need to chase away any coyotes. His sister, Sylvie, is all corgi. She can't sit still and has the attention span of a toddler. She constantly runs off to explore some new thing, and I have to reel her in like a sportfisherman with a blue marlin on the hook. Whereas Goji tends to be quiet and contemplative, Sylvie is brash and loud. She's a talker, and as I frequently remind her, "Yes, I hear you! People in China can hear you right now!" Both dogs love to read the local pee-mail.

As a dog dad, I know I'm not supposed to play favorites, but I admit Goji holds a special place in my heart. He came to me when I was in a bit of a rough spot personally and professionally. I had just lost another beloved dog who had been with me for fourteen years, and I really wasn't sure if I was ready to get another one. Fate conspired to bring Gojira and me together. The minute I laid eyes on him, I knew he was going home with me. As a puppy, he reminded me of Godzilla, so I named him after the King of the Monsters.

I was unsure what the future had in store. I didn't get commissioned until after twenty years of serving in the Navy. I was a chief warrant officer three (CWO3), and because of changes made to time-in-grade requirements, I wasn't sure if I was going to get another look at the promotion board before I was forced to retire. Normally, a second fail-to-select (FTS) was the kiss of death, and I wasn't sure how the time-in-grade changes affected my chances of being promoted. I had resigned myself to the idea that my Navy career was over, and that by March, 2015, I would be retired.

So I applied to a dual bachelor's and master's degree program in international relations and Pacific Rim studies at the University of California, San Diego. I was accepted to the program for the Fall 2014 term. Around that time, I adopted Goji. My command worked with me on my schedule, so I registered for classes and started attending them when my professional life took a turn. I was the weapons department head at Naval Special Warfare Group ONE Logistics Support Unit (LOGSU-1). Our mission was to provide arms, ammunition, and explosives logistics and material support to all active and reserve SEALS on the West Coast.

One day in September, I was inspecting the ammunition and weapons storage facilities out at the Mountain Warfare Training Facility near Campo, CA. It was late in the day when I wrapped up, and I decided to head straight home to Oceanside. No sooner had I walked through the door when my duty phone rang. It was our executive officer. I thought, "Just my luck. The one day I decide not to drive all the way back into the office before going home. He's going to ask me to drive all the way back to Coronado to update him on my findings." But he was calling simply to congratulate me on being selected for promotion to CWO4.

I wouldn't be retiring the following March. I called my detailer at Navy Personnel Command to see what my options were for my next duty station. I loved my job at LOGSU-1 and would have been thrilled to stay there, but my relief already had his orders to report in January. So I was surprised when my XO told me that rather than keeping me in the San Diego area, he needed me to go to Okinawa, Japan, where I would take over as the XO at the weapons station there. I had been to Okinawa before for brief stops at the naval facility at White Beach. I didn't know much about the place. I heard from others it was a mixed bag: they either loved or hated it, and there was no in-between.

It was one of the best jobs I ever had, in one of the best places I ever lived. If it were up to me, I would still be living in Japan. Life in Okinawa was idyllic. Gojira and I lived in a new three-bedroom house next to Sea Lake Zakimi. The community was mixed Japanese and American: native Okinawans, retirees from mainland Japan, American service members and their families, and some American military retirees. I loved the neighborhood because it was simultaneously familiar and foreign. Goji loved it because there was a dog park a block away. We played there every morning before I went to work.

I knew I wasn't going to be stationed there forever. When I first arrived, the thought was comforting. The stories from people who hated their time there made me wary of the place, so the fact that I could make an escape at some point kept my attitude positive as I packed up my household to move overseas. After a while in Okinawa, when I had grown to love the place, the thought of leaving was heartbreaking.

I tried to work out a deal with my detailer that allowed me to stay in Okinawa for an extra year. I told him I was willing to stay even longer if he needed me to, and I even suggested ships in Yokosuka or Sasebo as compromise. Anything that would let me spend more time in Okinawa. That's when I found out about Title 10 of the United States Code, which organizes the US Armed Forces. Title 10 caps the time in service of chief warrant officers to thirty years and sixty days. It was frustrating. If I had converted to a limited duty officer (LDO) earlier in my career, I would have been able to stay in as long as I wanted had I continued to be promoted on schedule. I had thought about converting but ruled it out because I liked the cachet of being a warrant officer. Had I known about Title 10 (in retrospect, it's something I should have known), I wouldn't have hesitated to convert. In any event, I was going to be statutorily retired, whether I liked it or not.

Credits and Commitments

Okinawa really spoiled me. It took fifteen minutes to commute to work. I went home to eat lunch and take Goji to the park. On days of really bad traffic, the commute took me twenty-five minutes, tops. That alone had a profound impact on my future life choices.

I have lived in North San Diego County off and on between duty stations and assignments for three decades now. Before I moved to Okinawa, I had been commuting to San Diego. Traffic in the mornings wasn't bad, but you never knew what the parking situation would be on base, especially at Naval Air Station North Island. It might take an hour to get through the gate and find a place to park. You never knew. And getting home in the evening, no matter what, was always soul crushing.

Before I had to leave Okinawa, I accepted a position working in the material readiness branch of Littoral Combat Ship Squadron One (LCSRON 1) as a combat systems inspector. I wasn't sure if I wanted to work for the government again right away. I thought I might want more free time to explore other options. But I also wasn't sure what my retirement income was going to be, and I had bills to pay and things I wanted to do, and that meant I needed income for a while.

I had been previously involved in the process of hiring civilian employees. It can take months to onboard someone. Been there, done that. I didn't want to commit to a job knowing I might bail on it to pursue some other opportunity after only a few months. If someone had done that to me, making me start the whole hiring process over again, I would have been justifiably peeved. I didn't want to do the same thing to someone else. But that's exactly what I did.

I blame Okinawa. And the San Diego traffic. One day on leave, I had to drive down to the Naval Station. I got stuck in traffic on I-5 somewhere between La Costa Ave and Leucadia Blvd, and I thought, "What kind of idiot does this every day?" And I realized that for the lasts twenty years, I had been that idiot. I decided then and there to tell the guys at LCSRON1 thanks but no thanks. I knew they would be angry and disappointed, but I had to do what was right for me.

I drove to the Joint Education Center at Camp Pendleton and visited a counselor at Park University. I had worked on completing my undergraduate studies for years in fits and starts. I had taken classes on base. I had taken classes onboard ships. I had taken correspondence courses. I had credits from military training. I realized that I might not have to balance a full-time job with taking classes part-time. I had the Post-9/11 GI Bill whose housing allowance helped offset the financial hit of committing to finishing my degree. I had taken classes at Park University in the mid-90s, and I re-enrolled for 2017.

My credits were all over the place. I had so many interests and never settled on a major. I had taken a bunch of classes in different departments but not enough units in one discipline to satisfy the requirements for graduation. I finally decided to major in management and computer information systems. It was in my wheelhouse and complemented the jobs I had done and the skills I had acquired in the Navy. I finally graduated in April, 2018, with more than 250 credits on the books. I could have saved myself a lot of time and grief if I had committed to a major, but I think there's benefit to education for education's sake, and I'm glad I learned about so many different things.

With my bachelor's degree, I joined the Master of Business for Veterans program at the USC in July of the same year. It was a journey decades in the making.

Destiny

On New Year's Day, 1973, the USC Trojans demolished the Ohio State University Buckeyes 42–17 in the Rose Bowl. I was nine years old. The sportscasters on TV kept talking about the "Men of Troy," which I liked to hear. I was teased a lot for having a weird name—Troy. But there on the TV were all the fans at the Rose Bowl cheering for guys with the same name. From that moment on, I wanted to be a Trojan and attend USC more than anything else in the world. As I got older, it became an impossible dream. I lost my way because I lacked the confidence to succeed.

As a youngster, school was easy for me. I was a straight-A student from elementary to junior high school. Then I got complacent. I was put in a talented and gifted (TAG) program after moving to California as a fourth grader in 1972. I'm sure the people who created the program had the best of intentions, but in my case, it made me entitled and lazy. If I could score a B by just cruising along, what was the point of putting in any effort? In high school, the work was more challenging, and my grades were impacted. My confidence suffered along with my academic performance. I made poor choices and worse friends. By the time I was a senior, my future appeared to be a dead-end job and community college if I was lucky. No way was I ever going to USC! I was accepted to Chapman College in Orange, CA. I enrolled for the fall and dropped out after one semester. I went to California State University, Fullerton, the next year and dropped out after one semester. Quitting became a habit, as did a lot of other self-destructive behaviors.

When I was growing up, my on-again, off-again stepfather was a real Dr. Jekyll and Mr. Hyde. He drank. He drank a lot. It was the only thing he was ever good at. On his good days, he was the coolest, kindest person. On his bad days he was angry and abusive. And he had a lot of bad days. Coming home in the evening after staying out with friends, and later, coming home from work, I walked through a minefield in my own home. I never knew what I was going to get when I walked through the door, and I never knew what was going to set him off.

It's hard to believe that I followed in the footsteps of someone so damaged, who I detested so much, but I learned since then that children and adolescents are sometimes more likely to identify with and emulate a powerful alcoholic parent because they learn that alcohol can make them feel powerful. Maybe that's what happened in my case.

After more years aimless and adrift after high school, I decided that a radical change in my life was necessary if I was ever going to live up to my potential and achieve anything meaningful. I'm reluctant to use the word "destiny," but if anyone was destined to join the Navy, it was me. Thanks to my grandfather Bill Padrick and his brother, my great-uncle Lyman Padrick, the sea had been calling to me for several years.

My grandpa Bill was a merchant seaman aboard the SS Heredia, a bulk stores carrier operated by United Fruit Company. He used to tell me stories of his trips across the Pacific from Portland, OR, to Shanghai and Hong Kong and the Philippines. There was something romantic about exotic ports overseas with different foods and cultures and smells and sights to see. Whenever I stayed at his house, he woke me up by hollering in Cantonese. It was the same spiel that had called him to breakfast aboard the SS Heredia after the ship picked up a cook in Hong Kong.

My great-uncle Lyman was also a sailor. He enlisted in the US Navy in 1936, and he was assigned to the Asiatic Fleet when the USS *Panay* (PR-5) was sunk on China's Yangtze River by Japanese aircraft in December, 1937. My great-uncle spent the bulk of his service aboard salvage ships doing repair and recovery work. He survived the war and was promoted to chief shipfitter.

He was also a master diver. After the war, he dove to the wrecks at Bikini Atoll following Operation Crossroads' Able and Baker nuclear weapon tests in 1946. Uncle Lyman was a larger-than-life figure, and the stories he told at his home in Lemon Grove captivated me. Most kids my age looked forward to visiting the zoo or Sea World when they visited San Diego; all I wanted to do was hang out with Uncle Lyman and listen to his sea stories. If being a navy chief meant being like him, it would be a real honor. When my grandpa Bill pinned my anchors at my promotion to chief petty officer, in 1995, I thought about how nice it would have been if Uncle Lyman had lived to see the day and do the honors himself. He passed away in 1984, two years before I enlisted.

I enlisted at the age of twenty-three. My first memories of recruit training are of waking up at 0300 hours to the sounds of crashing garbage cans and screaming drill instructors. Everyone was terrified and disoriented by the sudden shock. It was our first lesson, and an important one: bad things happen when you are most vulnerable, and you must overcome the initial shock, gather your wits, and do your duty.

The Navy changed my life. It became the one thing I had succeeded at that was challenging and required honest effort. Hard work and success turned out to be as habit-forming as mediocrity and laziness had been. The Navy also changed my worldview. Traveling the globe and living overseas grew my appreciation for all that we have in the United States, and it also gave me a deep and abiding respect for other cultures and their customs.

I owe my success to others. Leaders and mentors showed me the way and challenged and encouraged me to do my best, even when I believed I wasn't up to the task. I was given jobs I didn't want in places I didn't want to go to, but the tours I thought would be the worst turned out to be the most personally and professionally rewarding.

The military is like the ultimate team sport, because real success means working with others to achieve great things.

When I was promoted to chief petty officer, I was assigned to the staff of Destroyer Squadron 23 (DESRON 23). It was there I got a real education about what it meant to care for your people, and that caring for your people began with caring for yourself. Our commodore was Captain Barry Costello. He was ultimately promoted to vice admiral and command of the United States Third Fleet. Commodore Costello was one of the most inspirational men I ever met. An alumnus of Holy Cross University, he was a great example of the Jesuit ideal of the servant-leader. He had absolutely zero ego, except on the golf course and at the poker table, where his trash talking would shame the devil. The absence of ego made him a skilled negotiator. If you went to him with a solution to a problem, he never had to be right, and he never had to be the smartest guy in the room. If you could defend your position, that was that. If he disagreed with you or was going to overrule you, he had a persuasive way of talking you through the problem and the solution he had in mind, and you would walk away feeling as if his idea had been your idea all along.

The first conversation I had with him at my check-in interview set the tone not only for my tour on the staff but also for the remainder of my career. We talked at length about our backgrounds and interests. It was standard stuff at first. Then he asked me about my goals while I was with the squadron. I told him I wanted to learn as much as I could and work hard while I was there. I went so far as to tell him I was especially interested in learning more about Anti-submarine Warfare (ASW) since that was the primary mission of a destroyer squadron. In reality, I thought ASW was about as interesting as watching paint dry.

It was the commodore's turn to lay out his goals for me during my tour at DESRON 23, and what he told me was not what I expected to hear. He said, "Chief, I've got three priorities for you while you're here: Take care of yourself, take care of your family, and take care of your job, in that order. Take care of yourself. Physically, mentally, spiritually, education, whatever you need to do to be happy and productive and be your best. If you can't take care of yourself, you can't take care of anyone else. Next, take care of your family or your girlfriend or anyone else you care for. You can't do your job properly if you're always worried about the folks back home. If you take care of yourself and you take care of your family, then the job takes care of itself."

Doing those things was not as easy or quick as I thought they would be. As they say, old habits die hard. For a long time, I had felt like I was losing more ground than I was gaining. Two steps forward, one step back? More like one step forward, two steps back. I still struggled with the demon rum. I was a pretty high-functioning drunk, but I was a drunk.

The Blessing

On Thursday, June 17, 2010, I suffered a heart attack while I was on transfer leave between duty stations. I had just checked out of my previous command, USS *Halsey* (DDG-97), on Monday. My new command, USS *Chancellorsville* (CG-62), was getting ready to leave San Diego for Pearl Harbor to participate in the biennial Rim of the Pacific (RIMPAC) exercise. I still had some stuff in my old stateroom aboard which I needed to move into my new stateroom. I had been on sea duty for seven consecutive years and three deployments, so I took leave for a couple weeks, intending to officially report aboard *Chancellorsville* after it reached Hawaii.

I got up at around 0700 hours with the worst indigestion ever. Or so I thought. Thanks to a poor diet and too much alcohol consumption, indigestion and acid reflux were nothing new, but this was something else entirely. I ate some Tums and debated whether I should drive down to San Diego to move my stuff or just stay at home. The indigestion did not improve. I ate more Tums. The indigestion did not improve. By the time I had consumed a third of the bottle of Tums, I started to get concerned that it was something other than indigestion. Even though I didn't have the classic symptoms of a heart attack, I thought I might, in fact, be having a heart attack.

I debated calling 911, but I wasn't thinking clearly, and I was worried about getting stuck with a big bill should it turn out to be nothing and the Navy wouldn't cover it. I lived less than a mile from Tri-City Hospital in Oceanside, so I considered driving there, but again I was worried about getting stuck with a bill I couldn't pay. I decided my best bet was to drive myself to Naval Hospital Camp Pendleton. It was fourteen miles from my house, but I thought I could make it there without any trouble. I cannot stress this enough: driving fourteen miles when you are having a heart attack is insanely irresponsible!

I made it to the emergency room, and they asked me what the problem was, and I told them it was either the worst indigestion ever, or I was having a heart attack, maybe. When they asked me about my specific symptoms, they were also unsure what was going on with me, because I didn't exhibit the classic symptoms of a heart attack. Thankfully, they erred on the side of caution until they could rule out a heart attack. It was a strange morning. They did several tests, including sending me to radiology for X-rays and a CT scan, but nothing was conclusive.

Finally, the cardiologist told me I was being sent home, and he would see about getting me a portable heart monitor to wear to collect more data. He told me to sit tight, and he left to work on the discharge paperwork. Twenty minutes later, he returned, and I asked, "So Doc, am I good to go? Can I take off?"

"Actually, you're not going home today at all," he said. "We're trying to decide whether to send you to Balboa [Naval Medical Center, San Diego] by ambulance or by Life Flight."

"I thought I was going home?"

"No, your second enzyme test came back elevated," he replied, "and we need to send you down to the cath lab [cardiac catheterization] in San Diego. We don't have the capability up here, and we need to get you down there right away."

The flight down to San Diego was pleasant. I had been on dozens of helo rides, but never in such a good seat. The Life Flight helo had windows in the rear, so I enjoyed good views on the flight to Balboa, even though I was lying on my back on the gurney and looking at the world upside-down. Later, the invoice for the flight priced the twenty-minute joyride at $26,000, billed to taxpayers of course.

When we got to Balboa, they rolled me into the cath lab and placed a stent in my left anterior descending (LAD) artery. I was drugged but not unconscious, so I got to watch the whole procedure play on the monitor over my head. Absent the drugs, I would have been freaked out watching them poke around my heart. What I had blocking my artery they call the widow-maker. The next day, when I was still in the ICU, the cardiologist came in and said seven words I will never forget: "You can quit, or you can die."

I knew what he was talking about. The smoking. The drinking. Bingeing on junk food after a night of smoking and drinking. I can't count the number of times I had gotten "the lecture." Every time I went to a doctor: "Do you drink? Do you smoke? You should quit." When I went to a dentist: "Do you drink? Do you smoke? You should quit." We had general military training (GMT) sessions about the dangers of smoking and drinking.

I had heard it all a million times, and it never made a dent. But the way the cardiologist said those seven words finally hit the mark. His tone was neither angry nor condescending. It was matter of fact. His tone was that of someone who knew. He had probably said those same words to so many people who hadn't taken them to heart and died, that he was beyond the point of trying to convince anyone. I haven't had a drop of alcohol or a cigarette in thirteen years.

My heart attack was a blessing in disguise. I wouldn't be where I am today without it. I was young enough to survive it but old enough to know I couldn't put things off until someday, because someday might be too late. As soon as I was on the path to recovery, I went to a tattoo shop and got Tommy Trojan tattooed on my left shoulder. I believed it would motivate me to finally do the work necessary to attend USC. Because if I didn't do the work now, I would look stupid lying in a coffin with a Tommy Trojan tattoo on my shoulder with nothing to show for it.

Finding Identity Again

One scene in *The Hurt Locker* that nails the experience shared among many veterans, myself included, comes near the end of the film. The main character, played by Jeremy Renner, is an adrenaline junkie on a US Army Explosive Ordnance Disposal team, and he has returned home from deployment to Iraq. In the scene, he shops with his wife in a grocery store for the first time in more than a year. The pivotal shot shows him standing in the breakfast cereal aisle, paralyzed by all the choices confronting him. For me, and for many other vets, this was one of the strongest representations of being a veteran ever filmed because of the way it captured a certain common feeling. A lot of service members returning home from deployment, or service members retiring or separating from the military, experience the same feeling as they try to reintegrate into American civil society. Military society and culture present institutional and personal certainty and security, however flawed. The limited choices in military life can decrease the fear of making the wrong choices. For many veterans, American civil society, with all of its choices, is neither comforting nor liberating. It is often terrifying.

In the movie's next scene, Jeremy Renner's character returns to Iraq with a new team. As great as his love for home and family may have been, something pulls him back to the killing zone. Like the character, there are a lot of transitioning veterans who wonder, "Will I ever be important again?" or "Will my work ever matter like this again?" Each of us, regardless of rank or length of service, was somebody while we served. We were important people doing important things. Would we ever feel the same way once we left the service?

About the time I retired, Walmart did a "Salute to Veterans" promotion in their stores that highlighted the stories of veteran employees. One really stuck with me. This veteran, a Walmart store greeter, had piloted a B-24 Liberator with the Eighth Air Force and completed twenty-six bombing missions over Nazi Germany in WWII. And there he was now, greeting people and checking their receipts at Walmart. I thought, "This guy survived the war for... this? To end up at Walmart?" It was an unsettling thought. Going from B-24 pilot in his youth to Walmart greeter in his old age seemed like a real downgrade. Did the same fate await me? Was I going to go from being somebody to being a Walmart employee?

As my time in Okinawa and the Navy was winding down, the thought that I was going to lose my identity, and that I was no longer going to be somebody, weighed heavily on my thinking and decision making.

While I finished my undergraduate work at Park University, I looked at several graduate programs. USC and its MBV program were not among them. I wasn't aware of any graduate program tailored specifically for veterans. Even if I had known, I couldn't afford a program at a top-twenty business school like USC Marshall.

Then a weird thing happened. I was at home one day when my doorbell rang. There at my door was my neighbor, who happened to be part of the MBV Cohort V. He and his wife lived one street over. Our house numbers were identical. The delivery guys were constantly mixing up our houses. I had actually never met him before. He was looking for his delivery. He saw all the USC memorabilia in my house, and he asked if I went to USC. I told him I didn't, but I had always wanted to go and never had the chance. He told me about the MBV program and invited me to an information session. So even before I attended USC, the Trojan network was already a powerful resource!

The information session was held on campus. James Bogle did the bulk of the presentation, and he convinced me that the MBV was right for me, but I was still unsure if I was ready. I had doubts I could do it. Then Dr. Turrill shared his thoughts on the program, its history, and its future. He talked about the Trojan family and service to others. It made a pretty powerful impression, and I decided then and there that I was all in. When I heard Dr. Turrill's presentation, I understood that the only thing keeping me from doing anything was me, myself.

My experience in the MBV program was all that I had dreamt it would be and much, much more. The quality of instruction was exceptional—but I could have received quality instruction at any other number of business schools. The people and relationships that came with the MBV program made it special. The real blessing of the Trojan network is not the people you can call upon all over the world when you have a question or when you need assistance. The real blessing is the people all over the world who call you when they have a question or need help with something because you're a Trojan and they know they can rely upon a fellow Trojan.

The opportunity to bond with such a diverse group of veterans can't be found in other programs. Having a mixture of professionals who were long-separated from the service, or recently retired, or currently transitioning, or remaining on active duty or reserves; and who had varying experiences, from junior enlisted to field-grade officers, gathered in the same room and on the same team was the secret sauce that made the program such an enriching experience.

I wanted to try my hand at owning my own business, so after finishing the MBV program, I started a printing company that specialized in business-to-business promotions, specifically customer and employee recognition programs. It was a blessing and a curse. I loved the work, and I loved being my own boss. But the busier I got, the more difficult it was to decide if the business belonged to me or if I belonged to the business. To raise more cash to buy new equipment to expand my product lines, I started scouting government job listings for any fixed-term positions that would be suitable for a second job.

I found a position with the Commerce Department managing decennial census operations for North San Diego County. We were responsible for about 427,000 housing units and 1.2 million residents. The old hands at the office gave us a rundown on the operations and the challenges we could expect. But none of us could have been prepared for what it would be like conducting the census in the midst of a global pandemic.

When COVID hit in early 2020, I had 150 people on staff, and I was preparing to bring on another 1,400 for the major door-to-door canvassing operation in the summer. At first, it looked like we might weather the pandemic, but we were eventually forced to suspend operations. What we thought would be just a two-to-three-week delay turned into a hiatus that pushed the schedule back several months. I was able to keep a core staff, including my managers, but I had to furlough the rest with no idea if or when I could bring them back on board, or if they would even want to come back given the circumstances.

I went through a lot of crazy and challenging and even some scary things during my Navy career, but gathering the team and telling them we were shutting down indefinitely was the hardest thing I had to do in my professional life. These people had worked hard. Their reasons for being there were as varied as those of the people I served with. Some wanted to serve their country and their communities. Some were retirees looking to keep active. Some really needed the money, and for them the census meant paying the rent and putting food on the table.

When the census was finally finished, I was finished, too. I was tired of the printing business running me instead of the other way around. I missed out on things I wanted to do with friends and family because we had orders to fulfill. I had a small crew who were good workers, but it was tough to pay them a fair salary once wages began to rise. COVID made it really tough for a lot of businesses, and none were too eager to spend what little money they had on promotional materials.

I decided to sell the business and buy a plane. I finally figured out that I didn't need to prove anything to anyone anymore. Not to me, not to my family or friends, not to anybody. My time was my own, and what I really wanted to do was fly.

Aviation has been a lifelong passion. In my youth, I thought I might join the service as a naval aviator. I had taken flying lessons when I first moved to San Diego in 1992. I wish I would have stuck with it so many years ago, but it was an expensive hobby, and San Diego was an expensive place to live.

When I finished with the census and sold the business, I thought, "If not now, when?" I went back and finished ground school and took the written test. Because of my medical history, it was going to take some work to get my third-class medical certificate. Without it, I wouldn't be able to fly solo.

Flying suits me. I like the freedom, but it also requires skill, precision, attention to detail and adherence to procedure. It's liberating but comes with a good dose of discipline. I like to go over to the airfield and visit Bella, my 1972 Beechcraft Musketeer, to make sure she is washed and waxed and looking her best despite her age.

I have been working on my commercial pilot rating. It has been a challenge. I am the primary caregiver for my ailing mother, and it's hard to divide my time between caring for her and her home, caring for my home, and studying and practicing the maneuvers for the exam, known as the "checkride." But so far, I have stayed on top of everything. The goal is to complete my certified flight instructor rating so I can share my passion with others. It's a win-win-win proposition for me: I will get to fly, teach others to fly, and get paid for doing it. And the best part, I can do it (mostly) on my own schedule, so I will have plenty of time to spend with my pups and with a certain acacia tree in the park near my house.

"Ordinary People Who Experienced Some Pretty Extraordinary Things"

Troy Baisch reflects on telling his story

When Drs. Arámbula and Turrill first approached me about this project, I was pretty excited and eager to get to work and share my story. Eagerness turned to apprehension. In order to tell my story properly, I would have to share some things I wasn't so eager to open up about. I began to wonder if I had taken on a commitment that was too far outside my comfort zone.

Once I got over my initial apprehension, the real challenge became what to share and what to omit. I chose to omit a few details that are too raw or too personal to include. Either they didn't add anything to the story, or they were too complex and would take up too much space and time to relate properly. One story was both too personal and too long to tell properly: April 19, 1989, when forty-seven of my shipmates perished in an explosion in the center gun room of turret two. Even after all this time, I feel a bit ashamed. I feel lucky that we weren't all killed, but I think about what life might have held in store for the forty-seven had they lived.

It was important for me to share my story because there are still a lot of people who don't understand the challenges and the triumphs that veterans experience as they transition. Some of those people are civilians who have never served. Some are the family members of veterans who have no real idea of what their loved ones went through or are going through. A veteran's spouse, significant other, and/or children may not know the entire story.

But another group of people who don't understand the challenges of transitioning veterans are those service members still on active duty who may have no idea or may have serious misapprehensions about what the future holds for them. There is no one single story that is going to resonate with every reader. That's why a compilation of different experiences is important, so that readers find a story that resonates with them.

In participating with this project, I also wanted to dispel a few myths about veterans. We are often portrayed in media as damaged humans who have been so broken and devastated by our experiences that we can't function properly in civilized society; or as wackadoodle "vetbros" who are so "thank us for our service" gung-ho about the military and our superior patriotism; or as disgruntled, angry powder kegs who got a raw deal from "the gubmint." I'm not suggesting that those people don't exist. But they are not representative of the community of veterans I know. Most of us are ordinary people who experienced some pretty extraordinary things. I am grateful for each and every experience, good and bad. In some ways, the bad experiences were the best because they taught me that I am a lot more resilient than I would have ever thought possible, especially before I enlisted.

Some veterans talk about how the service is tough and wears you down. I spent the better part of my life in the service, so I admit that I don't have much to compare it to, but I think a lot of the challenges we chalk up to the service are just life's normal challenges dressed up in a spiffy uniform. There are some veterans who feel their challenges are greater than others', especially compared to the challenges of those who have never served. The point is: life is hard, no matter who you are or what profession you choose. Be kind to yourself and to others and give the people around you, and yourself, some grace.

Chapter 25

What I Left Behind
by
Ty Smith

Ty Smith is a decorated US Navy SEAL and accomplished technology entrepreneur. His twenty years of service included six combat tours to Iraq, the United Arab Emirates, and Afghanistan. He received military decorations including the Bronze Star with valor, the Joint Commendation Medal with valor, and the Combat Action Ribbon.

Ty began his naval career as a translator. After 9/11, he was selected for Basic Underwater Demolition/SEAL (BUD/S) training in Coronado, California, and he completed successful tours of duty with SEAL Team Eight and SEAL Team One.

During his time with ST-1, Ty's leadership of Echo platoon resulted in more than twenty valor awards and a Navy Unit Commendation for SEAL Team One.

Ty received a BA in organizational management from the University of Arizona Global Campus (UAGC) and a Master of Business for Veterans (MBV) degree from the University of Southern California.

He is currently the CEO of CommSafe AI, a venture capital-backed technology company.

A Protector of People

I was born and raised in East St. Louis, IL. Because of the trauma I endured in that place, growing up in that house in particular, I have not been ready to address how my experiences impacted me as an adult. But I'm ready now.

Over the course of a few years, I have had time to work on myself with the help of several exceptional professionals, my family, and my friends. And it's why I have come to realize that who I am today is a result of the experiences of the traumatized little boy inside me.

I felt for a very long time that I was inherently a protector of people, and that was why I became a Navy SEAL. But I was not looking deep enough. I know now that someone should have protected that little boy growing up in East St. Louis. As the little boy grew older, the fear he lived with constantly turned into his rage. Rage became focus and grit. His focus and grit made him believe that with sheer will, anything was possible.

As a Navy SEAL, I needed those emotions. And I was blessed that I did not lose my humanity during combat. Because I became a protector as a result of a longing to protect my adolescent self, hurting the innocent simply was not programmed in me. No wonder that when a twelve-year-old Ty Smith watched *Navy SEALs* starring Charlie Sheen, he was bitten by the bug. I remember that day like it was yesterday.

We lived in a quickly dying community in St. Clair County called Parkfield Terrace on the outskirts of ESTL. When we first moved into the neighborhood, I was seven years old. The neighborhood was charming and an upgrade from the neighborhoods where I spent the first six years of my life. Shortly after we moved in, so did drugs and violence. Within a few short years, Parkfield Terrace became one of the most dangerous neighborhoods in the county.

Most of the houses in Parkfield Terrace looked alike. We lived in a single-story, 1,200 sq. ft., three-bedroom white house with red trim. My little brother and I shared a bedroom. My late sister, Antinia, had her own room, and my mom and stepfather occupied the master. We had decently sized front- and backyards, like most of the Midwest. There was a driveway big enough for two cars that ran from the street to the backyard. Behind our house was a cornfield the length of the entire neighborhood.

I spent summers running through those stalks with my friends. Behind the cornfield were some woods where my friends and I played special forces.

One day, while my sister and little brother were playing in her room, and my stepfather was at work, my mom and I watched *Navy SEALs* for the first time. My mom was sitting on the couch, and I sat cross-legged on the floor with my shoulder leaned against her leg. That day, I fell in love for the first time. At the end of the movie, I looked over my shoulder and said, "Mama, that's what I wanna do when I grow up. I wanna be a Navy SEAL." My mom could have said anything. She gave the best response she could have given. She said, "Son, that would be the hardest thing you've ever done in your life, but if you want to do it, you can." She taught me a critical lesson that day.

What I saw those SEALs doing on screen was as real to me as any superhero I could imagine. And if I could be like those guys, I could protect myself and others like me. I knew it in my heart. To this day, I tell people that I made a few good decisions and ten times that number of bad ones, but the best decision I ever made was joining the US Navy.

Rage

Growing up as the son of a woman police officer in ESTL was hard. I got bullied a lot. Weekly beatings at the bus stop and around the neighborhood were common. We didn't have a lot of money, therefore my clothing was an easy target for bullies and everyone else. Kids are cruel.

Several of my aunts also lived in Parkfield Terrace with kids of their own. One day, one of my cousins argued with the boy next door, and he called her a name. My aunt and the boy's mom also got into an argument. The boy's mom sent him outside. I was hanging out in my aunt's house, and she called me up and sent me outside to face the boy. We met face to face in the front yard. I was a tough and confident seven-year-old kid. But when that boy caught me in a standing front guillotine choke, I started choking on the bubble gum I was chewing, and for the first time in my life, I thought I was going to die, and my spirit was broken.

I don't think it was the beating that really damaged me. It was the fact that I had to go home, where I was bullied by my stepfather. I could not go home and talk to my dad about what had happened, try to understand and grow from the experience, and keep my confidence intact. Instead, I went home to more trauma. A lot more.

Going to school and living at home were living nightmares. Things got better when my mom finally divorced my stepfather when I was thirteen. But the damage was done. Not only did I not know a healthy home, but I also did not have an example of what a successful parent or adult looked like. If I had a nickel for every time my stepfather told me I was stupid and would never amount to anything, I would have retired long ago. It took me until my early forties to forgive him. He was only projecting his own insecurity and trauma onto me.

Things did not get better until I was a sophomore in high school. By then I had been on the wrestling and football teams for two years, and I had discovered the strength and fire within me. I loved football but I sucked at it. On the wrestling team, I discovered the fighter, the killer, inside me. The protector in me really started to blossom.

One Saturday, after playing sandlot football with my friends and other boys from the neighborhood, one of them started in with the bullying, and just like that, I was finished tolerating bullies forever. Out with the fear, in with the rage. There was no thinking, talking, or arguing, there was only violence. I fought aggressively. I was completely unforgiving. In great shape, wrestling shape, I was relentless against the boy. I was no longer afraid. He paid for the fifteen years of fear that others made me feel. We fought on our feet and on the pavement, in the middle of the street, until two neighbors ran outside and stopped the fight. Everyone declared I won. I was different from that moment forward. By the time I joined the Navy at seventeen, I had a serious chip on my shoulder.

Forever Changed

The twenty years I spent in the Navy were an absolute dream, and I wouldn't change a single day. I would do it all over again, it was that good. During my time in the SEALs, I got exactly what I dreamed of and more. My combat experiences were epic, and the men and women I served alongside were the highest caliber of human beings I ever got to work with.

Seven years into my transition from the military, I am exploring the traumatic effects of having lost so many teammates in the Global War on Terror. I also have to acknowledge the times I myself almost met my maker, and how those experiences continue to affect me as a human being. The times I witnessed a child get hurt, those are the experiences that harmed me for good. I really look forward to talking to God about that some day.

I am learning how to break the habit of living in constant vigilance. Living in a hypervigilant state became my default when I was seven, and it sharpened over the course of a two-decade career as a warfighter. I never want to lose my killer instinct. I don't think I ever could. But I want to live a more normal life where I can be present for my friends and family.

We have all survived traumas. I can give you only a glimpse of a fraction of my own. I want to leave my history behind me so I can live a future of healing. My survival depends on it.

I am forever changed from combat. During my career as a SEAL, I deployed six times to Iraq, Afghanistan, and the United Arab Emirates. I experienced a tremendous amount of combat and loss. I survived more than a hundred gunfights, several green-on-blue attacks (in which our Iraqi or Afghani allies turned their weapons against us), numerous IED attacks, small arms fire ambushes, and the deaths of fifty teammates.

If I had not developed a vigilant state as a kid, it would have certainly developed early on in my naval career. I was a hammer and everything else was a nail. I couldn't drive down certain streets in the United States without feeling as if I was entering a chokepoint, which might come with deadly consequences like being blown up or ambushed.

I hadn't slept normally in several years before retiring from the military. I learned to live on a vampire shift, sleeping all day and hunting all night. When I did sleep, I was back at war. A few months into my transition, I was dreaming of gunfights. My wife would wake up to me violently moving in my sleep. She saw my trigger finger twitching in my sleep as I dreamed.

I became an emotional wreck. When something set me off, I immediately went to 15 on a 10-point scale. If someone brought fire against me, I returned the favor with a nuclear bomb. The Navy taught me to detach from the emotions I felt losing my friends one after the other and continue the mission. Not acknowledging the hurt taught me to ignore and override not just my own feelings but the feelings of others. I felt guilty for being alive when so many of my teammates, who were better than me in so many ways, were dead. I was angry and sad all the time, and I couldn't explain why I felt that way.

Then came the shame and remorse when I saw how badly I had treated people when I didn't understand my behavior. Realizing how I made others feel made me want to destroy myself. I'm a protector of people, even if it means I have to protect them from me.

I lost and gave away time I can never get back. My younger sister died at the age of thirty-one. I missed seeing most of her life. I never once thought that I wouldn't get the opportunity to make up that time with her. What hurts the most is that my sister and I grew apart because of the changes I underwent in the Navy. The person I grew to be pushed her away. She died peacefully in her sleep in 2015 while I trained in Fort Bragg, NC, with a special mission unit. Two years later, I lost my three-year-old nephew in a tragic accident. I never got a chance to be his uncle, and it hurts more than I can find words to explain.

No Time for Monsters

Depending on your job in the military, it was tough to find meaningful employment during the first years of your transition. Some veterans struggled to translate their military skillset into something valuable to the private sector. Also, for so many veterans, we had to learn how to communicate with civilians again. Often, veterans bounced from job to job for several years after transitioning from the military.

I learned that expanding my network and my tribe was critical, especially as an entrepreneur. "It's not what you know, but who you know" was one of the biggest lessons I learned. It was particularly challenging going from having the support system of the military to not having a professional safety net. I also struggled with wondering where all my friends had gone. My active-duty teammates were continuing with their military careers and deployments. Meanwhile, I was learning about human resources.

My friends were still living their purpose, and I was trying to figure out mine. Defining, or rather redefining, my purpose became a critical part of my transition. I came to understand that my purpose wasn't to be a SEAL. Protecting others had been my purpose all along, and my heart will always be on fire with it. That discovery allowed me to recenter my focus. I decided to build businesses that allowed me to protect others. I learned to think of my purpose as where I am balanced, my center. If I am not living my purpose, I am not on steady ground, regardless of how good everything else in my life appears.

It was really painful and confusing to discover that I no longer knew how to communicate with regular people like my family and friends. And it was after I stopped drinking alcohol that I discovered how poor my communication had become. In the military, there was very little room for negotiation when things needed to get done. There were orders, and you followed them. The military put little focus on critical communication skills like sympathy, empathy, and vulnerability. I felt as though I had been brainwashed into being an asshole.

It broke me to learn I had hurt those closest to me by behaving like a monster when a monster wasn't needed. When I discovered how much of a bully I had become, it hurt more than any pain I had felt in a long time. Shortly following that realization was the first time I almost took my own life. The protector had become the person everyone needed protection from.

I had to first recognize the problem. I was the problem. Then I had to work on myself, which included accepting the fact that I wasn't strong enough to do the work alone. I needed help from professionals like marriage and family therapists, counselors, psychiatrists, psychologists, executive coaches, and even a spiritual advisor. More than anything, I needed grace and patience from my loved ones. To get the help I needed from my loved ones, I had to do something I had stopped doing years before. I needed to talk to them. I needed to share with them my experiences and my feelings. I couldn't hold on to these things any longer. My family needed to understand the person I had become so that they could find their own empathy and grace. And I needed to tell them how sorry I was for the person I had become. Then I had to make my effort consistent and permanent.

What Have I Done to My Family?

One night in December, 2015, a few weeks after the death of my sister, and less than six months before I retired from the navy, I went out to dinner with my wife and son. It seemed like any other night. I wasn't smiling from ear to ear or anything like that, but I also didn't feel like anything was wrong. Not until midway through dinner, when I thought, suddenly, "Dear God, what did I just step on, and why did it make a clicking sound?"

Post-traumatic stress feels like trying to survive a mission of capturing or killing an enemy of the highest caliber, someone who is possibly more proficient at killing than I am. That's how I felt on the night I acknowledged that something was terribly wrong with me. I had arrived at what I thought was the end of my life. The stress was on par with being a one-man army in Afghanistan, stealthily infiltrating a packed enemy compound at night to assassinate a high-value target, but with no communication with friendlies and no quick response team to save me if things went wrong. "Dear God, what did I just step on, and why did it make a clicking sound?"—followed by a prayer. I was eating in a restaurant in San Diego when I realized just how much shit I was in. One wrong move, and boom, I would be dead.

I stood up from the table and kissed my wife goodbye, to her surprise. I looked at my son and told him I loved him. I walked out of the restaurant and called a yellow cab. I asked the driver to take me to Naval Medical Center. The cab dropped me off at the base's front gate, and I walked the rest of the way to the emergency room.

A very kind woman in a nurse's uniform sat behind the desk and asked me, "Sir, can I help you?" Not meeting her eyes, I said, "Yes, ma'am. I need help, and I need it right now." That was all I needed to say. She came around the nurse's station to accompany me. The command psychologist joined her. I didn't feel alone anymore. I don't know how I knew so surely that night that I would kill myself. I just knew. It came out of nowhere like an assassin.

At my worst, not every day felt terrible; in fact, I thought most days were good. But I wasn't myself. For the first few years of my transition, I had done what came naturally to me—I fought. I fought the PTS as if my life depended on it. But I was going about it all wrong, and it nearly killed me. I had to let the PTS have its way with me. I learned to acknowledge my feelings and understand them. I learned to be mindful of my emotional state, especially during times of discomfort, like when I was communicating with others. Because the more I fought, the harder the PTS hit back, and it always hit harder than I could. I had to learn to surrender and show myself grace. I had to understand that grief will decide when it is finished with me. In combat, service members lose something they can never get back. I had to make some serious admissions, not just to myself but to my loved ones. They had no idea who I had become. They had no idea about the scars I had received. When I left my family at eighteen, I was the timid skinny kid from ESTL. I had retired from the Navy as someone else. I retired as a professional killer who was really good at his job.

I tell the truth in hope there is healing on the other side. What I did to my relationships is what nearly killed me. I was arrogant, abrasive, misunderstanding, misunderstood, harsh, critical, hard to love, and hard on my loved ones. I was a monster. And I didn't recognize what I had become because I was so accustomed to being around other monsters.

I raised my children as if they had enlisted in the military. I was so hard on them, especially my son. As if my absences weren't painful enough, when I was present I wasn't the person they needed. I wasn't soft and gentle with my wife. I wasn't understanding with my son. Instead of simply being his father and trusted advisor, I treated him as though he was a SEAL training candidate. I really messed that up for him, and I regret it more than anything. When my daughter was little, I didn't talk to her enough.

I'm surprised my family didn't leave me. In my defense, I was going through so much. I had just lost my sister, and I was coming off two of the most action-packed combat deployments of my career. I lost a lot of friends on one of those deployments (twenty-plus in one mission). The anniversary of their deaths is August 6. I was also struggling with traumatic brain injury (TBI) from all the blasts I survived. I had a chemical imbalance in my brain that I wasn't nearly mindful enough about.

There were so many times that I hurt my loved ones so badly. I remember having drunken conversations with my family during which I was abrasive and rude. I fumbled my relationship with my little sister, selfishly and foolishly assuming I would be able to make up for lost time with her. I don't know if I will ever forgive myself, but I am trying. I don't know if I deserve the grace I want to give myself. I was supposed to be my sister's protector.

The most shame I felt in my life came four years ago. I remember it like it was yesterday. The memory is that painful. My wife was pregnant with our four-year-old daughter. All nine of my aunts flew from ESTL to San Diego to be with us for the baby shower. They were so excited that I was having another kid. I think they all knew that my daughter was sent by God to save me.

On the day of the baby shower, my mom and aunts were late. I was so stressed out about giving my wife a great experience that I took my anger out on my mom and aunts when they showed up. I remember their looks of hurt and disappointment as I criticized them in the garage before they walked into the house to join the party. My mom told my aunts it was ok, go inside, and she would stay with me. And she did. She stood there and listened as I talked down to her. I will never forget the look on her face and her posture. She was hurt and disappointed in me, and she should have been. My words were heavy on her. She wanted to carry some of the burden I was carrying. My mom and aunts were a few minutes late to the shower because they were preparing a surprise for my wife.

I can't believe I treated them like that. They had given me so much. I'm still so ashamed of my behavior. For the life of me, I can't tell you why I behaved that way. I don't understand it. That night I came close to blowing my brains out. I very seriously considered it.

It was never PTS itself that caused me to want to hurt myself; it was how I treated others as a result of PTS.

Other instances here and there have stayed with me. A few years ago, I was smoking cigars with one of my best friends when he told me, very matter-of-factly, that I get really mean when I drink. Not only did his words go right to the center of me, but my own realization struck me hard. If this large and powerful, violent and protective man thought I was mean, then oh my goodness, how much of a monster must I be in the eyes of my family and friends? I had drunk enough alcohol for three lifetimes. Again, I was ashamed. I kept imagining my precious little ones looking up at me in fear. I should have been their father, friend, trusted advisor, their safety and security. In so many ways, I failed to provide for them. I don't know if I will ever defeat this demon. Maybe I will never make up for those times but going forward I can try to be there for them and their children, now that my son has two beautiful children of his own. In the years I have spent explaining my feelings of sorrow, shame, remorse, and regret, my family and friends have all told me that they forgive me, and I believe them, but I am struggling to forgive myself.

Willpower and physical strength had nothing to do with my ability to live with PTS and start to thrive. I simply had to surrender to the very human feelings I felt following a lifetime of trauma. My childhood traumas by themselves, let alone what I survived as a special operations commando, were enough to drive almost anyone crazy with paranoia and anxiety. I had lost faith in humanity for the most part.

In surrendering, I learned to show others and myself much more grace. I had done so much damage over the years that it was impossible to fix it all, so I needed to forgive myself. And I have done more than apologize; I have worked to better myself, so I don't repeat my mistakes. I enlisted the help of professionals. I worked weekly with a psychologist, a psychiatrist, a marriage-family therapist, a spiritual counselor, and an executive coach. I gained a better understanding of my family's perspective. I will likely use antidepressant medication for the remainder of my life. That's OK. I would not trade my experiences for anything, but I have been forever changed by them, and now I need help so that I can be a better chief to my tribe. I want to be purposeful and have more positive rather than negative interactions with my family, especially with my wife and kids.

Additionally, I focused on building relationships with people outside the veteran community. My civilian friends helped me to communicate better and made me aware of how my presence and energy affected others. They taught me that things work much differently in the private sector. For example, things don't happen quickly in the civilian workforce. I used to get so frustrated when I didn't hear back from someone I emailed. I just didn't understand that in the private sector, people rarely take anything as seriously as we were taught in the military to take everything, and for good reason. What we did in the military was very high risk, whereas nothing in the private sector, in corporate environments to be specific, is even close to high risk. I had to stop taking myself so seriously and stop taking everything around me so personally.

In doing so, I have learned so much about myself. For instance, when I was younger, I hated school because I wasn't taught to value education. I was told how stupid I was and that I wouldn't amount to anything, let alone be a man. But I learned that I actually love education. I don't just love it, I thirst for it, and I am really good at learning due to my focus and tenacity, which I learned in the military. Realizing just how little I know in the grand scheme of life has made me more curious than ever.

I am more than a Navy SEAL. I'm a father, husband, academic, entrepreneur, teacher, and artist. But in order to see those things in myself, I had to be honest and admit that I had become some other things too. I had developed characteristics that would not benefit me in the next chapter of my life. I had become a kind of introvert and recluse—I didn't want to engage with you if you weren't in the military or a veteran. I considered it a waste of time to communicate with people I thought wouldn't understand me. And how could they? They had no idea what it felt like. They hadn't lived my life. What was I supposed to say to them? That I can't stop seeing the four-inch hole blown through a little girl's leg, and the look on her face? That she was standing with a group of Afghanis waiting to vote, when the Taliban threw a grenade?

I was a buzzkill in many ways. For everything my wife and kids suggested we do for fun, I found the potential dangers and risks. Either I shot down their ideas or reluctantly participated without enjoyment.

I was incredibly insecure, and that insecurity started early. When I was in boot camp, I had my first heartbreak. My son's mom, my high school sweetheart, cheated on me shortly after I left. It hurt in so many ways, and there was so much I didn't understand about the situation. I don't blame her now. We were so young and silly. We both had so many more mistakes ahead of us, not just with each another, but in general. I failed to realize how she must have felt when I left her behind in ESTL. She must have been so scared and felt so insecure about the future. And maybe she didn't know any better because of her own childhood experiences and traumas. Eventually, our broken hearts healed. We matured and got on the same page. I'm so happy that she and I are still close friends. She gives me strength and advice all the time. More importantly, she gives me grace when I feel at rock bottom because of my failures as a father. She forgave me a long time ago, too.

Moderation Is Not for Cowards

I learned a lot of lessons throughout my military career that have served me well in this new chapter. My business is still operating following the COVID-19 pandemic that closed many businesses, including some of my competitors. We are even thriving during a global economic crisis. Only the persistent and tenacious win at entrepreneurship. Determination and an unshakeable faith are the most essential qualities.

My experiences have taught me to walk by faith in God. I have survived when I shouldn't have and only because a higher power protected me. I see God everywhere, and that is what keeps me going, considering I also see ugliness and sometimes I feel helpless.

I developed an indomitable spirit in the military, and I never quit. I learned servant leadership at a level I could not have learned anywhere else in the world. In peace and war, my leaders performed some of the most courageous and selfless actions I have ever seen. Putting others before myself is my default, and my family taught me how. During Thanksgiving and Christmas, my grandmother would open our door to the homeless, so that they had shelter and hot food on the holidays.

The military taught me about unconventional warfare, and I have incorporated those lessons into my role as a tech CEO. My training made me an expert at intelligence gathering and asset/target/source development. In my current role, I use that training for market intelligence gathering and customer development. I use my skills to map organizations and build trust and relationships within them so that they feel comfortable working with my company. In the military, we called that key leader engagement.

Lastly, the military taught me to build teams like nobody's business. I learned how to build tribes capable of accomplishing any mission, and since leaving the military I have learned how to lead them better.

Originally, I wanted to transition into the FBI at the end of my naval career and continue as a special operator on the hostage rescue team (HRT). But when I got into graduate business school at USC, I distanced myself from the idea. I realized just how psychologically and emotionally challenged I had become as a result of being a special operator And I really fell for entrepreneurship while studying at USC.

Entrepreneurship at this level is the hardest thing ever. Many days it feels as though getting shot at for a living would be easier. But I have grown both personally and professionally from my entrepreneurial experience.

I decided that I needed to leave alcohol behind for the most part. Alcohol enabled and even motivated me to hurt myself and others. And because I am a person of extremes who needs to spend more time in the middle, quitting alcohol altogether didn't seem right. Instead, I had to do the opposite of what the SEAL ethos tells us. Anything worth doing is not always worth overdoing, and moderation is not for cowards.

I needed to live near my SEAL community, so my wife and I built a life on Coronado Island, San Diego. When I need to spend time in that environment, to heal when I am hurting, and when I need it, it's critically important that I listen to those feelings and have access to my teammates. I would be in real trouble if we moved someplace that kept me from being around my SEALs. I was also concerned that I wouldn't be able to mentor (protect) my sailors from what I had gone through during my transition. That alone is extremely important to me. I will have to fight for my boys' lives for the rest of their lives too. I'm never out of the fight. That means I can never stop growing as an individual.

I learned to be more empathetic, patient, vulnerable, mindful, gentle, and emotionally aware. As a leader and tech CEO, I needed to grow because real leaders are never afraid to admit what they don't know. Authentic leaders are only concerned with individual, personal, and professional growth and proficiency and the well-being of their tribe, which requires the leader to be constant in his/her personal and professional growth efforts. My employees count on me to be more than their boss, and they need me to be empathetic, vulnerable, and firm when needed. They also need me to be their mentor and their trusted advisor.

I don't have to be a military man at home. It's not what my family needs, even considering the safety and security which they depend on me to provide. Most of the safety and security they need from me comes from my ability to be consistent in my efforts toward growth and the correction of my past behaviors.

I have grown from not understanding the value of an education to becoming an educator. Five years ago, I had the opportunity to serve as a board member at Ashford University (now part of the University of Arizona Global Campus), working on the Academics and Student Affairs Committee. I spent three years with some of the brightest and most dedicated educators. In 2022, the UAGC awarded me with an honorary doctorate. I have accepted that I'm a teacher at heart, and I will always mentor and teach, especially the veteran community.

But I am still learning to delegate and trust, even after so many years in the military as a senior enlisted leader. If I become a casualty to stress, I do myself and my tribe no good.

USC's MBV program was uniquely valuable to me during my transition from military service. I never really felt alone during the program. It taught me to think more critically than I had before. I was forced to think critically about myself, which was scary. I'm so glad I committed and never stopped doing it. I'm much more introspective now. Just as other people have their own insecurities, which they project onto me, I also do that to others. And I can control that.

The MBV program motivated me to leave behind the things I no longer needed. I learned business at a professional level, developed a strong entrepreneurial spirit, and learned entrepreneurship from excellent professors who walked the path before me. One of them is now a majority investor in my company. Talk about full-circle market feedback telling me how much I have grown, learned, and how far I have come.

What I Left Behind

Alcohol was killing me. As a substantial part of military culture, I had to leave it behind as I transitioned from the military. When I was at my lowest, and in so much interpersonal trouble, I considering taking my own life, and alcohol was at the center of my days. I used alcohol as a coping mechanism to my detriment. We were taught for so long to ignore and override our feelings and drown them in alcohol. It's a deadly lesson.

I learned to leave my ego in bed most mornings, too. It's a strange feeling, but I feel like the more successful I become, the stronger I become, the more intelligent and accomplished I become, then the more control I gain over my ego. I feel like I have nothing else to prove to myself or to anyone else. It's all beginning to make sense to me now. I'm grateful because everything does, in fact, happen for a reason. I am the person I am today because of what I survived. If I had not been repeatedly humbled as a child and throughout my military career, maybe I wouldn't be so happy, fulfilled, and in search of humility as my success grows. I have been blessed with so much, not just with what I need, but with everything I want.

Fighting isn't my default anymore. Understanding someone's pain and insecurity is more important to me now. I now listen to understand as opposed to listening to respond.

There has been a tremendous change in how I view what it means to be a man, and not just a man, but a badass. I grew up in hard, old-school Middle America. I was taught that men don't cry. That's bullshit. Not acknowledging my pain, not allowing myself to cry and mourn the loss of my friends, was nearly fatal to me. I remember sitting in a restaurant alone one day while taking a break from work, and I couldn't stop thinking about my friend Kevin "Juicy" Houston, who died on August 6th, 2011, in Afghanistan. I didn't try to stop the emotion that rushed out of me right there in the middle of that restaurant. I no longer cared if I cried, because I really missed my friend. Responding to heartbreak doesn't make me weak. It makes me human.

Although I retired from the Navy in 2016 at the age of thirty-eight, I'm still working on transitioning from military service. I spent more time in the military than I spent outside of it. I needed more time to learn to live as a normal person, because I hadn't done so for twenty years.

Transitioning from the military is a journey that none of us can complete overnight. Society shouldn't expect it from us, and we shouldn't expect it from ourselves. Transitioning from the military in and of itself is traumatic. It takes a long time to recover. I experienced more in the service than I could have ever asked for, and my personal growth and learning benefited exponentially. My life is amazing as a result. But I came to understand that my service in the Global War on Terror came with a price I will be paying for the rest of my life. I accept the cost. As part of accepting it, I'm up for the work and the journey ahead of me. I know that life will continue to reward me.

"Be a Human First"

Ty Smith reflects on telling his story

It has taken me multiple decades to understand these words from *The Power of Myth* by Joseph Campbell: "All of the Gods, all of the heavens, all of the hells are within you."

I focus on sharing my truth with readers and purging bad karma. As a result, I found peace and forgiveness for the most important person I needed to forgive—myself. My work will create more light-workers, spreaders of hope, and pour positive energy into a world currently tilting closer toward hell, not heaven.

Out of everything you can be, be a human first. Love thy neighbor. God is preparing you for something. Know your purpose and surrender to God's will for your life. Flow. It is good. God is good. Build a partnership with your ego, or your ego will control you to your own detriment. Gratitude is everything. I can't stress that enough. Rarely take yourself seriously. Success takes a village. Grow your tribe and be the chief in your own special and unique way. There is no one like you. When God blesses you with success, never lose sight of the fact that your village and God got you there.

Chapter 26

And the Beat Goes On
by
Vanessa Bolognese

Vanessa Bolognese is a US Army veteran who served for four years including two tours of combat in Operation Iraqi Freedom. As a combat medic for the brigade personal security detachment (PSD) and the dedicated combat logistics patrol, she coordinated with foreign dignitaries and high-ranking officials in the development and execution of emergency medical treatment and evacuation. She achieved the rank of sergeant, and she received three Army Commendation Medals, an Army Achievement Medal, and a Good Conduct Medal.

She obtained a BS in nursing (BSN) from West Coast University, where she received the Jarvis Scholarship for outstanding clinical expertise. She went on to gain her Master of Business for Veterans (MBV) degree from the University of Southern California.

Vanessa works as a registered nurse. Currently, she is a quality consultant for veterans' healthcare at VA San Diego, where she manages the delivery of quality and safe patient care facility-wide.

Generations

I have a photo taken when I was about five years old. It's me with my mother, grandmother, great-grandmother, and great-great grandmother. Five generations in one photo is pretty incredible. But it isn't just a frozen moment of smiles, it's a narrative that encompasses perseverance, pain, and ultimately triumph over the adversities each woman faced, from humble beginnings in rural Oklahoma to the challenges of single motherhood. The photo is a testament to the resilience and strength passed down through generations. Beautiful and rugged women, they lived through unimaginable struggles, but in the photo they are beaming and gleeful. There's no doubt that giving birth and becoming a mom is one of the most difficult things a person can do, but when I was growing up in the '90s, the media often reduced women to one-dimensional roles. They were either passive caregivers or glamorous professionals. There is so much more complexity to a woman that often goes unrecognized and undervalued, even today. My beloved multi-generational photo shows the many facets of what a woman is, has been, and can be. My story is not just about a woman growing from an awkward girl into a soldier, but it's about the gift of life, courage, and the tenacity those women gave me.

I was born in Nevada in 1986 (what the kids call the late 1900s). My mother and biological father had me several years after they got married. When I was three, it became apparent to my mom that my father was on his way to being a life-long criminal. Terrified for my safety, she divorced him and relocated us to California where my grandmother and great-grandmother lived. The three of them, all single women, did everything to take care of me. My biological father was hardly in the picture. I saw him two to three times a year, then not at all for decades. He provided no financial support whatsoever. This time period was dizzying for me. I never quite knew what the next day would look like. It was also an adventure. I stayed busy rollerblading with Mom, baking with Grandma, and gardening with Great-Grandma. They were still working: Mom managed a gas station, Grandma was an in-home caregiver, Great-Grandma cleaned houses, and they took me along to work. It could not have been easy or convenient in the least, especially since I was rather mischievous. I wrote my ABCs on Mom's office walls at the SmogPros, and when I was five, Grandma discovered that I had folded one of her clients into a "human taco" by pushing the buttons on his powered hospital bed. After getting him settled, she laughed and said, "That girl's gonna be a nurse!"

Things changed when my mother met my stepdad. They married and we moved into a spacious house in a peaceful neighborhood in Southern California. My school was a short walk down a residential street. My brother was born two years later, and we all lived at that house until I moved out at eighteen. We became a nuclear family, and I quickly embraced it by dropping the "step" from stepdad without much thought. My dad's guidance kept me grounded and focused on my ambitions and achievements that might have seemed beyond reach in a less stable environment. He was a natural at being a father—he paid the bills, helped me with homework, and planned family vacations meticulously to ensure we experienced new places and cultures together. And yes, he wore the quintessential dad attire: sneakers, graphic T-shirts, and shorts, emblems of his dedication to the unglamorous but essential roles of fatherhood.

Likely a result of my colorful early years, I was a weird kid. I was hyper, talkative, and I had an odd sense of humor. It didn't help that I was also a little chubby. Because I stood out, I caught a lot of bullying from students and teachers alike, unfortunately. In second grade, my teacher would count down from three when I talked, and everyone would turn and shout, in unison, "Shut up, Vanessa!" That teacher egged the bullying on all year. One time they put me in a literal box for an entire day. Another teacher asked the class to guess how big my brain was. One ounce was the only reasonable answer, the teacher said. I got the message that I was inferior to others and deserved to be humiliated. I went home crying a lot. In fourth grade, after months of being threatened by a boy who said that he and his friends were going to jump me, it was time to let him know that I was incredibly dangerous. I told him that I would chop his legs off with my nunchucks. I was almost expelled for that.

My sixth-grade teacher, Mrs. Arguello, saw my weirdness and energy as strengths. She helped me apply those strengths toward stimulating projects and experiences. I did considerably better in school that year as a result. She also called me a natural beauty, which I had not heard from anyone outside my family.

In my opinion, overcoming challenges like bullying when you are a child hinges on having strong role models in your life. My biggest role model was my mother. She was a bit of a firecracker who never shied away from confrontation. Her assertive and unapologetic approach taught me valuable lessons about standing up for myself. She equipped me with sharp comebacks, like "I can lose weight, but you'll always be ugly."

But my upbringing lacked a certain gentleness. My parents, themselves products of no-nonsense families, were often stern and focused on toughness before tenderness. Their tough love left me craving softer, more compassionate support. Lucky for me, the YMCA after-school program, free from schoolyard hierarchies, introduced me to Crystal, who was a grade ahead of me and went to a different school. Crystal had the confidence and self-assuredness I lacked. She was bold, tenacious, hilarious, and above all, compassionate. Things began to slowly improve after meeting Crystal. In high school, I occupied a weird social limbo between band and drama nerds, the problem kids, and the jocks. Having a single identity was kind of boring (wonder why!) so I liked embracing my multifaceted persona, which allowed me to connect with diverse groups.

I felt like a fish out of water living in easy, sunny Southern California. I knew there was something else out there, more to the world. What I was living wasn't the norm. Every morning on AM 640— "More stimulating talk radio"—I heard chatter of disaster, major weather events, racism, poverty, war—so much outside my reality. It made sense that my curiosity grew. With the challenges I faced growing up, and a lingering lack of a sense of self, the idea of stepping into the unknown became less daunting and more exhilarating. I started to see the potential in venturing beyond the familiar and seeking out new opportunities.

In 2005, when I was almost nineteen years old, after a semester of dreadfully uninspiring community college, when the country was well into the invasion of Iraq, I joined the Army. I wish I could say I joined for some incredibly patriotic reason, but my boyfriend broke up with me to join the Marine Corps. I was so upset, and I craved a major change. The breakup was the catalyst that made me finally consider the military as a legitimate option. It had piqued my interest for some time. I chatted up the recruiters on my high school campus and watched a bootcamp graduation in complete awe. I was struck by how I saw them. They were so separate from us, as if they had unlocked a secret world together and spoke a different language now. Also they all looked like Greek gods. As I talked about the possibility of joining, the adults in my life became preoccupied with the amazing benefits that came along with serving. At that age, I didn't care about benefits. What even are benefits? I just wanted to get out of my lackluster reality and enter their secret world.

When I walked into the recruiter's office, I wasn't aiming for anything too rugged. I wanted to enter the healthcare profession. My desire was likely the result of my mother's constant reminder that "You can't send the old folks to China!" She meant that shipping off elderly, sick Americans to outsource their care was not going to happen in the near future. Healthcare had job security. I didn't always listen to my mother, but I did hear her. That day at the office, the Navy recruiter had left for the day, but the Army recruiter was eager to talk.

From Boot Camp to Iraq

The recruiter told me that women weren't sent to the front lines. "As a healthcare specialist, you'll be assigned to a military hospital in Germany or somewhere like that," he said. Enamored with the possibility of living and working in a distant land and having the world at my fingertips, I swore into the Army. Two weeks later, I found myself enjoying luxury linens at the LAX Westin where they put us up for three days before we shipped out. "Nice, this is going to be a breeze," I thought. The reality was the Californian who had never seen snow was on her way to Missouri for bootcamp in the dead of winter.

The next year of my life was like the movie *Private Benjamin*, where Goldie Hawn naively joins the Army after her rich fiancé dies at the altar. A recruiter dazzles her with stories of the incredible opportunities for world travel and adventure in the US Army. She signs up, overlooking the fact that she was joining the actual military.

Basic training was awful. Yes, it was cold. Yes, I was tired. But the emotional turmoil was worse. Grown adults belittled us and played mind games. I will never forget watching a heavier recruit being forced to do push-ups with a Snickers bar on the ground. Each time he pushed, he had to shout, "I'm a worthless fat fuck," and they egged him on until he was in pure agony, a puddle of exhaustion, tears, and embarrassment. Recruits were being aggressively belittled like this for anything that could be considered a flaw—their body sizes, big or small, their hair color, hair texture, skin, teeth, you name it. If there was something about you that stood out, they would jump all over it, shouting the worst, most vile things I had ever heard spoken to a person. The drill sergeants wanted to break spirits. My heart ached for those being picked on.

I called my mom at the start of the program, when they still allowed phone calls, and told her the Army wasn't for me, and I wanted to go home. True to the family ethos of tough love, she said, "If you quit now, then what?" I really didn't have an answer. At that moment, something solidified. I realized she had a point, and I didn't want to know what "then what" looked like. I wanted to do hard things. Plus, the idea of being an Army chick was pretty freaking cool, so I pressed on.

I avoided the attacks at first, but things abruptly changed. A drill sergeant, Sergeant X, began hitting, punching, and kicking me, all totally unprovoked. It's how I would be woken up, it's how he finished his shake downs after the firing range, it's how he let me know at any given time who was in charge. Sometimes he hit me so hard that I buckled to the ground breathless, while everyone was turned to stone with fear and stood and stared ahead. Was this what being a tough Army chick meant? Then I was sexually assaulted. One might say, "That's enough, tell leadership!" But when you are that young, naive, already being abused by leadership, and in front of everyone, who do you tell? It didn't help that the person who sexually assaulted me was Sergeant X's favorite star recruit.

My reasons for staying quiet were validated when two recruits went to another drill sergeant to report what happened. I was pressured into confirming the reports, which made everything worse. The physical abuse heightened, psychological abuse began, and I was ousted from my team, sent to another platoon, and nothing happened to my assaulter, the star recruit. Word got out about why I was transferred, and my peers turned on me. One day, Sergeant X ordered my former team to circle around me and take turns calling me awful names.

The only thing that kept me going was the Sunday church service. It was a way to leave the shadow of my abuse and feel hope. I collected a rosary from each service as a token of getting through another week and as a countdown to graduation. A week before graduating, Sergeant X found me alone and leaned in close to say, "One day you'll understand why I did all of this to you. You'll be a stronger person because of it." I left that place thinking he might be right. In order to succeed in the Army, I had to stay in line, respect authority at all costs, and no matter what happened to me, drive on. The truth I have come to know is that I was already tough enough.

Later, I heard through the grapevine that the entire basic training leadership was investigated for several issues. I never followed up on it as I wanted to move on. I simply accepted there were despicable people in the world, and that they were worse when they held positions of power. These days, people seem to be more comfortable coming out against abuse, and I can only hope that things have gotten better for young recruits. The last thing we need are patriots who end up resenting their leaders.

After bootcamp, I went into advanced individualized training (AIT) at Fort Sam Houston, TX, to officially become a combat medic— not a "healthcare specialist," which had so beautifully rolled off the tongue of that savvy recruiter. Tomato, tomato. The experience was far less traumatic but no less difficult. Many more sleepless nights and physically demanding challenges, but they made sense. The emotional turmoil had seemingly ended, since these drill sergeants were tough but not abusive. With good scores and behavior, I earned personal time off, even off-post. I actually enjoyed this part of my training, and I made friends.

I was very lucky to be stationed in Hawaii after my training, and some of my Private Benjamin dreams were fulfilled. It was an incredible place to be a young adult on my own for the first time. I wasted no time and quickly assimilated into island life. I fell in love with an infantryman and got married in a small Hawaiian court at twenty. It turns out wartime love stories are pretty common, as this had all been predicted by the recruiter.

After months of galivanting in paradise, reality sank in when I had to leave to support Operation Iraqi Freedom. I also had to leave my new husband because he was in a different brigade. I cried a lot on the trip to northern Iraq, but I knew it was my sworn duty to defend my country, and the time to do so was now. Once we landed, I buried my sadness and put on a brave face.

My unit was a medical company within a support battalion, and our mission was staffing and running a field clinic, or aid station, on base. However, shortly after arriving, I was chosen to serve as one of two medics for the brigade personal security detachment (PSD), a team of grunts and artillery men. On a regular basis, we would be running missions, often risky ones, outside the confines of a protected base. I flashed back to when the recruiter assured me that women weren't sent to the front line. I should have registered that a couple of things about this war were different. First, a "front line" was outdated. Second, we were trying to be sensitive to the Muslim community and respect their beliefs and customs that prohibited men from seeing or touching women who weren't family. Because the only dual-gender combat role at the time was a medic, many teams chose to have female medics to search and treat the local women. The other PSD medic was also female. We shared a container housing unit (CHU) near the guys where the brigade headquarters was located.

I felt safer with the PSD away from my company. In fact, the aid station turned out to be full of daytime talk show-worthy drama— affairs, baby daddies, girls fighting over guys, guys questing for girls, contraband, finding innovative ways to get high, and the like. Entertaining for sure, but not a great way to live. If you get enough young people, relocate them away from their families, house them all together, and give them relatively mundane tasks, some wild things happen. The big difference between my company and the PSD were the missions we ran outside the wire. Many were unplanned. Among my team there had to be a mutual reliance for safety and security.

Early on, I learned a valuable lesson about how close I could get to any of my team members when I became friends with one of the artillery men. One night, my new friend went missing, and around midnight our head sergeant barged into the CHU, sure that he would find him with me. He didn't find my friend, and I had no idea where he was. I was pretty upset by the sergeant's assumption. Everyone knew I was married, and I had never considered my relationship with the artillery man as anything other than platonic. "Perception is reality," the sergeant told me. That statement became a major life lesson. I have used it as a tool for my success, and it has posed considerable challenges. It's a heavy burden to manage how others see you. Trying to control that narrative can be exhausting and isolating. It has been a constant balancing act between maintaining my authenticity and navigating the expectations and judgments of others.

My assignment with the PSD was full of the stuff of US Army commercials. We ran missions all over the country, almost daily, flying across the warzone in choppers, riding in armored Humvee convoys, firing heavy weapons, crouch-running through fields, high-fiving elite military leaders, meeting and dining with foreign dignitaries, etc.

Our mission was crucial. We brought local Iraqi and American leaders together to strategize on rebuilding Iraq. Which included efforts to develop Iraqi military and police forces and bridge relations among diverse regions and religions. It felt like we were at the heart of the war, not pursuing insurgents but fostering solutions and empowering local leaders. Helping facilitate those reformative discussions while ensuring the safety and security of our team was profoundly rewarding. It came with constant stress, as that year was a particularly dangerous time in northern Iraq, the bloodiest on record since 2004. Someone I knew or had just met was killed or injured almost weekly, and it went on for months. Near the end of deployment, we lost fourteen guys in one day. I woke up each morning feeling an intense eagerness to return to bed again that night because it would mean that I survived another day and was closer to going home.

But when I did return home from deployment, the tragedy continued. A PSD vehicle, with three guys and the medic who replaced me, was attacked with a roadside bomb. They were all killed while traveling the same roads I had in the last fifteen months.

Coming from the lush landscapes of Hawaii, Iraq offered an eye-opening experience. The closest place I could compare it to was Death Valley, but unlike the mostly uninhabited Californian desert, there were many large cities in Iraq, full of people living their lives in harsh conditions. The weather went from intensely hot to very cold, and it even snowed! Powerful winds and sandstorms were common, and the wet season was an additional small trauma. Walking in the rain felt like wearing moon boots made of mud pies, and we tracked mud all over our personal belongings and spaces everywhere.

Daily life for the Iraqis encompassed surviving a war in addition to the extreme conditions of the environment. Infrastructure was rudimentary, but some signs of modernization were visible, mostly in the form of precarious power lines haphazardly strung above dwellings. Children playing soccer was one of the few scenes of recreation. Having this kind of experience at just twenty years old, while my friends back home were enjoying their carefree youths, really highlighted the major differences in our worlds. I felt immense pride knowing we were genuinely trying to help these people. And I was hit over the head with gratitude. I still pause when I'm enjoying simple pleasures, such as getting fresh air on a tranquil walk, drinking ice water, getting cozy on a comfy couch, taking a hot shower on demand, and especially when using an above-ground, porcelain toilet. It's incredible that these things are considered typical for the modern American, but they are luxuries I don't take for granted anymore.

The first deployment had one more casualty—my young marriage began to fall apart. The beginning of the end came months into deployment when I found out that my husband was essentially living life as a single man. Part of me wondered what had I expected? Were a nineteen- and twenty-year-old going to live happily ever after through the incredible stress of multiple deployments? Then we stayed married for another three years.

Deployment was one of the most intense, confusing, lonely, terrifying, and pivotal periods of my life. Here it all is on paper, summed up in a few short paragraphs. It feels surreal and reductive.

I haven't completely unpacked my experiences. I have spent a lot of time and energy trying to move forward. By the end of the first deployment, there wasn't much healing to do because I was still in the thick of it. However, I knew inevitably that I needed to confront the full impact my service had on me.

Bona Fide Soldier

Six months after returning home from my first deployment, I volunteered to leave paradise again. I switched brigades and headed back to Iraq to join my estranged husband. My patriotism was booming. I was invigorated by the incredible work we did during my first deployment and determined to not let the lives we lost have been lost in vain. I was desperate to do what I could. The opportunity to fix things with my husband was also on my mind.

After all I had been through, I was a bona fide soldier. I didn't know how to be a soldier without being boots on the ground.

Even though my dedicated combat logistics patrol team ran more than 200 missions resupplying smaller bases and camps, my second deployment wasn't nearly as active, or badass, as the first. We had a lot of downtime, which most people spent playing video games, working out, sleeping, or getting into trouble reminiscent of the aid station on my first deployment. I simply could not waste my time being bored in the desert, so I enrolled in virtual community college, began correspondence courses for promotion, and studied for the promotion board. After several months of earning college credits and maxing out my physical fitness test and every other item on the promotion list, I was promoted to sergeant by an all-male infantry board, which I was told was a first for them. The infantry first sergeant on the board said, "You march better than half my men!" Writing it out now makes me wonder if it was meant inappropriately.

During my second deployment, the Army was still figuring out how to deploy mixed-gender units in a warzone. Our infantry leadership wanted to be progressive, so they assigned me and my husband to the same company. He was my CHU roommate. One would assume this was a good way to deter cheating. Making matters more complicated, neither of us had lived with a partner before. Trust was broken, we had no time to reintegrate, and we had to go through the risk and stress of war, which meant the odds of having a successful relationship were practically nonexistent.

It was a long deployment. My husband turned out to be a jealous man. I was isolated, especially since there weren't many females at the headquarters, and they had their own troubles. One of the only friends I had was an Iraqi woman, one of our translators. I took her under my wing when I noticed she received unwelcome attention, and we ended up having a lot of good times together. She was beautiful and sweet, fun to talk to, a wonderful spirit of joy and kindness, with an infectious smile. She brought me trinkets and treats from the city, and we covered so many topics talking during our lengthy missions—fashion, religion, family, pop culture, the war, and what her job meant for her and her family. She put herself and her family in great danger working with the Americans, but her family was proud of her, and she was determined to help her country rebuild. Her friendship during that very lonely time remains a cherished memory.

The missions were monotonous, but I still enjoyed my role as the only medic on the team. In addition to running missions and my other responsibilities—like packing aid bags, teaching various medical classes, developing emergency medical evacuation plans—the guys visited my CHU at all hours to report their sprains, strains, sniffles, fevers, aches and pains, and I took care of them. My husband was not thrilled. I staffed an aid station, which was my first experience in a real clinic. Everything I practiced before then was field medicine, the "rub some dirt in it" approach. This got the wheels turning in my head about my future in medicine, and it's when I seriously started considered nursing for a career.

After returning to Hawaii, my husband and I tried again to make things work, but it got worse than ever, and I found myself contemplating my life without him. While we probably (definitely) should not have gotten married, the legal reality meant we (or, I) was forced to think about the deeper significance of marriage. Nevertheless, after a string of unsettling events, any reserves of hope, patience, and compassion were depleted, and the divorce was finalized later that year.

It was a peculiar thing to be in an intense relationship during such an unconventional and pivotal time in my life. It felt like we had beamed to a different planet, turned into action hero versions of ourselves, adventured in paradise, kicked enemy butt, went through fear and grief together, and lived a young love story through it all. We were the only ones who understood what the other was going through, because we were going through it together. Saying goodbye to that bond was unimaginably hard and, admittedly, it did not happen quickly. Despite the pain the relationship caused, it was fortunate that I didn't have to go through those experiences alone.

I thought about extending my time in the Army, especially since I had become pretty dang good at being a soldier. I wanted to see what working in a hospital was like, so I asked the re-enlistment officer if such a placement was possible, and I got a big ol' "no can do" in return. No guarantees could be made. So, after twenty-seven months in combat, I reached my end time of service, filed for divorce, and without any further ties to Hawaii, I moved back home to live in my parent's house.

Press Forward

At twenty-three years old, I had already lived what felt like a lifetime of exploring far away lands, getting married, fighting in a war, and getting divorced. After having a hard time grasping onto a sense of self,

I had finally built an identity as a soldier, a medic, a sergeant, a wife, and a Hawaiian resident. Back in California, it felt like I had to tuck it all away. It was time to rebuild myself.

I studied nursing in order to work in the emergency department, which made the most sense considering my experience. On the first day of community college, in one of my competitive prerequisite classes, the professor did something genius. The classroom overflowed with last-minute sign-ups hoping to get a spot in the course. During his introduction, the professor talked on and on about how hard the class would be, the coursework's large time commitment, and he shared the high dropout rates. Hearing this, many people filed out of the room, but I wasn't going anywhere. I had done hard things—how hard could Anatomy and Physiology be? I was just about to roll my eyes at the professor's ego when he finished the class with a poignant statement: "The only way to certain failure is to take yourself out of the game," he said. "Congratulations. Those of you who stayed are admitted to the course." That moment resonated with what my life had looked like up to that point, and it reinforced the strength, wisdom, and confidence that was growing within me. I wanted to press forward no matter how intimidating or impossible my goal seemed.

I lived a fast life the next several years while waiting tables and going to school full-time. Memories of waking up eager to go back to bed on my first deployment fueled a determination to live my life and never turn down an opportunity for fun or adventure. It was a challenge to balance it all. I hardly slept, as I really didn't see the point if I still had the energy. A friend asked how I was seemingly everywhere doing everything all the time, and I told her, "I don't want to look back on any given calendar day and ask myself what I did and the answer to be 'nothing.'"

I got very close to a wonderfully kind, bubbly, and downright fun friend, who I now call a best friend. I moved out of my parent's house and into an apartment with her. She was also working toward her undergraduate degree, and she became another great role model—comfortable in her own skin, smart, determined, close to her family, and always willing to help others. Becoming best friends was easy through baking dates, late night study sessions, bonding over boy drama, and building our young lives together.

We partied our platform heels right off, and while the scene was fun at first, my social drinking evolved into something more compulsive. For my friend, drinking was a good time, but for me it grew into a craving, and I fixated on when and where I would get my next cocktail. It was hard for her to recognize what was happening to me, as it was early enough in life that it wasn't obviously problematic, since most twenty-somethings partied similarly. My drinking had become a way to cope. My unresolved fear, anxiety, and an overall sense of confusion inside me was drowned out with loud music, strong drinks, and a good dance floor. Over time I pushed the envelope to a dangerous degree. Some of it was self-sabotage: I didn't believe that I deserved a promising life, the life I was building. I wanted to see how far into oblivion I could go. Thankfully, I came out relatively unscathed.

I made a ton of other really great friends by leveraging how I understood the universal desire to be seen, heard, and valued. I recognized and fulfilled those desires in others, and in doing so I learned to foster genuine and lasting connections. I used to call it "manipulation for good," but it was more like strategic empathy, a way to build and strengthen bonds based on mutual respect and understanding.

I graduated with a BS in nursing in 2014 and passed the RN licensure exam. Then I volunteered to go on a week-long medical outreach mission to rural Nicaragua, where the accommodations would be minimal, but the impact would be great. Intrigued, three of my nursing peers volunteered with me, which caught the attention of our college, and the college ended up making a large donation to the cause. Over four very busy days at various field clinics, we newly minted nurses triaged and helped treat about 1,100 locals, providing them with vital diabetic and blood pressure medications, antibiotics, and antiparasitics. We aided in performing dental work, and we educated people about infection control, public health, nutrition and dental hygiene. It was a moving experience, and it showed me that with the right support, and by fostering a spirit of cooperation, I could help others, and myself.

From April until October that year, I applied to every open new grad and entry-level nursing position within a seventy-five-mile radius. I even begrudgingly applied to a medical-surgical position (a typical floor nurse role) at a small hospital even though I really wanted to work in emergency medicine. I figured if I could get some experience under my belt, I could go for the job I actually wanted later. That hospital called me back but asked if, considering my field experience, I wanted to interview for the emergency department (ED) instead. I interviewed and was hired on-the-spot.

The hospital was in Gardena, and much of the facility's emergency transport came from Compton, so we saw it all in the ED. It was an insane ride—many very sick, very poor people, and physical assaults against the staff. The intensity was reminiscent of my Army days. My team was stellar, with some major players worthy of combat themselves, with their grace under pressure and can-do attitudes.

In nursing school, they did not teach us that while you may want to give people the best care, the healthcare system will often prevent you from doing so. Private healthcare is extremely complex and poses many challenges, including access, continuity of care, and affordability of tests and procedures. Prioritizing the feeding and housing of their families, or other demands, the people we treated often did not seek routine healthcare and used the ED for emergencies and issues that weren't urgent, both. Their chronic ailments had worsened without routine healthcare, so we saw really, really sick people, all the time. When it was time to discharge them, it was up to them to arrange follow-up appointments, something that most of them didn't have the time, means, or experience to do. It was heart-wrenching to see the same faces in this doom spiral, knowing that the gap between emergency care and ongoing health management was only widening. Stressed because my patients had to navigate health systems as complex as their health issues, and tired of the constant exposure to violence, both potential and real, I worried my mental health was being affected, and my optimism wore thin.

In all aspects of life, a positive outlook fuels resilience and motivation, while a realistic perspective ensures that we are not surprised or tolerating abuse. The balance enables us to pursue meaningful change. So I needed to go somewhere I could genuinely make a difference instead of only managing to keep myself, and my patients, above water. Luck and timing were on my side when a hiring manager from the Veterans Health Administration (VA) called me about an application I placed. I had applied to many job openings at the VA trying to get my foot in the door. I wanted to not only serve my brothers and sisters, but as a long-standing patient at VA, I knew they were on the right side of healthcare, especially after what I was seeing at work at the ED. The hiring manager told me the position I had applied for months ago was no longer open, but I would be a good fit for another position—in the emergency department! I interviewed and got the job.

I moved into the cutest little loft apartment in West Hollywood and began what I thought the rest of my career would look like. It really felt like the VA job aligned perfectly with my experience, both as a combat medic and as an ED nurse. I was well prepared for the challenges of serving a special population, but most importantly, I could work with confidence knowing the VA's access to care, case management, and affordability was unlike anywhere else. Plus, the VA wasn't the area's primary emergency provider, so the volume and criticality of patients was generally lower and more manageable. While I was hired for the less-desirable night shift, I liked the job and made the most of it. (Blackout curtains helped a lot.) After learning the ropes and finally feeling like I could dig my heels in, I found that I had a knack and passion for quality and efficiency. I started taking on extra duties, for instance becoming the VA's shared governance secretary and a Women's Health Champion. One of my biggest contributions was facilitating the first-ever charting tool for use on women veterans seeking emergency care, since nothing in the nursing template even mentioned the existence of female anatomy. My leaders took notice of my influence and work ethic and promoted me to charge nurse, where I helped improve real-time quality and efficiency for the unit.

One night in 2018, while out with work friends, one of them started raving about a relatively new program for former and current high-performing military leaders at the University of Southern California, Marshall School of Business. He said I should check it out. I told him I was never going back to school because I had a good thing going and didn't want the stress. Also, the idea of potentially ripping open traumas from the Army was not something I was ready for. The more he talked about it, though, the more it sounded like something to consider, at the very least to set myself apart from entry-level nursing. After talking to graduates and learning more about the program, I applied, interviewed, and was accepted at USC.

The ten months of the Master of Business for Veterans (MBV) program went by like lightning. It was intimidating at first, learning so many new concepts, being surrounded by, and working closely with so many members of the military. I was pushing my boundaries and returning to an identity I had spent almost a decade separating from. The program's leadership was well-equipped with the skills and experience to delicately address and respond to my kind of apprehension, and they helped me tremendously. They were not only veterans themselves, but they were incredible students of the intricacies of reintegration and how they differ from person to person.

Transition can be even more complicated for women veterans, and early on, it was made clear they understood this and were dedicated to providing active support. The program director was the father of two adult women who had educated him through the years on the complexities of existing in male-dominated environments, which I greatly appreciated. After some uncertainty, I felt strongly supported and safe in the program, which helped me begin to reframe my experiences in the military. Safety, security, and empowerment continued to build with the support of strong role models, challenging and meaningful course materials, and most importantly, the family mentality of the Trojan network, a main feature of the program.

We built a new family through many group projects, school events, happy hours after class, and going out on the town, all while leaning into our similarities as service members and veterans. I couldn't avoid being awkward and reserved most of the time, but when graduation came around, I had created lifelong bonds with my classmates.

Business Is Booming

As a bedside nurse, I had a narrow view of how healthcare systems operated on a business level. It first occurred to me that a hospital was a business when I worked in Gardena. We had to itemize each supply used on a patient (for billing). Charting a certain way affected government funding and reimbursements. So I had some curiosity about how a nurse could use her business knowledge. It's not very common for a nurse to go to business school, and I must have piqued the interest of MBV admissions when I started my application essay with "Healthcare is a business and business is booming."

Six months after graduating from USC, I applied to a nursing supervisor position, also known as house supervisor, for VA San Diego. This was a big deal, since it was about a three-rung leap on the traditional nursing leadership ladder. A nursing supervisor is the healthcare system administrative liaison who provides 24/7 operational leadership and real-time healthcare planning and logistics, including patient flow, while ensuring safe staffing. They supervise the entire facility's clinical staff and critical events and regularly report all statuses to the executive leadership team. There is often only one of them on duty at a given time. That kind of scope and responsibility meant the role would be demanding and come with substantial visibility, a major transition from being a charge nurse on a single unit. Considering my experience and recommendations from mentors and peers, I was ready for the challenge. Taking this kind of leap was a tactic often repeated in the MBV program. "Aim high and use your resources!" To prepare, I cold-messaged a few people on LinkedIn for advice, interviewed people who had either worked the role or were currently in the role, reworked my resume many times to ensure my experience spoke to the position, and practiced interview questions. The interview process was intense. In my second interview with the executive leadership team, I was white-knuckling the arms of the chair and caught the chief nurse staring at my hands. I thought I was done for.

Well, I got the job and relocated to beautiful San Diego. Two months into the job, the COVID-19 pandemic hit. We were all trying to navigate a scary new reality, but being so new and highly visible was more pressure than I could bear. It was an honor to lead and make important decisions that ensured the safety of the staff and patients, but because everyone was chronically stressed, it felt as though I became the scapegoat for their fear and frustration. The complaints were often legitimate, since maximizing operations during the ever-changing COVID-19 guidance meant we stretched resources beyond the norm. My passion for and dedication to quality had to be compromised, as I was working fourteen-hour shifts every day of the week, on weekends and holidays, plus extra shifts to cover absences. I lived to work and planned my life around when I could sleep, since I rotated through day and night shifts. I had no time or energy for professional growth or quality improvement projects, and I certainly had no time for meaningful self-care. While I took pride in my work and was determined to do it well, the chaos took a toll on me and those around me.

The only way I could climb out of the well of darkness I was in was to take a step back and turn to individual therapy, thankfully provided by the VA. My decision to seek help was personal and challenging and forced me to confront deeply-rooted fears and vulnerabilities I had long ignored. Therapy turned into a lifeline. While I have made a ton of progress with managing my current lifestyle, unpacking past traumas will take much longer.

With no end in sight for the COVID-19 chaos, after two years in the supervisor role I was practically begging to go back to Iraq. Time to find a new position better aligned with my interest in quality and need for balance. The perfect opportunity happened to fall into my lap when I received an email about a job opening for a quality consultant with the Quality Management service at the same VA facility. I had heard so much about consultants in the MBV program. Because I was familiar with the role and liked what I knew, I was ready for the opportunity and went for it. I got the job, and it's where I work now. Because the position is all about leadership, strategy, networking, and influence, and is usually reserved for nurses with masters' degrees, graduate school has really paid off. I am still pretty visible in my role, but I am no longer the perceived enemy. Instead, I am a resource that provides big-picture perspectives and assistance in quality, and I have a significant impact on care delivery. Plus regular hours, flexible working environments, and weekends and holidays off, the job has been a major breath of fresh air. I could not be happier in any other professional role at this point in my life, and I am so grateful to continue serving veterans in a way that aligns perfectly with my skills, experience, and dedication to quality.

Self-Care

I have always felt that it is a privilege to help others for a profession, and although I initially pursued nursing for economic reasons, my passion for helping others became a core motivation. Whether through work or volunteer efforts, like with Habitat for Humanity and American Red Cross, I have long been drawn to helping those in need. I got really good at meeting other people's needs. Then I realized it was a great distraction from addressing my own. Since my life is no longer swirling with chaos, I have had the time, energy, and wisdom to focus on strengthening my own foundations. I am close to returning to helping others in meaningful ways outside of my professional duties as a nurse.

After finally finding an identity by building one, and after finding a fulfilling and balanced job, my current life with my husband and dog in a peaceful home by the sea might appear boring compared to the more adventurous chapters of my past. I embrace this phase wholeheartedly. To reach this peaceful, warm, and loving plateau, the path I trekked was steep and filled with numerous challenges. With the support of my friends and family, I am here, and here is where I have space to heal. We hope to raise a family now, adding another link to the chain of strong women who raised me. Through it all, the beat goes on.

"The Catalyst I Needed"

Vanessa Bolognese reflects on telling her story

I have a newfound respect for writers of autobiographies. While I only had a chapter to write, exposing myself in words was much harder than I expected. Not only did I have to constantly stop myself from thinking that my story was one big complain-fest, but I also had to weigh potential outcomes for revealing certain details of how my life has unfolded. Legal implications aside, laying it all out could mean unnecessarily hurting people in my life. I worked hard through many rewrites to get my main points across without hurting anyone. While I chose to hold back in some ways, addressing sexual abuse was terrifying and liberating. It allowed me to continue writing my narrative feeling empowered. I watched the text grow beyond some of my lowest points.

While sexual abuse might be considered the most sensitive, I experienced other kinds and degrees of abuse which I have no doubt resonate with many. I hope sharing those experiences can help others feel less alone, and I want to encourage them to begin or continue their healing journey with renewed strength and determination knowing that in the darkest moments, there will always be a path toward light and healing.

I agreed to be part of this project because of the MBV program. It reunites veterans and creates parallels between professional networks and military camaraderie, which is massively influential, and I will always help support the mission any way I can. MBV wasn't just a graduate degree to me, it was the catalyst I needed to begin healing some past traumas and again become part of something bigger than myself.

Finally, as I reached the end of the chapter, I reflected on where my life of service has taken me and where I am in my healing journey now. Serving others while fostering a sense of community not only improves physical and mental health but has always been the force that binds us together, making us stronger and more resilient.

Having lived this timeless truth, and being in a stronger place mentally, I feel more compelled than ever to reconnect with and find more ways to contribute to my community as the next step in my journey toward total self-actualization.

Chapter 27

Some Insights and Reflections

New Narratives

This collection of twenty-five veteran narratives reveals a complex interplay of resilience, transformation, and the ongoing challenges that veterans face. The stories highlight the universal themes of overcoming adversity, finding new purpose, and the critical role that support systems play in navigating the transition from military to civilian life. However, accurately summarizing twenty-five unique personal stories is a humbling task. We will share with you our insights and summary comments, but we encourage you to reach your own insights and conclusions about the veterans and their life experiences.

By reading, listening to, and comprehending veterans' stories told by veterans themselves, we can better understand the sacrifices they make, and we can develop more effective ways to support them and assist in their reintegration to non-military cultures. Rather than having the burden of their transitions fall solely on the veterans themselves, sharing their stories may allow for a two-way reintegration process, including with their families, communities, and the broader culture.

We see some of this shared responsibility in many organizations and economic sectors that create job programs especially for veterans; in social services that deal with a broad array of veteran issues; and in specialized higher education programs, such as the MBV, that address veterans' transition challenges in the context of other veterans. Such programs succeed insofar as they form a collaborative and supportive community that enhances (or enables) the opportunity for transformational change and personal growth.

The general strategy of this book was to let veterans speak for themselves. Typically, we see only a limited, external view of veterans and the lives they lead, and these points of view are often stereotypical, with little contribution from and understanding for the actual lives of veterans. While major stereotypes of brokenness and heroism dominate the discourse about veterans, those definitions are too narrowly defined, and they don't speak to the experiences of the many veterans who may not fall stereotypically within either category, who go about their daily lives like other private citizens.

The twenty-five veterans who contributed to this book as co-authors are graduates of the MBV program at the Marshall School of Business at USC. Many who originally volunteered found the challenge of confronting and sharing their lives and experiences greater than they expected, and they dropped away over the course of the writing process.

In their writing, and at our biweekly meetings, the authors mentioned that they enjoyed being with other veterans from the MBV program and committing to the completion of the book together. In some respects, getting together regularly as a group became as important as the writing process.

Call to Service

Most civilians, unless they have family members in the military, now or in the past, know very little about the experiences of veterans. The lack of veteran-told stories reinforces the various stereotypes that surround the demographic, especially the often-implicit bias of blame and rejection forced on veterans because of the role they play in international conflicts and national defense.

One of the most important values and intentions that is not visible to the casual observer is the veterans' call to service. There are many motivations to join the military, as evidenced by our authors, but the service motive is often suppressed in the wider discussion. For those who have not served in the armed forces, this call to service may seem strange and unusual. Several of the authors expressed that part of their motivation for sharing their story was to provide benefit and assistance to other veterans who might be facing the same challenges of transitioning and reintegration.

It has been said that the military knows how to do leadership right.[3] Within the stories, our veterans demonstrated their leadership in action through their contributions to the project and to their collaborators. They supported one another, offered feedback, and maintained a collective focus on the goal of the book. Their ability to lead by example, fostering a spirit of camaraderie and mutual respect, underscores the broader leadership qualities they carry into their civilian lives.

This group of veteran co-authors exemplify how individuals can leverage their military experiences to inspire, mentor, and lead in new contexts, reinforcing the idea that the leadership skills honed in the military are invaluable in civilian life. Their commitment to this project and their ongoing dedication to service and community highlight the multifaceted roles veterans play as leaders and contributors to society. Many of the stories highlight the veterans' leadership in their lives, careers, and communities.

Demographics

The twenty-five authors included in this book share among them a total of forty-two graduate degrees, demonstrating a strong commitment to personal development, and readiness to take on new challenges and improve their skills and achievements, and a great respect for higher education as an intervention strategy for personal growth. Many of our authors describe their graduate experience as "transformational," contributing to their personal and professional growth. Many have used their programs to shape their careers differently, changing direction from what they thought they would do, into totally new areas of interest, including entrepreneurship.

Among the four branches of military service represented here, there are ten officers and fifteen enlisted members. This proportion closely represents the proportion of officers and enlisted within the MBV program itself.

---

[3] Barbara Kellerman, *Professionalizing Leadership,* Oxford University Press, 2018, pp. 54-55.

About eighty-five percent of officers attain a bachelor's degree before active duty, but as little as eight percent of enlisted personnel have a degree before entering the service. Many service members earn college credits while on active duty, so the number of enlisted with a degree increases during and shortly after active duty. A bachelor's degree is required for entrance into the MBV program, so the fact that almost sixty percent of the program is enlisted service veterans means the program attracts an ambitious group who see higher education as a path to success and personal growth.

The MBV program reduces both the impact of rank on the social structure of the classroom and on the concept of competition between students. In the case of this book, there was little influence of rank on the choice to participate.

Additionally, the book collects stories from eight women and seventeen men. Women make up about seventeen to twenty percent of active military, and within the MBV program, there is typically a similar proportion in the student cohort. Therefore, women are more represented here than in either the armed services or in the MBV program.

The greater representation of women is significant for several reasons. First, it provides diverse perspectives, as women often have different experiences in the military, which include challenges related to gender dynamics, leadership roles, acceptance, and work-life balance. By highlighting the voices of female veterans, the book provides a richer and more comprehensive understanding of what it means to transition from military to civilian life. This representation challenges common stereotypes about women in the military and veterans who are women, demonstrating that they are not only integral to military operations but also deeply engaged in the process of reflection, healing, and growth after service.

For current and future service members, seeing women well-represented can be inspiring and affirming, showing that their unique experiences matter and encouraging them to share their own stories. Moreover, by including a significant number of women's voices, the book broadens the scope of veteran issues under discussion, addressing topics such as balancing parenthood with deployment, navigating gender dynamics, and managing healthcare needs specific to women. This higher representation of women ultimately enriches the narrative and contributes to a more inclusive and accurate portrayal of military service and veteran life.

On the Page

We found several themes that are critical to understanding the authors' stories.

First, ***overcoming adversity and showing resilience*** is major among the writing. Throughout their experiences and transitions, the veterans have demonstrated their capabilities to adapt and overcome adversity, including poor childhood environments like abuse, family disruption, and personal health problems. Personal resilience and drive, and problem-solving skills, are characteristic of many of the authors. Their stories demonstrate their strong resolve, especially in pursuit of continued education and personal growth. Their ability to adapt to changing circumstances and difficult problems, rather than to quit a difficult task or situation, and their courage to bounce back from setbacks and failures, were often times remarkable.

Second, ***service and sacrifice***. A recurring theme is the profound sense of service and sacrifice the authors feel, whether for military service, community service, or post-military careers that focus on giving back to society. This theme intertwines with themes of personal sacrifice, in which the veterans often prioritize the welfare of others before their own, reflecting deep-rooted values instilled in them during their service. Multiple stories emphasize a commitment to taking care of one's family. Also, veterans feel a drive to give back in different ways, such as by creating new organizations that address the needs of other veterans. Almost all our authors have a list of service contributions and commitments that is very impressive, many of which relate to veteran issues.

Third, ***the search for purpose and identity in civilian life*** is often quite different and more difficult than military purpose and one's military identity. Both issues are mentioned frequently by our veteran alums. One's purpose in the military is typically clear, through both the mission and one's role in the mission. Military identity is presented via the uniform, rank, and unit designation, which is visible to everyone. A search for purpose and identity is integrated within the military experience. The military provides what's missing within civilian environments, namely camaraderie and belonging.

New students in the MBV program recognize that everyone in the room has a military background, which instantaneously produces a cohesive group attitude and culture that benefits from their shared military values, norms, and experience. Cohorts are happily surprised when the camaraderie they miss from their military days returns during the MBV program. Veterans in their search of greater purpose and new identities can use higher education as a way to structure their searches. Graduate programs open up new possibilities for finding new purpose and developing new identities, including the opportunity to develop and establish one's own business and organization. For example, the entrepreneurial success of former Navy SEAL Ty Smith, a member of MBV Cohort III, was an early breakthrough in reimagining one's future career.

The fourth major theme is one of **transformation and growth** brought about by military experience as well as advanced education. Many of our authors experienced a difficult childhood or homelife and found the military experience itself transformational in developing discipline, structure, purpose, and a sense of belonging to a larger group.

Looking at the challenges of transitioning out of the military, many of our authors found that higher education was transformational to their sense of self and their aspirations. In programs that emphasize the development of leadership skills as well as creative abilities, many of our alumni found new opportunities and came away empowered. In addition, informed by both their military service and higher education, many veterans experienced a greater sense of leadership and effectiveness in their current careers.

The fifth theme was **hope and inspiration**. Our authors are often introspective about their past challenges and future aspirations, and they maintain a hopeful outlook towards continuous personal growth. The focus on hope and inspiration acknowledges the hardships and challenges while emphasizing the possibilities of transformational successes. Higher education helps them commit to life-long learning and personal development. This theme reflects a sense of optimism about the future and the capability of veterans to grow and meet new challenges.

The sixth major theme was *the impact of storytelling on healing and cathartic outcomes* as expressed by our authors, primarily in their reflections. Seeking authenticity, self-awareness, and self-understanding through confrontation with oneself, along with deep and consistent reflection and introspection, created some outcomes for our authors that both surprised them and created a deep sense of meaning, satisfaction, and release of long-term injury. Our authors found themselves peeling off layer after layer of hidden injuries by going deeper with their introspection, with a wider scope of discovery and understanding.

There are two aspects of storytelling pertinent to this project: first is the content of the story itself, and second is the actual process of writing. The act of writing can lead to healing, growth, and self-acceptance. Included in the writing process was the veterans' participation in sharing their stories with others, which allowed them to claim their stories as their own, which is often one of the most difficult aspects of authentic storytelling. In this process, veterans not only tapped into the creative act of storytelling, but in sharing their stories, also experienced the crucial support of their families, communities, and fellow veterans.

In the Group

While the process discussion meetings were instrumental in fostering growth and healing, several common pitfalls emerged that challenged the authors along their journey.

Self-doubt about the value of their story. Many authors initially struggled with the belief that their personal stories were not "good enough" to be part of a project like this. Some felt that their experiences did not measure up to the more dramatic or intense stories they had heard from others. A sense of inadequacy sometimes made it difficult for them to start writing or to fully commit to the process.

The innate desire to compare stories. Another common challenge was the tendency to compare their stories to those of their peers. The inclination of authors to compare often led to feelings of insecurity and self-doubt, with some questioning the significance of their experiences. This pitfall was particularly challenging because it could undermine the confidence needed to tell their story authentically and without reservation.

Doubts about the process and outcome. Throughout the process, some authors questioned whether their stories would ultimately resonate with readers or make a meaningful contribution to the book. These doubts could create resistance, especially regarding the reflective process, as some authors were unsure if their efforts would lead to a satisfying outcome. This skepticism sometimes made it difficult to push through the more challenging aspects of writing and sharing their narratives.

Recognizing these pitfalls became an important part of the group's journey. By addressing these challenges openly in the process meetings, the authors were able to support each other, reaffirm the value of every story, and reinforce the importance of authenticity and self-acceptance. The shared understanding that every story had its unique value helped the authors overcome these hurdles, leading to a richer and more diverse collection of narratives.

For some, the groups were a way to confront and process painful memories; while for others, it was an opportunity to reflect on personal growth and achievements. The act of writing allowed some to find peace, forgiveness, and a deeper understanding and acceptance of themselves.

It became apparent during our Saturday small group meetings that there were some authors who came not so much to talk about the content of their story, but to interact with the other veterans and to share the process of insight and healing together. Also, getting together with other veteran alumni of the MBV program became its own attraction for many of the members of the discussion groups.

Veteran Reflections on Storytelling

During our biweekly meetings with the authors, it became clear that reflecting on their stories was not only insightful but an essential part of the storytelling process. Reflecting on the journey of writing added significant depth to each author's story and contributed to their sense of completion, offering new insights into experiences.

By inviting the authors to reflect on their writing processes, we encouraged them to engage in different levels of reflection. Their reflections not only describe the challenges and emotions they encountered but also analyze the impact of these experiences and, in some cases, critically examine how their stories relate to broader themes of identity, service, and resilience.

Reflection is a process often used in educational and personal development settings to deepen understanding and foster personal growth. Reflection requires deep introspection into one's history, or into any topic one wants to address or understand. Reflection can create awareness, understanding, assessment, and potentially change.

The reflections revealed a wide range of emotions and insights, from the therapeutic effects of writing to the challenges of confronting painful memories. Below are some of the key themes that emerged from their reflections, illustrated through direct quotes from the authors.

Expression and Healing

Authors found the process of writing their stories to be like a form of therapy, allowing them to process emotions and experiences that had been buried for years. The act of putting their thoughts on paper helped them heal and find peace with their pasts.

- "Sitting with those emotions and writing about them as to where I am now was very therapeutic."
- "Remarkably healing!"
- "Retelling the story is incredibly cathartic . . . it ripped off scar tissue that I'd built up to deal with the past."
- "The process of writing my story has acted as a sort of therapy."
- "Profound realization that only now I'm in a place of peace to let this flow freely."
- "Facilitates a focus of self-healing."

Challenges and Difficulties

Writing about personal experiences, especially painful ones, was not easy. Authors struggled with starting their stories, being honest and direct, and ultimately finishing the process. However, despite these difficulties, the authors recognized the value in confronting their pasts. There were also many comments of how difficult it was to get started, be painfully honest and direct, and to finish.

- "It's difficult to talk about oneself."

- "There will never be a good time to write painful stories."
- "I realized in the venture there will never be a good time to write painful stories."
- "Most of us are ordinary people who got to experience some pretty extraordinary things."

Personal Growth and Connection

The process of introspection and storytelling led to significant personal growth. Writing their stories helped the authors understand themselves better and facilitated a deeper connection with their own experiences and with others.

- "Transitioning my thoughts to a document has always proved to be cathartic."
- "Introspection and storytelling as vehicles for personal growth and connection."
- "Reflection during challenging times will help create the most growth and change for the better."
- "Self-discovery and acceptance; adaptability, directness, and problem-solving."
- "The process of storytelling and structured recollection is profound to where I find myself in life."
- "I have submitted to this personal journey to become a better me."
- "Telling my story allowed me to lower my walls and share deeply personal parts of myself. I feel completely unburdened as a result."

Community and Support

The collaborative nature of the project fostered a strong sense of community. Many found comfort and strength in reconnecting with their MBV alumni and sharing their stories, knowing they were not alone in their struggles.

- "The best part of this storytelling was reconnecting with MBV alumni and sharing our stories."
- "I am grateful to have so many incredible people with me in my corner to reinforce that I don't have to go through this alone, and neither do you."

- "In sharing our narrative, we not only process our own experience, but we contribute to a collective understanding and empathy."
- "Positive aspects of weekly meetings. Getting closer through caring."
- "Armed with more confidence, humility, dedication and perseverance because of everyone here (co-authors)."
- "Serving others and fostering a sense of community not just to improve physical and mental health."

Hope and Inspiration

Despite the hardships and challenges they faced, authors expressed a hopeful outlook towards continuous growth and contribution. They emphasized the possibilities of transformation and the importance of lifelong learning and personal development.

- "I believe in the power of continuous growth and the possibility of a better future."
- "Our stories can help bridge the gap between veterans and the broader community. Participating in the storytelling project has been a transformational experience."
- "Whatever challenging obstacles you are experiencing, you realize you are not alone."
- "Resiliency, family support, embracing change, and seeking growth and transformation."

Ownership

While the process of writing and reflecting on their stories was challenging, the result was a powerful tool for healing and growth. Authors spoke of how "owning" their story was difficult but ultimately enhanced their self-confidence and provided a sense of completion and peace with their history and themselves.

The reflection process itself allowed the authors to accept their unique lives and experiences more fully. While it sometimes involved personal judgment, it also offered a greater sense of acceptance and comfort with the outcome. The act of reflecting on and writing about their lives helped these veterans to achieve a healthier reconciliation with themselves and, often, with their families, friends, and community.

In conclusion, the reflections shared by the authors not only provide insight into the process of writing but also highlight the profound impact that storytelling can have on personal growth, healing, and connection. By confronting their pasts with honesty and vulnerability, these veterans have found a sense of completion, self-confidence, and peace. Their reflections are a testament to the transformative power of storytelling—a process that not only heals the individual but also helps bridge the gap between veterans and the broader community.

Afterword

The Process

Heriberto "Eddie" Arámbula, PhD, MBV Cohort IV, became our storytelling guide and champion, and my partner in writing and producing this book. Eddie's storytelling knowledge and approach became the overarching process-guidance for our work of collecting and combining the personal life stories of our veterans, who additionally became our co-authors.

Eddie and I communicated frequently after he left the MBV Program in 2017. I had written him a letter of support to begin his PhD program in Educational Leadership at Texas State University in San Marcos, TX. For his dissertation, "Story of Re-creating Home: Learning as a Tool for Healing Post Military Service," he authored an autoethnographic account exploring the healing process of Gold Star families—those who have lost family members in combat. This work led Eddie to recognize the power of personal storytelling as a path to healing and learning for both veterans and their families in post-military service. He also identified the critical role of community support in this process. Eddie was passionate about teaching, researching, and developing programs to help veterans navigate reintegration and achieve their educational goals.

Eddie's focus on storytelling as a healing process had a broader background and grew from the outcome of a significant battle in the Iraq War. He was a member of the Second Battalion, Fifth Cavalry regiment (2/5 CAV) of the First Cavalry Division out of Fort Cavazos, TX, formerly Fort Hood. The division was ambushed in Sadr City in Baghdad on April 4, 2004. Eddie counted among the troops sent to rescue the ambushed patrol. During the battle, eight soldiers lost their lives and another sixty were wounded. This unit and this fight became the subjects of Martha Raddatz's book, *The Long Road Home*, published in 2007 by Penguin Books.

Throughout the course of a full year, Eddie and I held small group meetings with our veteran co-authors to discuss and share the process and struggles of writing an honest and direct nonfictional story about oneself. During the week, we read, discussed, and provided feedback on the various drafts that were turned in. Then on the weekend, meeting with other veterans over a video call, we encouraged those who wanted or needed to, to share their progress, seek feedback, or share their successes and challenges and provide support for each other. The Zoom meeting structure accommodated authors from across the country, from the east coast, mid-west, west coast, Pacific northwest, and Hawaii.

Soon, it became apparent that the process of holding discussion meetings was taking on a life of its own and helping many of the co-authors dig deep within themselves and experience healing or catharsis from sharing their vulnerabilities. This evolution was apparent in the various reflection pieces which each author wrote at the end of their story. Rather than simply sharing one's own story and insight in order to help other veterans successfully manage their transitions, the writing and storytelling process became a deeply insightful and growthful process for the writers themselves, potentially healing deep personal wounds often hidden within their backgrounds as well as their military experiences.

In creating a supportive environment, the authors were able to choose what to share and what to withhold, making the act of sharing and discussing personal vulnerabilities a commitment to honesty and integrity, despite the risk. The creation of this safe space was crucial, allowing the co-authors to participate fully in the peer-to-peer support meetings, where they found encouragement and understanding from others who had faced similar challenges.

While we didn't begin the Saturday process-meetings until about midway through the writing experience, it became clear to us that this technique, as a new addition to our writing process, was an effective intervention for completing the stories. The safe space cultivated within these meetings provided the trust and security necessary for authors to confront difficult memories and experiences, facilitating both personal growth and the healing process.

After beginning with about fifty volunteers, twenty-five ended up contributing to the project. A number of our alums dropped out, as the challenge of the personal story became too great. In some cases, the resistance to digging deep within one's own life experience, both the positive and negative, became not only challenging but painful.

Some of our co-authors—veterans and MBV alumni all—who said they could commit to writing and sharing their stories were unable to either start or finish because of a strong sense of personal humility that prevented them from talking or writing about themselves. This value, shared among an entire group of veterans, was the most apparent value of any group with which I had interacted in the classroom during my time spent in higher education.

Transitions in Leadership

Late in my career, I led the development of the MBV program and spent six years as the academic director and leadership instructor. It was a satisfying, growthful, and unique experience, for me personally. Typically, after an assignment such as this one, you move on to the next assignment at some point. In my case, I was moving on to retirement. I had failed a couple of times at retiring, but after six years as the academic director and leadership instructor, and forty-seven years in total at USC, it was time for me to let go!

For *Transitions in Leadership*, it seemed appropriate and somewhat demanding to memorialize the program and my experience with the veterans. I began writing that book on the evening of Ty Smith's military retirement ceremony in San Diego in May, 2016. Ty was in Cohort III, our third year of the MBV Program.

Even though his cohort had graduated a week earlier, and everyone had departed, Ty invited them to join him for the ceremony and celebration. I assumed it was merely a courtesy invitation. But as I looked around at the guests, I saw that forty percent of Cohort III had shown up for Ty! In almost fifty years of university work in business schools, I had never seen this depth and strength of connection and shared membership. From the moment the student veterans entered that first class session and saw that everyone else in the room was also a veteran, the MBV program replaced the loss of camaraderie, which is one of the biggest losses for veterans upon transitioning out of active duty.

After working on the previous book for two years, I invited some students to contribute short personal stories that would be added to each chapter. Forty-four individuals enrolled in Cohorts I through VI (my last class) contributed, and I included their stories in order to provide brief glimpses into the lives of some of our veterans.

Volume 2

In 2020, I began thinking about a second book with a theme—"where are they now"—written by the veterans themselves. I was very curious about what happens to our alumni a few years after graduating from the program. I thought my curiosity might provide some insight into the impact of the program on the lives of our alumni, which in a small way might add some value to evaluating the program's effectiveness. It might also provide a different sort of narrative about the lives of veterans who are not solely categorized as either broken or heroes, as so often is the case of the media stereotypes of veterans. This would be a real view into the actual lives of veterans who have pursued graduate education as an intervention into their personal lives and careers.

I drafted a proposal for the book focused on purpose, participation, and personal storytelling. I turned to Ty Smith to give feedback on the proposal. I knew I would get a straight response from Ty, as I always did. Ty has the capability of responding quickly and congruently from both an emotional and analytical perspective on these types of issues and questions. Having shared my proposal with Ty, I asked for his response and approval.

"Do you think we should do a book about our veterans' transitions into new careers, reintegration, successes, challenges, and personal growth, including the impact of the MBV program, using their personal stories?" I wrote to him. "Do you think we can obtain enough volunteers to make this book worthy of the effort and topic, and should this be done in the first person, in their own voices?"

Ty's response was immediate: "No! It's not that we should do such a book, we must do it!"

Robert B. Turrill, PhD

About the Authors

Dr. Robert B. Turrill (left) received his MBA and PhD in applied behavioral science for management from the Anderson School of Management at UCLA. A recipient of multiple teaching awards, he has developed courses in leadership, organizational design, team design, and interpersonal relationships. He is professor emeritus of clinical management and organization at the Marshall School of Business at USC and helped found the Master of Business for Veterans program, whose board of executive partners he serves on. In 2020, he published the first volume of *Transitions in Leadership*, documenting the first years of the MBV program. He served in the US Army from 1957-63.

Dr. Heriberto Arámbula, a Pat Tillman Scholar, combat veteran, and public pedagogue, served in the US Army and Coast Guard, achieving the rank of E-5. He holds a PhD in educational and community leadership from Texas State University, where he researched US military veterans' trauma and social integration following their service. He is the founder of the Journal of Interactive Veteran Experiences.

www.ingramcontent.com/pod-product-compliance
Lightning Source LLC
Chambersburg PA
CBHW010935120626
46554CB00007B/2476

9798992586824